T0336397

Solving Climate Change

A guide for learners and leaders

Online at: https://doi.org/10.1088/978-0-7503-4032-8

Solving Climate Change

A guide for learners and leaders

Jonathan Koomey
Koomey Analytics, Bay Area, CA, USA

Ian Monroe
Stanford University, Stanford, CA, USA

IOP Publishing, Bristol, UK

ISBN   978-0-7503-4032-8 (ebook)
ISBN   978-0-7503-4030-4 (print)
ISBN   978-0-7503-4033-5 (myPrint)
ISBN   978-0-7503-4031-1 (mobi)

DOI   10.1088/978-0-7503-4032-8

Version: 20221201

IOP ebooks

British Library Cataloguing-in-Publication Data: A catalogue record for this book is available from the British Library.

Published by IOP Publishing, wholly owned by The Institute of Physics, London

IOP Publishing, No.2 The Distillery, Glassfields, Avon Street, Bristol, BS2 0GR, UK

US Office: IOP Publishing, Inc., 190 North Independence Mall West, Suite 601, Philadelphia, PA 19106, USA

Image credit: Violet Kitchen 2022

From Jon: For my friend and colleague Florentin Krause, who started me down the path that led to this book.

From Ian: To Kai for his joy and curiosity, Susan for her resilience and support, my family for championing sustainability long before it was cool, and friends everywhere working on a better climate future for all of us.

Contents

Foreword

It's warming. It's us. We're sure. It's bad. We can fix it.

I first heard these words[1] on a sunny winter day at Stanford. Jon Krosnick was speaking at a memorial service[2] celebrating the life of Steve Schneider, who first inspired me to study climate impacts and solutions. I have used this 'climate haiku' (if you don't count syllables too carefully) ever since: in my teaching, research[3], writing[4], and on a footnoted protest sign[5]. Research[6] shows that understanding these key points drives support for strong climate policy.

In *Solving Climate Change*, Jonathan Koomey and Ian Monroe have focused on the most critical piece of the climate haiku: 'We can fix it', adding rightly, 'But we'd better hurry!'

This book is a valuable guide for students and practitioners to understand the principles and crunch their way through the numbers behind how to 'fix it'—how to stop global warming and start healing the damages it has caused.

Drawing from Jonathan and Ian's extensive experience in practice, and in teaching these concepts to generations of students, *Solving Climate Change* offers clear guidance for how to eliminate emissions by using fewer fossil fuels (decarbonize), as well as practical milestones to aim for along the path to ending fossil fuels altogether.

After decades of inaction, the authors urge us to 'move as quickly as we can' to retire polluting cars, factories, and ideologies. Rather than trying to map out the one perfect path to a fossil-free world (which doesn't exist), it's far better to focus on meeting our needs with the fossil-free options that are ready today. While a few of us can specialize in researching how to decarbonize the tricky last 10%–20%, almost all of us should instead be focused on putting what already works to work. This book will help you do that.

A key skill you'll develop is to use systems thinking to understand where fossil greenhouse gases come from, what's needed to eliminate them, and how to get started. Combining tools such as emissions inventories and scenarios will help you prioritize, so that you focus on the most effective actions to reduce emissions fast and fairly. You'll also understand how to align incentives and mobilize money to put those actions into practice. (Fortunately, since climate action offers so many co-benefits for health and equality, this work can go faster than many think!)

As the authors point out, we need many more climate solvers who can build simple spreadsheet models to quickly inform effective decarbonization decisions everywhere. The approach in this book will help you roll up your sleeves and get to

---

[1] https://www.kimnicholas.com/climate-science-101.html
[2] https://stanforddaily.com/2010/12/01/memorial-for-stephen-schneider-set-for-sunday/
[3] https://journals.plos.org/plosone/article?id=10.1371/journal.pone.0218305
[4] https://www.penguinrandomhouse.com/books/665274/under-the-sky-we-make-by-kimberly-nicholas-phd/
[5] https://twitter.com/billmckibben/status/808791393569243140?s=20&t=WDkZSn7ptnFE9en2qqHCPA
[6] https://journals.plos.org/plosone/article?id=10.1371/journal.pone.0118489

work, leading the way to make your home, school, company, industry, or city fossil-free.

Alongside the unassailable case from physics for why fossil fuels must be left in the ground, Jonathan and Ian make a compelling moral case for elevating truth, and against continuing to rely on the fossil fuel industry to make the changes needed to avoid climate catastrophe.

We are alive at an absolutely critical moment for humanity and life on planet Earth. Our task is to lead a transformation from our current world with a destabilizing climate and a fraying web of life, to a world where we succeed at halting warming, then go beyond net zero to climate positive, while benefitting people and nature.

We desperately need more citizens armed with the urgency, clarity, and skills to help build the bridge between these two worlds. This book can help you build that bridge. Thank you for picking it up. You are essential to doing the work that the world needs right now!

Kimberly Nicholas, PhD
Associate Professor of Sustainability Science, Lund University,
Sweden
Author of *Under the Sky We Make: How to Be Human in a Warming World*

Preface

There's one issue that will define the contours of this century more dramatically than any other, and that is the urgent threat of a changing climate.
—Barack Obama

We've both been working on understanding and scaling climate solutions for most of our lives. Our paths first crossed while teaching courses on energy systems and climate solutions at Stanford University. The ensuing collaboration led us to write this book.

Jon first studied climate change in his graduate program at the UC Berkeley's Energy and Resources group in the mid-1980s, under interdisciplinary Professors John Holdren, John Harte, Mark Christensen, and Anthony Fisher. At the same time, he joined the Energy and Environment Division at Lawrence Berkeley National Laboratory (LBNL) and began to collaborate with energy systems thinkers such as Mark D Levine, James E McMahon, Ashok Gadgil, Joe Eto, Ed Kahn, Arthur H Rosenfeld, and, most importantly for his climate journey, Florentin Krause.

Florentin invited Jon to work with him on two climate-related projects, one of which led to the first comprehensive analysis of a 2 °C warming limit, published in 1989 [1, 2]. He's been working on climate solutions ever since, until 2003 doing research full time at LBNL, after that as a visiting professor at Stanford (twice), at Yale University (once), and at UC Berkeley (once). He was also a researcher at Stanford and a visiting lecturer at Stanford's Graduate School of Business.

Ian's climate journey started by witnessing elevated threats from drought and wildfires around his family's small farm, which led to an international career in climate solution science and engineering, while co-founding a climate technology startup (Oroeco) and a climate investing firm (Etho Capital). Ian has taught climate solution courses in the Earth Systems Program at Stanford for over a decade, and he has worked in over 30 countries, including advising on climate accounting and decarbonization strategies for United Nations programs, governments, Fortune 500 companies, startups, nonprofits, and over $100 billion in investor assets. Ian's childhood climate nightmares came true when his farm and surrounding community were leveled by wildfire in 2017.

When we started working on this issue, climate change was mostly discussed as an issue for future generations, while now hardly a week goes by without new climate-related disasters in the news. Now even those who continue to misunderstand the certainties of climate science have a hard time denying the realities of unprecedented events around them. Renewable energy, electric vehicles, and efficiency technologies were niche products a couple decades ago, now they are readily available and becoming the cheapest options in many parts of the world. Nearly every major government, company, and investor now acknowledges the grave dangers posed by a warming world while taking steps to mitigate risks and shift towards climate solutions, with many adopting 'Net Zero by 2050' targets for eliminating climate pollution.

Figure 0.1. The Redwood Valley wildfire on October 9, 2017, burned through Ian's family farm in the middle of the night, fueled by drought and hurricane-force 'diablo' winds. Over 1000 people lost homes that night, including over a dozen in Ian's family, and nine people died from the flames. Wildfires are increasing in frequency and intensity around the world due to climate change, which will continue until we turn the Earth climate positive. Copyright Nathan Chance Franck, 2017.

Much has changed from when we started working on this book a couple years ago. The United States has now passed comprehensive climate legislation for the first time, joining Europe, China, and most of the world's other major economies to subsidize dozens of different climate solution technologies to tipping points that can catalyze scale. Russia's invasion of Ukraine has also highlighted the geopolitical and economic risks of fossil fuel dependency, driving record demand for electric vehicles, renewables, and energy efficiency technologies. Real climate progress is finally being made, in many cases spurred on by tragic events and suffering, but there are still major gaps between rhetoric and realities, and global emissions continue to rise. As we'll discuss in more detail 'Net Zero by 2050' is likely at least a decade too late to reach Paris Agreement targets, and it is still far from the climate-positive targets needed to stop and reverse the climate damages from which millions of people are already suffering.

There is also increasing awareness that our climate crisis is not primarily an environmental problem, it is first and foremost a human problem, and a problem enmeshed in tremendous social and economic inequity. While we're both scientists and technologists at heart, our climate solution work around the world has highlighted that climate inequities are everywhere, and are often the roots of the problem. Denying the links that climate pollution and climate impacts have to structural classism, racism, and colonialism is as wrong as denying the fundamentals of climate science. Climate solutions that fail to incorporate climate justice considerations are often not real or saleable solutions. Understanding and addressing climate justice issues is essential for scaling technologies and policies that move climate in the right direction, while solving climate is also essential for achieving every one of the United Nations Sustainable Development Goals (SDGs).

Climate change is the biggest collective challenge humanity has ever faced. The problem is difficult in part because temperature changes are driven by changes in *cumulative* greenhouse gas emissions. The longer we wait to fix the problem, the more difficult fixing it becomes, and the faster our *rate of change* needs to be to stabilize global temperatures. It is more complicated still because almost every human activity generates greenhouse gas emissions, so the needed *scope* of change is unlike anything that has come before.

We're already committed to some warming from past emissions of greenhouse gases, so as Professor John Holdren has said for years, humanity's choices for response are threefold: mitigation, adaptation, and suffering. The world will do some of each of these in coming decades, but how much of each we do is up to us.

The more mitigation we do, the less adaptation and suffering we'll impose on our descendants. Adaptation is hard to do unless you know exactly what's going to happen, and the current path we're on is likely to be so disruptive that we will have a hard time adapting fast enough. In addition, ecosystems can't adapt as rapidly as we're currently changing the climate, and there's little we can do for them except mitigate as rapidly as we can. Suffering, of course, is what remains if we fail.

For all these reasons, the main focus of this book is on climate mitigation (reducing emissions), although we touch on aspects of climate adaptation and resilience as well. Preserving a stable climate will require unprecedented and rapid changes around the globe, but most people have no idea just how big and fast these changes need to be.

Another reason the climate problem is complicated is because the fossil fuel industry and other large polluters have had more than a century to rig systems in their favor. Most technical treatments of climate solutions ignore or gloss over this reality, but we make it central to our analysis.

Stoddard *et al* [3] addresses this question head on in their 2021 article 'Three decades of climate mitigation: why haven't we bent the global emissions curve?' That article shows the power of status quo interests to frame the debate and stymie progress on an important problem, using myriad techniques to delay climate action. Related to that power is deep-seated (but often invisible) corruption and regulatory capture enabled by the vast revenues of the most powerful industry in human history (about $5T US/year at last count). Perverse incentives from small but powerful groups that profit from climate pollution continue to spread disinformation, deliberately obscuring climate truths from the public while blocking policy progress. Confronting this corruption is central to stabilizing the climate, a theme to which we return later in the book.

This book grew out of a class we developed and taught at Stanford University's Earth Systems Program in 2017 and 2018 titled 'Implementing Climate Solutions at Scale'. It had as its primary goal teaching students quantitative and qualitative methods for understanding the rate and scope of change needed to truly face the climate challenge. Each student (or group of students) produced and presented a project for a state, region, or country that achieved net-zero emissions by the middle of the twenty-first century, assessing potential emissions reductions in different sectors, based on the resources and characteristics of the geographic unit under study.

Our class didn't focus on estimating total costs of emissions reductions, as that is a complicated task not well suited to a one-quarter class (the students had their hands full with the emissions reductions side of the equation). We did encourage students to use cost-effectiveness assessments in choosing emissions reduction options, we just didn't require that they add up all the costs and benefits, because that would have been too much to ask for this introductory class.

Our intended audience

This book introduces tools for understanding how to reduce emissions rapidly, hitting net-zero emissions no later than 2040, then pushing beyond to remove carbon from the atmosphere and reduce carbon dioxide concentrations to those of the stable climate in which human civilization developed. We've written this book first and foremost as a resource for professors and advanced undergraduate and graduate students ready to teach and learn the key analytical tools needed to create a climate-positive world. We hope self-motivated students, investors, entrepreneurs, policy-makers, climate solution practitioners, and others who want to learn these skills on their own will also find this book useful.

What will readers learn?

This book goes beyond our original courses to provide a more comprehensive framework for solving climate change than we've found elsewhere. We include an overview of climate solution technologies, as well as the analytical tools necessary to identify solutions that really work. We also explore what's needed to align incentives, mobilize money, and elevate truth in climate conversations, key pillars of climate action that are often overlooked by techno-centric discussions of global emissions reductions.

Some of the tools we cover include:

- Emissions inventories and country-wide energy balances.
- Decomposition tools to understand historic and future scenarios.
- Technology costing studies for emissions reductions.
- Climate 'wedges'.
- Life-cycle assessment (LCA) methodologies.
- Financial modeling tools that integrate and assess sustainability.
- Whole-system, clean-slate integrated design.
- Social and environmental tradeoff analysis.
- Climate adaptation and resilience strategies.

Learning by doing only happens if you DO, so the book focuses on how to put these tools and associated data to work in student projects.

The structure of the book

As explained in chapter 2, we organize the book around the overarching goal of achieving net-zero emissions by 2040, making the world climate positive thereafter. We present three key high-level ideas for stabilizing the climate: ending fossil fuels, minimizing non-fossil emissions, and creating a climate-positive biosphere. We then elaborate on eight pillars of climate stabilization, laid out in detail in chapters 4–11.

This book is divided into thirteen chapters and seven appendices:

Chapter 1—Introduction to the climate problem (short form): The continued orderly development of human civilization requires that the world reduce greenhouse gas emissions to zero as fast as possible, starting now. This chapter explains the evidence and rationale for that point of view, in a shorter form than in appendix A.

Chapter 2—Introduction to climate solutions: This chapter offers a comprehensive vision of how to stabilize the climate and describes the eight pillars of climate action, which we use to structure the book.

Chapter 3—Tools of the trade: This chapter summarizes how to create emissions reduction plans using emissions inventories, then how to turn those inventories into projections of future emissions for different scenarios.

Chapter 4—Electrify (almost) everything: This chapter draws upon the latest research and field experience to describe how to electrify virtually all energy end-uses, which is the single most important measure to end fossil fuels.

Chapter 5—Decarbonize electricity: This chapter describes the technologies, policies, and business practices needed to move electricity generation to zero emissions, while increasing total generation three- to four-fold in coming decades.

Chapter 6—Minimize non-fossil emissions: The often neglected non-fossil emissions really need to be tackled for climate stabilization, and addressing the short-lived non-fossil pollutants is a way to slow increases in global warming in the near term.

Chapter 7—Efficiency and optimization: This chapter focuses on efficiency in the broadest sense, that of optimizing energy, capital, and materials flows while reducing emissions to zero.

Chapter 8—Remove carbon: This chapter discusses the prospects and pitfalls of technologies that promise to remove carbon dioxide from the atmosphere.

Chapter 9—Align incentives: We'll need to re-design markets and re-align incentives throughout society to reflect the goal of getting to zero emissions as soon as possible.

Chapter 10—Mobilize money: Solving the climate problem means replacing high-emissions infrastructure with capital and technology. Financial innovation can help smooth that transition.

Chapter 11—Elevate truth: Elevating truth and fighting disinformation, apathy, and cynicism are critical to a successful transition to a climate-positive society.

Chapter 12—Bringing it all together: Planning for a climate-positive future involves complex tradeoffs of timing, logistics, and economics. This chapter introduces students to some of those tradeoffs and how to think rigorously about them.

Chapter 13—Truly facing the climate challenge: This concluding chapter describes reasons for hope in the face of a truly daunting societal problem, and describes the best practices for solving and discussing the climate problem.

Appendix A—Introduction to the climate problem (long form): The continued orderly development of human civilization requires that the world reduce greenhouse gas emissions to zero as fast as possible, starting now. This chapter explains the evidence and rationale for that point of view, including more graphs and detail than in chapter 1.

Appendix B—Detailed example on modeling capital stock growth and turnover: This appendix presents an example you can copy for your own model of capital stocks.

Appendix C—Why much existing fossil capital will need to retire: This appendix explains why we can't stabilize the climate without retiring existing fossil capital.

Appendix D—Expanded Kaya decomposition: This appendix presents the expanded Kaya decomposition shown in chapter 3.

Appendix E—Proper treatment of primary energy: This appendix explains why some conventions for calculating primary energy are misleading when the world is transitioning to non-combustion generation.

Appendix F—Revenues from top fossil fuel companies and the tobacco industry: This appendix documents revenues from fossil fuel and tobacco industries in 2019.

Appendix G—The effect of carbon prices on existing coal-fired electricity generation and retail gasoline prices: This appendix shows the effect of a \$10/tonne carbon dioxide charge on electricity generation and retail gasoline prices to illustrate why carbon taxes are far more effective at changing the utility sector generation mix than in affecting small consumer behavior.

The material in each chapter supports a lecture or two in our class. We incorporate figures for each topic area into power point slides for teachers to incorporate into their own lectures as they see fit. We also create spreadsheets showing example emissions scenarios from which students can learn. Each chapter gives suggested readings. References and footnotes lead to more detailed sources for those who want to learn more. At the beginning of each chapter is a summary of key conclusions to introduce what readers will learn. We also post a template and some advice about student projects at http://www.solveclimate.org, alongside a collection of climate solution action checklists for government, company, investor, donor, and community decision makers.

How to use this book

Professors who want to build a course around this book can use the material in each chapter as the basis for lectures, supplemented by examples drawn from their own experience.

General readers can read the book straight through. Those in a hurry should read chapters 1 and 2 then the summary at the beginning of each succeeding chapter. They can conclude by reading chapters 12 and 13.

How to contact us

We encourage readers to contact us with their comments, ideas, and thoughts about how the book could be improved for a second edition. You can contact Jon via http://www.koomey.com, Ian via LinkedIn, and both of us via http://www.solveclimate.org.

Jonathan Koomey and Ian Monroe

Bay Area, CA, and Redwood Valley, CA, USA

30 August 2022

We are stealing the future, selling it in the present, and calling it GDP.

—Paul Hawken

References

[1] Krause F, Bach W and Koomey J 1989 *From Warming Fate to Warming Limit: Benchmarks to a Global Climate Convention* (El Cerrito, CA: International Project for Sustainable Energy Paths) http://www.mediafire.com/file/pzwrsyo1j89axzd/Warmingfatetowarminglimitbook.pdf

[2] Krause F, Bach W and Koomey J G 1992 *Energy Policy in the Greenhouse* (New York: Wiley)

[3] Stoddard I *et al* 2021 Three decades of climate mitigation: why haven't we bent the global emissions curve? *Annu. Rev. Environ. Resour.* **46** 653–89

Acknowledgement

Climate change is an interdisciplinary challenge, and few universities have the faculty or inclination to address this problem inside or outside of traditional disciplinary departments. There are more and more exceptions to this rule, but most universities have in the past not addressed climate change in a wholistic fashion because of its interdisciplinary nature. We are grateful to Professor Pamela Matson, formerly Dean of Earth Sciences at Stanford University, as well as Deana Fabbro-Johnston from Stanford's Earth Systems Program, who both supported our interdisciplinary class and allowed it to reach its current state of development.

We'd also like to thank the students who've taken this class for helping us develop our arguments and for challenging us when we weren't clear about important issues.

Kimberly Nicholas has our gratitude for writing a lovely foreword that summarizes the essence of the book in a compact and compelling form.

Jon and Ian are both grateful to their families for their support throughout the long process of writing this book.

We are also thankful for others along the way who have supported our intellectual development. One of the most important of these for both Jon and Ian is Gil Masters, Professor Emeritus at Stanford University who has been for decades an inspirational pioneer in clean energy and green building education.

Jon is grateful to his friend and colleague Zachary Schmidt, whose research, data analysis, and programming skills have contributed in many ways to the tables, graphs, and appendices in this book.

Many friends and colleagues have helped and supported our writing and teaching efforts, and we are lucky to have them as part of our intellectual community. These supporters gave comments, suggested ideas, gave guest lectures, and supplied quotations about the book. For these and other contributions, we are grateful. These supporters include (in alphabetical order by last name): Allison Archambault, Drew Baglino, Drianne Benner, Bruce Biewald, Lukas Biewald, Carl Blumstein, Adam Brandt, Marilyn Brown, Karl Burkart, Ben Caldecott, Chris Calwell, Mark Campanale, Ralph Cavanagh, Dan Chu, Moya Connelly, Danny Cullenward, Laura D'Asaro, Noah Deich, Amanda Denney, Tim Duane, Georges Dyer, Delphine Eyraud, Chris Field, Leslie Field, Nathan Chance Franck, Benjamin Franta, Amberjae Freeman, Allison Gacad, Gil Friend, Ashok Gadgil, Michel Gelobter, Ashley Geo, Margot Gerritsen, Kenneth Gillingham, Justin Gillis, Peter Gleick, David Goldstein, Eban Goodstein, Deborah Gordon, Saul Griffith, Arnulf Grübler, Genevieve Guenther, Juliana Gutierrez, Corwin Hardham, Chante Harris, John Harte, Hal Harvey, Karl Hausker, Thomas Hayden, Gang He, R Paul Herman, Richard Hirsh, Katie Hoffman, John Holdren, Nate Hultman, Holmes Hummel, Soh Young In, Stacy Jackson, Mark Z Jacobson, Dan Kammen, Bob Keefe, Julie Kennedy, Min Kim, Ryder Kimball, Paul Komor, Florentin Krause, Saumya Krishna, David Kroodsma, Bård Lahn, Skip Laitner, Dan Lashof, Doug Duckjun Lee, Michael Lepech, Mark Levine, Alvin Lin, Dawn Lippert, Scott Loarie, Amory B Lovins, Katherine Mach, Lisa Mandle, Eric Masanet,

Kieren Mayers, Andrew McAllister, Ed McCullough, James E McMahon, Michael Mann, Vineet Mehta, Evan Mills, Robert Munro Monarch, Rebekah Moses, Dustin Mulvaney, Gregory Nemet, Eliza Nemser, Jenna Nicholas, Wendy Ong, Robbie Orvis, Richard Plevin, Robert Perkowitz, Sian Proctor, Jean Ann Ramey, Richard Reiss, Lisa Renstrom, Joe Romm, Abe Schneider, Gia Schneider, Alicia Seiger, Gianluca Signorelli, Susan Su, Joel Swisher, Auriane Tang-Subtil, Melissa Vallejo, David Victor, Diana Vorsatz, Amy Guy Wagner, Gernot Wagner, Vance Wagner, Rose Wang, Michael Webber, Bill Weihl, John Weyant, Sheldon Whitehouse, Austin Whitman, Jim Williams, and Eva Woo.

Author biographies

Jonathan Koomey

Jonathan Koomey is a researcher, author, lecturer, and entrepreneur who is one of the leading international experts on the economics of climate solutions and the energy and environmental effects of information technology. Dr Koomey holds MS and PhD degrees from the Energy and Resources Group at the University of California at Berkeley, and an AB in History and Science from Harvard University. He is the author or coauthor of ten books and more than two hundred articles and reports on energy efficiency and supply-side energy technologies, energy economics, energy policy, environmental externalities, and global climate change. He has also published extensively on critical thinking skills. *Solving Climate Change: A Guide for Learners and Leaders* is his latest book.

For more on Dr Koomey's research, writing, and accomplishments go to Koomey.com.

Ian Monroe

With a career spanning climate science, technology, policy, and finance, Ian Monroe has taught at Stanford University for over a decade and worked on climate challenges in over 30 countries. A pioneer in climate solution investing, Ian is the CIO of Etho Capital, which runs the ETHO ETF and some of the world's first and best performing deeply decarbonized index strategies. Working directly with family offices and institutional investors, Etho has helped decarbonize over $100 billion in assets, and Ian is now helping move even more capital into climate solutions by co-founding the new Climate+Positive Investing Alliance (C+PIA). Ian is also a Founder of Oroeco, a pioneering personal climate solution technology platform, and Ian has advised the United Nations, World Bank IFC, governments, companies, investors, and the creation of several international sustainability standards, including Climate Neutral and Science Based Targets (SBTi). Beyond his work and graduate degrees in Earth Systems science from Stanford and the University of Oxford Artificial Intelligence Programme, Ian has also been educated by climate-fueled droughts and wildfires on his family's small farm in Mendocino County, California.

IOP Publishing

Solving Climate Change
A guide for learners and leaders
Jonathan Koomey and Ian Monroe

Chapter 1

Introduction to the climate problem (short form)

Facts are stubborn things; and whatever may be our wishes, our inclinations, or the dictates of our passions, they cannot alter the state of facts and evidence.
—John Adams

Chapter overview

- Following Nicholas [4], we summarize the current state of knowledge about the climate problem in a few phrases:
  - It's warming.
  - It's us.
  - We're sure.
  - It's bad.
  - We can fix it (but we'd better hurry).
- These conclusions are based on some of the most well-established principles in physical science, corroborated by the work of thousands of scientists and many lines of independent evidence, measurements, and analysis. Every reputable academy of science, the United Nations, and the vast majority of the world's national governments and multinational corporations agree with these core climate truths.
- Keeping Earth's average temperature from rising much more than 1.5 °C from pre-industrial levels will require immediate, rapid, and sustained greenhouse gas emissions reductions. It will also almost certainly require the removal of carbon pollution from the atmosphere.
- Society should aim to hit 'net zero' greenhouse gas emissions globally by 2040 at the latest, sooner if possible, and every year thereafter should be 'net climate positive' until enough climate pollution is removed to reverse climate damages.
- Governments, companies, investors, and communities that can move faster should do so, and their early action will drive deployment-related cost reductions associated with learning-by-doing and economies of scale that will benefit the entire world.

doi:10.1088/978-0-7503-4032-8ch1

1.1 Introduction

Climate change is a complex issue, but the basic outlines of the problem are well known. In her book *Under the Sky We Make* sustainability scientist and author Kimberly Nicholas summarizes the climate issue in a few simple phrases [4]:

- It's warming.
- It's us.
- We're sure.
- It's bad.
- We can fix it.

Michael E Webber, a professor of energy resources at the University of Texas at Austin, rightly adds: 'But we'd better hurry'[1].

This chapter explains in a compact way why aggressive climate action is urgent, using Nicholas's framing (with Webber's addition) to structure the discussion of climate solutions. We provide a more elaborate version of these arguments in appendix A, with lots more graphs, which you can use as a resource when challenged on some of these key points by purveyors of misinformation. However, given the overwhelming scientific evidence pointing to the causes and magnitude of our climate crisis, much of the honest debate has shifted to *how* we should solve our climate challenges and *who* should be responsible for doing the solving. With that said, it's important to understand the basics of these core climate truths, as these facts underpin the urgency of action.

1.2 It's warming

Scientists have used many methods to estimate global average temperatures. Reliable and comprehensive direct measurements began around 1850, but proxy methods, using tree rings, ratios of isotopes, and other techniques, can yield estimates going back thousands or even millions of years. Of course, the further back we look, the more complicated it is to do accurate assessments, but even the proxy estimates are based on well-established principles of physical science.

Figure 1.1 shows a summary of the instrumental temperature record since 1850, created by researchers at the University of Oxford [5] and continually updated. These data combine four well accepted time-series of temperatures into what they call a **climate change index**. Such temperature data are expressed relative to a base year (or an average over a set of years).

This graph is expressed as a change in temperature relative to a 1850–1900 baseline, which is one way to characterize what we call a **pre-industrial baseline**. The industrial revolution, which was the beginning of large-scale exploitation of fossil fuels, started in the late 1700s, but didn't really gather steam until the late 1800s. That fact combined with widespread instrumental temperature data only

---

[1] https://twitter.com/MichaelEWebber/status/1335950979334893572?s=20

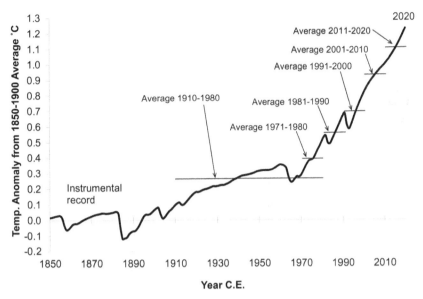

Figure 1.1. Global average temperatures, 1850–2020. Source: Based on data through 2020 from https://globalwarmingindex.org, first assessed and presented in [5].

extending back to about 1850 have led scientists to use the 1850–1900 baseline in the most recent research.

Temperature changes in the 1800s were mostly driven by volcanoes and other natural forces. The effect of Krakatoa in 1883 was particularly dramatic, with volcanic ash cooling the Earth by more than 0.1 °C over a year or two. Temperatures in the 1900s were also affected by such events but increased by more than 1 °C by 2020 due mainly to increasing concentrations of greenhouse gases during this period (see below). Each decade since the 1970s has been hotter than the last, and almost all of the hottest years ever recorded have occurred in the past two decades[2].

Warming since pre-industrial times has pushed the Earth out of the comfortable and stable temperature range in which human civilization developed [6, 7]. The *rate* of temperature change in the last century (as well as the rate of change in the underlying drivers of temperature change) is also much more rapid than humanity and the Earth have experienced in thousands of years, which is another reason for concern, as we discuss below.

This warming is reflected in other indicators. Northern Hemisphere summer sea-ice extent has reached historically low levels in recent years, while Antarctic sea ice saw significant declines in the early to mid-twentieth century [8] with Antarctic sea-ice extent falling below long-term averages from 2015 onwards[3]. Sea levels keep increasing as land-based glaciers melt [9] and a warmer ocean expands [10, 11].

---

[2] https://climate.copernicus.eu/copernicus-globally-seven-hottest-years-record-were-last-seven

Sea level rise has also been accelerating in recent years [12, 13], at the same time as global average upper ocean heat content has increased rapidly [14]. Finally, satellite measurements of Earth's energy balance have confirmed that greenhouse gases are trapping heat, just as scientists expected [15].

It's important to understand that **global warming** does *not* mean that everywhere is breaking temperature records all the time. Natural cycles and fluctuations still exist, and places can still experience record cold temperatures on a given day. Many places are experiencing extremes in both directions from climate change, particularly the **weakening of the jet stream** in the Northern Hemisphere linked to melting Arctic sea ice [16], which can allow hot air to move further north and cold air to move farther south (contributing to localized **polar vortex** cold snaps). However, despite localized variability, average annual and daily high temperatures are increasing almost everywhere we look.

We have even longer-term data for historical temperatures going back millions of years, as shown in Burke *et al* [17]. The uncertainties grow as we move back in time, but it is clear Earth was a lot cooler (−3 °C to −5 °C from our 1850 to 1900 baseline) for most of the 300 000 years before the present, with kilometers-deep ice sheets covering much of Earth's land masses.

Temperatures about 3 million years ago were about at about current levels, and before that (50–60 million years ago) temperatures were much hotter (10 °C to 15 °C above 1850–1900). Sea levels in that warmer period were more than one hundred meters higher than they are today [18].

Multiple independent lines of evidence are consistent with a rapidly warming Earth. The world's foremost climate science authority, the **United Nations Intergovernmental Panel on Climate Change** (IPCC) concluded in 2021 [19] that 'global surface temperature has increased faster since 1970 than in any other 50 year period over at least the last 2000 years'.

1.3 It's us

We know why average temperature have increased over the past century: emissions of **greenhouse gases** (GHGs) related to human activity have increased substantially since pre-industrial times, leading to increasing **concentrations** of these gases in the atmosphere. GHGs such as **carbon dioxide** (CO_2) trap energy from the Sun that otherwise would radiate to space, acting like a blanket that has been getting thicker for more than two centuries. To first approximation, warming to date is 100% driven by human activities.

Global temperature changes have historically corresponded very closely with atmospheric GHG concentrations [19]. The basic science behind how greenhouse gases warm the Earth through the **greenhouse effect** has been understood since the 1800s, and pre-industrial levels of atmospheric GHGs have been instrumental in keeping Earth warm enough for life to evolve. But while GHG levels have naturally

---

[3] https://nsidc.org/data/seaice_index/

fluctuated over millennia, the rapid rise in greenhouse gases since 1900, driven by human activities, is unprecedented in Earth's history.

Figure 1.2 shows man-made emissions of all greenhouse gases from 1850 to 2020, which have increased rapidly and exponentially. There are four major categories of greenhouse gases: carbon dioxide (CO_2), methane (CH_4), nitrous oxide (N_2O), and other gases (mostly what are called 'F-gases' that contain fluorine). The non-CO_2 gases are converted to what's called **CO_2 equivalent** (CO_2e), which is the equivalent amount of CO_2 that would result in the same warming effect as emissions of these other gases over a 100 year period. This technique allows us to approximate the total warming effect for all gases over time.

Total greenhouse gas equivalent emissions have increased at a rate of 2.2%/year from 1850 to 2020, with the emissions from fossil energy use growing at a 3.1%/year

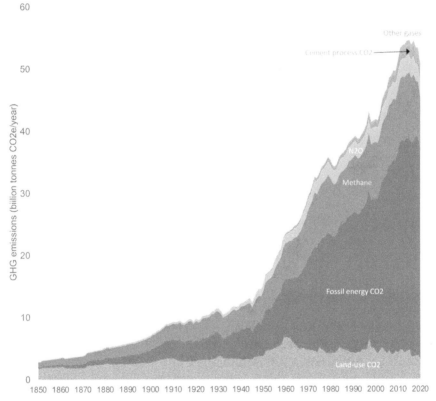

Figure 1.2. Greenhouse gas emissions expressed as CO_2 equivalent 1850 to 2020. Source: Fossil, cement, and net land-use CO_2 emissions 1850 to 1958 from CDIAC archives: https://cdiac.ess-dive.lbl.gov/trends/emis/ tre_glob_2013.html. Fossil, cement, and net land-use CO_2 emissions 1959 to 2020 from [20]. Fossil CO_2 emissions include combustion from flaring, solids, liquids, and gases. Methane, N_2O, and F-gas emissions from 1850 to 2019 from PIK, taken from https://www.climatewatchdata.org/data-explorer/historical-emissions. Global warming potential values for methane and N_2O adjusted to reflect AR6 100 year values in table 2.1 (PIK data use AR4 values as described in [21]). Emissions of methane, N_2O, and F-gases for 2020 estimated assuming emissions stay at 2019 levels (just like they did during the 2009 recession).

rate over that period. There have been periods of modest declines, but the overall upward trend has been inexorable.

The most important warming gas is carbon dioxide, which comes mainly from combustion of fossil fuels for energy, but also from deforestation and production processes for cement, steel, aluminum, and other materials. The burning of fossil fuels and land-use changes are the main sources adding CO_2 to the biosphere. The ocean and land have historically taken up about half of these emissions, but what remains goes into the atmosphere and stays there for centuries. As these emissions continue over time, the concentrations of CO_2 in the atmosphere go up, **climate feedback loops** (such as wildfires and melting permafrost) can further accelerate GHG emissions and associated warming.

Scientists have different methods to assess past concentrations of trace gases over time. One way is to drill for ice cores and extract and analyze air samples from bubbles in the ice. For more recent times (since 1959) we have detailed direct measurements of atmospheric concentrations from the observatory on Mauna Loa in Hawaii and other locations.

Over the last 800 000 years, Earth's atmosphere never held more than about 300 **parts per million** of CO_2 [22] corresponding to significantly cooler temperatures than in the Holocene (the twelve thousand year period before recent rapid human-caused warming) as shown in Burke *et al* [17]. In 2020 we hit 414 parts per million, and in May of 2022 we hit 421 parts per million. The increase over historical levels has occurred mostly in the span of about one and a half centuries. When it comes to CO_2 concentrations, we've moved rapidly into uncharted territory [23], driven by the increasing CO_2 emissions shown in figure 1.2.

Total CO_2 equivalent concentrations (including all warming agents) reached just over 500 parts per million (ppm) in 2020[4]. For comparison, concentrations in 1850 were about 305 ppm. That means current concentrations are now about 1.6 times pre-industrial levels.

The IPCC in 2021 tallied the main drivers of warming to the 2010 to 2019 period, relative to the 1850 to 1900 period [19]. Those results are shown in figure 1.3. Higher positive numbers mean more warming effect, and negative numbers mean that factor causes cooling, mainly by reflecting sunlight back to space rather than trapping it as heat.

The most important driver of increases in global surface temperatures since pre-industrial times has been increasing concentrations of carbon dioxide, mostly from combustion of fossil fuels, with a significant additional contribution from changes in land-use patterns (such as deforestation). Next is methane, driven by land-use changes, energy production, and agriculture, followed by non-methane volatile organic compounds and carbon monoxide, halogenated gases (such as the chlorofluorocarbons being phased out under the Montreal Protocol for ozone depleting chemicals, plus some others), nitrous oxides (also mainly from agriculture), black carbon, and airplane contrails. The cooling agents are sulfur dioxide

---

[4] https://gml.noaa.gov/aggi/aggi.html

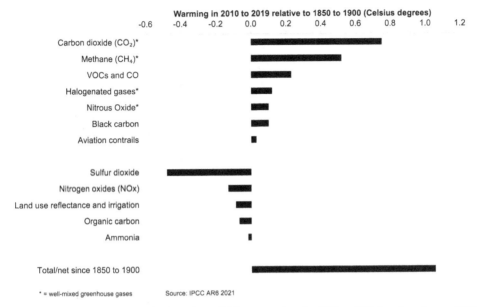

Figure 1.3. Contributors to warming in 2010 to 2019 compared to the 1850 to 1900 baseline. Source: ICPP Working Group I, Summary for Policy Makers, Sixth Assessment Report [24].

aerosols (small particles that reflect sunlight), nitrogen oxides, land-use reflectance and irrigation, organic carbon, and ammonia.

The cooling effects just about cancel out the effect of other factors than CO_2 and half of the warming associated with methane, but that doesn't mean that we can ignore the other warming agents. As we start to phase out fossil fuels, the cooling effects from the aerosols will also be reduced, so we'll need to compensate by also rapidly minimizing the shorter-lived warming agents such as methane, some fluorinated gases, and black carbon.

1.4 We're sure

Uncertainties always exist in science, but when we confirm scientific findings by multiple, independent lines of evidence, we call them 'facts'. There may still be uncertainty about some details, but the preponderance of the evidence points towards these facts accurately describing how the physical world operates. Every credible scientific organization now agrees that burning fossil fuels and other human activities are almost entirely responsible for current climate change, based on decades of research from thousands of scientists. Human-caused climate change is now as much of a settled scientific fact as gravity.

The US National Academy of Sciences [25], which is not known for its wild speculation, concluded in 2010:

A strong, credible body of scientific evidence shows that climate change is occurring, is caused largely by human activities, and poses significant risks for a broad range of human and natural systems....

Some scientific conclusions or theories have been so thoroughly examined and tested, and supported by so many independent observations and results, that their likelihood of subsequently being found to be wrong is vanishingly small. Such conclusions and theories are then regarded as settled facts. This is the case for the conclusions that the Earth system is warming and that much of this warming is very likely due to human activities.

The academies of science for 80 other countries (including China, India, Russia, Germany, Japan, Brazil, and the UK) have released comparable statements about the science of climate [26]. The IPCC, the global scientific body charged with investigating this issue, stated with uncharacteristic bluntness in 2007, that 'warming of the climate system is unequivocal, as is now evident from observations of increases in global average air and ocean temperatures, widespread melting of snow and ice and rising global average sea level' [27]. IPCC reports in 2013 [28] and 2021 [19] were even more emphatic, as was the World Meteorological Association in 2021 [29].

That greenhouse gases warm the Earth is a finding based on some of the most well-established principles in physical science, as well as extensive measurements. The largest historical source of warming is carbon dioxide, and we know for a fact that the increase in carbon dioxide concentrations after the industrial revolution (particularly after the mid-1800s) was fueled predominantly by human activity.

One reason that we know that fossil fuels and land-use changes are the cause of the measured increases in CO_2 concentrations is because the total carbon emitted since the dawn of the industrial age is about twice as large as the total amount that remains in the atmosphere nowadays (the other half was absorbed by the oceans and land-based biota), and there are no other known sources of carbon that could account for such an increase in the atmosphere's CO_2 content.

In addition, scientists can measure the prevalence of different isotopes of carbon in the atmosphere to understand the sources of CO_2. Carbon from recent biological activity (such as deforestation) contains a different mix of carbon isotopes than carbon from fossil fuels and other geological sources, and scientists have measured changes in the concentrations of those isotopes in the atmosphere that are consistent with the additional carbon coming from human-linked sources [30]. So it's virtually certain that humans are the cause of elevated CO_2, as well as causing similar increases in the past two centuries in concentrations of methane, nitrous oxides, and other GHGs [19].

These changes are warming the planet, as shown in figures 1.1 and 1.3, and as predicted with surprising accuracy as early as 1975 in *Science* [31] and in 1982 by scientists at Exxon [32]. The best current estimates are that global average surface temperatures have increased 1.1 °C–1.2 °C since 1900.

The physical science results supporting these conclusions go back at least a century to the Swedish scientist Svante Arrhenius, who calculated in 1896 the first climate sensitivity of 4 °C to 5 °C for a doubling of carbon dioxide concentrations. Arrhenius's analysis was supported by earlier measurements made by Eunice Foote and John Tyndall that demonstrated the heat trapping abilities of CO_2 [33]. The idea that greenhouse gases could warm the Earth is not a new one, and in fact the first informed speculation about this topic was by the mathematician Joseph Fourier in the 1820s [34].

These concerns are also validated by actual measurements of the climate system that corroborate theoretical predictions [35]. Satellite measurements show, for example, that as greenhouse gases have built up in the atmosphere, the heat emitted from the Earth has been declining at wavelengths that exactly correspond to those absorbed by various GHGs [36]. Similar measurements show that thermal radiation back to Earth's surface from the atmosphere, which we'd expect to increase if greenhouse gases trap heat, has been increasing exactly as we thought it would [37]. And the amount of heat stored in the oceans over the past few decades has been rising rapidly, which is consistent with a warming planet [24, 38, 39][5].

We can also examine data on some key indicators of warming, which by most accounts are changing at rates equaling or exceeding our worst-case predictions of just a few years ago [19, 40, 41]. These measurements are one of the main reasons why scientists are so alarmed about humanity's effect on the planet's temperature.

In summarizing these and other measurements, the IPCC [27] concluded in 2007 that 'observational evidence from all continents and most oceans shows that many natural systems are being affected by regional climate changes, particularly temperature increases'. By 2021, the IPCC [24] was even more explicit: 'Human-induced climate change is already affecting many weather and climate extremes in every region across the globe. Evidence of observed changes in extremes such as heat waves, heavy precipitation, droughts, and tropical cyclones, and, in particular, their attribution to human influence, has strengthened since AR5 [in 2013]'.

The observations cited above are powerful evidence for a warming world, but they are not the only ones [19]. Glaciers are melting [9, 11], wildfires are more frequent and more destructive [42, 43], agricultural growing seasons are changing [44], agricultural productivity is down [45], wildlife are altering their long-established patterns [46], and billions of people are (or will soon be) experiencing new heat extremes [47, 48]. Something's clearly happening to the climate, and these indicators are consistent with what the last one and a half centuries of science has been saying about this problem all along.

1.5 It's bad

If we continue on our current trajectory, we'll push Earth's climate into uncharted territory [49]. Figure 1.4 shows the result from one widely cited 'current trends continued' case (also known as SSP-2) created around 2015 or so [50]. For comparison, we also plot the instrumental record from figure 1.1 and the current best estimates for temperatures over the past twelve thousand years, from [6, 7].

---

[5] Also see https://www.climate.gov/news-features/understanding-climate/climate-change-ocean-heat-content.

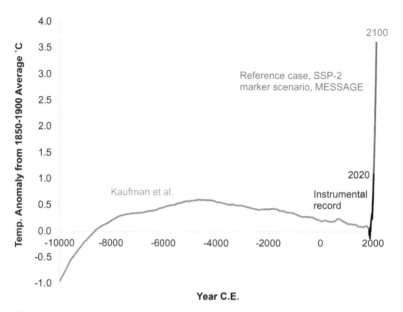

Figure 1.4. Historical temperatures contrasted with the reference-case projection from a prominent 2017 study. Source: Kaufman *et al* [6] for the past 12 000 years [6], https://globalwarmingindex.org for the instrumental record [5], and Fricko *et al* [50] for the reference-case, SSP-2 marker scenario [59].

That emissions trajectory for greenhouse gas emissions (and thus, global temperatures) results in an increase of about 3.6 °C above pre-industrial levels by 2100. This result is typical for assessments of a 'current trends continued' path circa 2015, which fall in the range of 3 °C to 4 °C above pre-industrial times. In the past few years, with significant climate action in some major countries, the 'current trends continued' case has improved even more [51–54] to more like 2.5 °C to 3.5 °C.[6]

In 2012 when one of us (Koomey) wrote a book assessing the options for reducing emissions [55], the current path at that time implied an increase of about 5 °C by 2100. Since 2012 we've made great progress in bringing down the cost of clean technology, we've shut down many coal plants, we've implemented and strengthened policies to reduce emissions, and we've realized that some of the more dire projections circa 2010 were based on overestimates of exploitable reserves of coal [56]. Of course, there is great uncertainty in any projection into the future, but we need a benchmark against which to measure progress, and 2.5 °C to 3.5 °C by 2100 is as good an estimate as any.

That improvement is small comfort, however. Even our current trajectory would be a disaster for the Earth and most other species [49, 57]. It would raise Earth's average temperature to a level not seen for about four million years [17], and to do so with unprecedented speed.

The most important direct effect of increasing temperatures is stress on natural and human systems [49]. Greenhouse gases keep more energy in the climate system,

---

[6] Also see https://thebreakthrough.org/issues/energy/3c-world

and that energy must go somewhere. Where it goes is into extreme rainfall and temperature events, which will become increasingly difficult or impossible for humans to manage, and ever more devastating for natural systems (whose ability to adapt is even more limited) [58].

Figure 1.5 illustrates how a warming climate 'loads the dice' and makes extreme temperature events more likely and the extremes more extreme, just as more moisture in the air makes high precipitation events more likely (and oddly enough, makes droughts in some places more likely and more intense as well) [24, 48, 60–64]. It also makes wildfire much more likely [43].

This concern is not a theoretical one. In a NASA report published in 2012, James Hansen and his colleagues examined distributions of measured temperatures for the period 1951 to 1980, comparing them to those from the 1980s, the 1990s, and the 2000s [65]. In each succeeding decade, the shifting of the distribution to towards higher temperatures became more pronounced as the climate warmed.

As we showed in figure 1.4, the current global environment is accustomed to a relatively narrow temperature range, one that has prevailed for thousands of years. Ecosystems can sometimes migrate slowly (over many millennia, not decades or centuries), but that migration is limited by geography and other constraints. For example, a forest biome can gradually move up the mountainside as the climate warms (soil and geography permitting), but once it reaches the peak there's nowhere else to go, and extinction is the result. On our current path, thousands of plant and animal species that have existed for eons will be driven to extinction in the span of a century or so [66–68].

Humans and their support systems are also vulnerable to a warming climate. Heat waves will become ever more frequent [69], and people in locations without

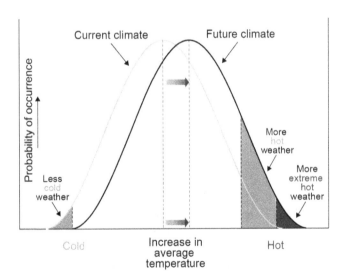

Figure 1.5. Increasing temperatures load the dice and make extreme temperatures more likely (and the extremes more extreme). Source: Adapted from a graph made by the University of Arizona, Southwest Climate Change Network, http://www.southwestclimatechange.org. Reproduced with permission from [55].

air conditioning will either add it (which will worsen climate change), suffer, or even die (as tens of thousands of Europeans did during the heat wave of 2003). Our wastewater treatment and water supply systems are designed to handle current conditions but will be difficult and expensive to adapt to a warming world's rapidly rising sea level and increasingly intense rainstorms. Wildfires will become more frequent and more intense [43]. Pollen and its associated respiratory effects will worsen [70]. Climate change will also increase risks of cross-species viral transmission [71, 72].

With even small increases in sea level, low lying coastal areas will become increasingly vulnerable to storm surges, putting millions of lives at risk, particularly in the developing world. Those areas will also suffer from increased saltwater intrusion into groundwater supplies. A 'current trends continued' path implies more than 0.5 m rise in sea level by 2100 [24, 73–75], which would represent significant challenges to human society, and sea level rise will continue for centuries, barring substantial carbon removal from the atmosphere.

There are also indirect effects. One of the most important is an increase in the acidity of the oceans, caused by more dissolved CO_2 (which creates carbonic acid). This development will make life increasingly difficult for many types of aquatic life, with rates of acidification proceeding more rapidly than at any time in the past 65 million years [76]. The acidification effect (plus the increase in ocean temperatures) means that coral reefs will likely be a thing of the past by the end of the twenty-first century, and will pose increasing challenges to marine life of all types [77, 78]. It also is one reason why schemes such as those proposed to inject particles into the atmosphere to cool the Earth are ultimately chimerical—as long as more CO_2 dissolves into the oceans, the acidification effect will intensify, and just reflecting more sunlight won't fix it.

Remember also that the climate sensitivity measures the *average* temperature change for a doubling of greenhouse gas concentrations. Changes at the Earth's poles have been [79] and will be much larger (that's just how the system works). The most likely case for climate sensitivity combined with the reference-case emissions forecast would ultimately lead to an ice-free planet Earth and sea level rises much bigger than even recent projections. It also means that large releases of carbon trapped in the permafrost and in methane hydrates beneath the ocean floor are much more likely, and that would amplify the warming effect.

Perhaps the most worrying aspect of climate change is the unknowable but non-zero probability that pushing the Earth's climate out of its recent equilibrium might lead to 'tipping points', discontinuous change, and catastrophic disruptions of weather and climate [80–82]. One example that has preoccupied scientists for decades is the possibility that melting ice could disrupt the Gulf Stream in the North Atlantic [83], and there has been evidence of a weakening of those currents in recent decades [84–86]. The rapid release of carbon and methane from melting permafrost and methane from warming oceans are two more. The probability and consequences from catastrophic events are inherently *unknowable*, and that reality disrupts the standard benefit–cost model for assessing the economics of climate mitigation [82, 87–90].

1.6 We can fix it (but we'd better hurry)

Let's start by defining what we mean by 'fixing' climate change. Solving climate change starts with stabilizing global surface temperatures at the lowest possible level above pre-industrial times, and that means *reducing emissions to zero as soon as possible* [91].

Many climate action plans now focus on getting climate pollution to **net zero**, the point where GHG emissions are reduced far enough that any remaining remissions are counterbalanced by GHG emission removals from natural and engineered systems [92]. Net zero by 2050 (or sooner) targets have now been adopted by many governments, as well as many of the world's largest companies and investors, catalyzed by goals of the United Nation's 2015 **Paris Agreement**.

Unfortunately, net zero is not enough. Even if Earth gets to net zero GHG emissions tomorrow, without additional action we're still locked into at decades of additional warming, and a new climate equilibrium that leaves the world worse off than it is today. Droughts, floods, heat waves, wildfires, rising seas, crop failures, and other climate-linked disasters will continue to get worse. Many more lives and livelihoods will be lost, and some once thriving places may become unlivable.

To truly solve climate change, we must view net zero as a transition point rather than an end goal. We should aim instead to get our planet to net **climate positive**, sometimes also called **carbon negative**, where there is a net removal of carbon from the atmosphere every year. Our collective goal should be to make Earth climate positive until we have returned the climate to a state that's as close as possible to the pre-industrial range for which human civilization and the ecosystems around us have evolved. We should also aim to repair the damages caused by climate change as much as we can and compensate those who have suffered when damages can't be repaired. Fixing climate change requires our best efforts to not just stop the bleeding, but also heal the wounds.

This doesn't mean that we shouldn't aim to reach (and surpass) net zero as soon as possible, but net zero by 2050 is not soon enough, as we discuss below. Because of the long residence time of the most important greenhouse gases in the atmosphere, warming is to first approximation proportional to *cumulative* emissions [93, 94]. That means every molecule emitted matters, and that if we stop emitting greenhouse gases, warming will eventually stop [95, 96]. This makes the climate problem different from other types of air pollution, which generally stay in the atmosphere for a much shorter time than do emissions of carbon dioxide, nitrous oxide, and many of the other gases.

The importance of cumulative emissions means we have no time to lose in reducing our emissions, a fact that is not widely enough appreciated. We've already dithered for more than three decades as emissions kept increasing, and every day we delay getting to climate positive makes the situation worse [91].

The good news is that we already have almost all the climate technologies we need, and these solution technologies are improving every day. The same technologies that can get us to net zero can propel us beyond to climate positive and a regenerative climate future. The biggest climate challenges are sociological, political, and economic rather than technological. Society has plenty of money to pay for the transition but shifting this money out of climate pollution and into solutions requires navigating powerful entrenched interests, perverse incentives, bureaucratic barriers, and outdated ways of thinking. Solving climate change will also require much better communications about our climate impacts and solutions, engaging all levels of society in shifting to a climate-positive economy. We can do it, but we need to decide to do it. We'll get into specific recommendations in the following chapters.

1.6.1 What is a warming limit?

The warming limit approach has its origins in the realization that stabilizing the climate at a certain temperature to minimize climate risks (e.g. a warming limit of 1.5 °C or 2 °C above pre-industrial times) implies a particular emissions budget, which represents the total cumulative greenhouse gas emissions compatible with that temperature goal [97]. That budget also implies a set of emissions pathways that are well defined and tightly constrained (particularly now that we've squandered the past three decades by not reducing emissions). This risk-minimization approach, which can also be described as 'working toward a goal', also involves assessing the cost effectiveness of different paths for meeting the normatively determined target [55].

A warming limit is more than just a number (or a goal to be agreed on in international negotiations). It embodies a way of thinking about the climate problem that yields real insights [98]. A warming limit is also a value choice that is informed by science. It should not be presented as solely a scientific 'finding', but as a value judgment that reflects our assessment of societal risks and our preferences for addressing them.

The warming limit approach was first suggested for externalities more generally in [99] and was explored (then quickly dismissed) by the climate economist William Nordhaus [100, 101]. It had its first fully developed incarnation in 1989 in [1] (which was subsequently republished in 1992 [2]). It was developed further in [102] and [103], and served as the basis for the International Energy Agency's analysis of climate options in 2010, 2011, 2012, and 2020 [104–107].

This way of thinking was developed as a counterpoint to the prevailing 'benefit–cost' approach favored by the economics community, and it has many advantages [98]. It encapsulates our knowledge from the latest climate models on how cumulative emissions affect global temperatures, placing the focus squarely on how to stabilize those temperatures. It places the most important value judgment up-front, embodied in the normatively determined warming limit, instead of burying key value judgments in economic model parameters or in ostensibly scientifically chosen concepts such as the discount rate. It gives clear guidance for the rate of emissions reductions required to meet the chosen warming limit, thus allowing us to determine if we're 'on track' for meeting the temperature goal and allowing us to adjust course if we're not hitting those near-term guideposts.

The warming limit approach also allows us to estimate the costs of delaying action or excluding certain mitigation options and provides an analytical basis for discussions about equitably allocating the emissions budget. Finally, instead of pretending that we can calculate an 'optimal' technology path based on guesses at mitigation and damage cost curves decades hence, it relegates economic analysis to the important but less grandiose role of comparing the cost effectiveness of currently available options for meeting near-term emissions goals [98].

The warming limit approach shows that delaying action is costly, required emissions reductions are rapid, and most proved reserves of fossil fuels will need to stay in the ground, and large amounts of carbon will need to be removed from the atmosphere if we're to stabilize the climate and repair the damage. These ideas are familiar to some, but many still don't realize that they follow directly from the warming limit framing:

- Delaying emissions reductions forecloses options and makes achieving climate stabilization much more difficult [108]. 'Wait and see' for the climate problem is foolish and irresponsible, which is obvious when considering cumulative emissions under a warming limit. The more fossil infrastructure we build now, the faster we'll have to reduce emissions later. If energy technologies changed as fast as computers there could be justification for 'wait and see' in some circumstances, but they don't, so it's a moot point.
- Absolute global emissions will need to turn down immediately and approach climate-positive territory in the next few decades [109] if we're to have a good chance to keep global temperatures 'well below 2 °C', as many in the scientific community advocate [110]. The emissions pathways given the current carbon budgets are tightly constrained. Even if the climate sensitivity is at the lowest end of the range included in IPCC reports (1.5 °C), that only buys us another decade in the time of emissions peak [111], which indicates that the findings on emissions pathways are robust, even in the face of uncertainties in climate sensitivity.
- The rate of emissions reductions, which is a number that can be measured, is one way to assess whether the world is on track to meet the requirements of a particular warming limit. We know what we need to be doing to succeed, and if we don't meet the tight time constraints imposed by that cumulative emissions budget in one year, we need to do more the next year, and the next, and the next. It's a way of holding policy makers' proverbial feet to the fire.
- The concept of 'stranded fossil fuel assets' that can't be burned, popularized by Bill McKibben [112] and Al Gore [113], follows directly from the warming limit framing. In fact, Krause *et al*'s 1989 book, *Energy Policy in the Greenhouse* [1] had a chapter titled 'How much fossil fuel can still be burned?', so the idea of stranded assets is not a new insight (but it is a profound one).

1.6.2 An evolution in thinking

The warming limit approach continues to be helpful in building the case for urgent action on climate [114–116] but its original policy usefulness rested on the over-arching assumption that there was still time left to address the crisis. As time has passed and climate damages continue to mount it has become clear to us that the world is running out of time [117], and given uncertainties in estimating the carbon budget [118, 119], it is in our judgment better to focus on concrete emissions reductions goals instead of worrying too much about how much carbon budget is left.

The IPCC completed its report on scenarios based on a 1.5 °C warming limit in 2018 [110]. That study found, after reviewing many studies that meet the 1.5 °C warming limit, that cutting global greenhouse gas emissions at least in half from 2020 to 2030 (with subsequent reductions to follow at a similar pace) is a critically important milestone for success. This rough rule of thumb, which was enshrined as a 'carbon law' by Johan Rockstrom and his colleagues in a seminal article in *Science* in 2017 [109], describes the minimum level of effort needed to hit a 1.5 °C warming limit. Rockstrom's 'law' implies halving of absolute emissions in each decade starting in 2020, reaching close to net zero emissions by 2050.

When talking about rapid emissions reductions it is customary to refer to rates of change as a percentage of base year emissions, instead of often-used exponential rates of decline [55]. This convention is used because rates of decline reach astronomical levels in percentage terms as emissions approach zero, and percentages of a base year maintain an intuitive physical meaning throughout the analysis period. With this convention, a 5%/year decline in emissions relative to 2020 would result in a halving of annual emissions by 2030 (5%/year × 10 years).

A strong case can be made for an even more aggressive goal than implied by Rockstrom *et al* achieving net zero global emissions by 2040. That goal implies a rate of emissions reduction of 5%/year relative to 2020 continuing to 2040. We explain the rationale for such an aggressive goal in the next chapter.

When evaluating long-term goals such as these, it's important not to place too much emphasis on precise numbers, particularly many decades hence. It is valuable, however, to look at broader lessons from such scenarios, and the most important lesson is the rapid rate of change embodied in all scenarios that put a 1.5 °C warming limit in reach. We'll need to build zero emissions energy and industrial process technologies at high rates and retire existing high-emissions capital on a rapid schedule.

Achieving such speedy emissions reductions will require unprecedented changes in how the global economy generates value. In coming decades, many processes in our economy will need to be re-evaluated and re-designed from scratch to minimize or eliminate emissions.

1.6.3 Can it be done?

The question of whether a modern society can achieve such rapid reductions is one we'll explore in the rest of this book, but it cannot be answered precisely by modeling or analysis. We'll only actually know how far we can reduce emissions once we start trying to do so in earnest, and we really haven't started yet.

Every tenth of a degree matters. If we overshoot 1.5 °C, so be it, but 1.6 °C is much better than 1.7 °C, which is much better than 1.8 °C, and by getting to a climate-positive state we can start to bring temperatures back down and reduce damages from temporarily overshooting climate targets. Even if we are daunted by the challenge of keeping warming below 1.5 °C, we need to try, and we need to act as quickly as we can.

For climate change, 'moving in the right direction' isn't enough, as Solomon Goldstein-Rose points out in his excellent book *The 100% Solution*:

> There's a lot of rhetoric about 'moving in the right direction' on climate change. But because of its difference from other issues (impacts being caused not by each year's emission but by cumulative emissions until we start removing them from the atmosphere), there's not really such a thing as 'moving in the right direction.' Climate change impacts get exponentially worse until we solve the problem 100%. That's why it is so much scarier and more urgent than other problems...

> The idea of 'doing what we can' is dangerous when it comes to climate change because it implicitly accepts that the maximum viable action is less than the minimum needed action.

'Can it be done?' is therefore the wrong question, and worrying about feasibility is the wrong framing. The right question is 'How can we change society to do what is necessary?'

Nobody knows what's likely or even possible until we start down the path of aggressively reducing emissions by deploying technology, capital, communications, and institutional innovations at the requisite scale. If we choose to do so, many things will become possible that wouldn't be possible if we didn't.

Feasibility also depends on context, and on what we are willing to pay and prioritize to minimize risks. What if we finally decide (as we should) that it's a real emergency (like World War II)? In that case we'd make every effort to fix the problem, and what would be possible then is far beyond what we could imagine today.

It is therefore a mistake for analysts to impose an informal feasibility judgment when considering a problem such as this one, and instead we should aim for what we think is the best outcome from a risk-minimization perspective, and if we don't quite get there, then we'll have to deal with the consequences. But if we aim too low, we might miss possibilities that we'd otherwise be able to capture.

History shows that under the right conditions, societies and industries can move quickly. In the beginning of World War II, the US retooled much of its heavy industry over the span of about 6 months [120], and some other nations engineered similarly rapid change. We now have some technology advantages over industrial firms of that era [121], especially information technology, which is our 'ace in the hole' [122].

We also know that existing technology offers opportunities to reduce emissions substantially right now, and maximizing immediate emissions reductions is where we should be focused, not on worrying about whether we'll be able to get to zero emissions by 2040. Do the obvious things: shut down fossil fuel power plants (starting with coal), mandate electrification where possible, install much more wind, solar, and energy storage, and deploy all other existing emissions reduction technology at scale everywhere we can. Our choices now create our options later because deployment drives costs down, creating new opportunities for reducing emissions elsewhere.

1.6.4 Keeping carbon in the ground

Another way to think about how fast emissions need to come down is by comparing projected fossil fuel consumption to the latest estimates of proved reserves of fossil fuels. In geology parlance, proved reserves are those stocks of fossil fuels that are known to exist with high confidence and that can be extracted using current technology at current prices. Resources are fossil deposits known with less confidence and/or are not extractable using current technology and current prices.

The first bar in figure 1.6 shows proven fossil reserves for 2018 from [123]. They total about 900 billion tonnes of carbon (not carbon dioxide). The second bar shows emissions from a widely cited reference case [50] from 2015, which shows about 1300 billion tonnes of carbon combusted from 2020 to 2100. That result implies that a small fraction of the remaining resources (which are more than tenfold bigger than reserves) would need to be converted to reserves through exploration or using new technology to meet that demand.

The more important bar for our narrative, however, is the one for the low energy demand case [59], which is a scenario that would keep global temperatures from increasing no more than 1.5 °C from pre-industrial times. In that scenario, the world can burn less than one tenth of the fossil carbon implied in the reference case, and only one eighth of the proved reserves. That means we'll either need to keep almost 90% of global fossil reserves in the ground unburned or identify some way to sequester the carbon from burning it, which will be a heavy lift at the required scale.

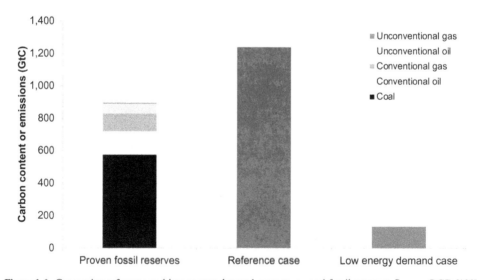

Figure 1.6. Comparing reference and low energy demand cases to proved fossil reserves. Source: BGR [123], Fricko *et al* [50], Grubler *et al* [59].

These exact numbers are dependent on some key assumptions in this scenario, but the main point is the same for all aggressive mitigation scenarios that keep the 1.5 °C warming limit in reach: *We'll need to keep a substantial fraction of proved fossil reserves in the ground or find another safe way to sequester that carbon.*

1.6.5 The economics of climate action

In the early years of studying the economics of climate, economists were concerned about the economic costs of moving too quickly [100, 101, 124–126]. As the problem became better understood, a different picture emerged, indicating that the benefits of at least modest climate action should significantly exceed costs from the societal perspective. Subsequent analyses showed that substantially reducing emissions would come at gross societal costs of *at most* a few percent of GDP in coming decades, resulting in the loss of roughly a single year's growth in GDP [127–134]. Further, these analyses showed that delaying action had serious downside risks, as discussed below.

Over time, economic assessments became more sophisticated and started to include important factors that the earliest simple models omitted—factors whose inclusion generally showed that achieving lower emissions would be cheaper, easier, and more beneficial for society than the earlier assessments indicated [135–139]. Models often omitted air quality and other health co-benefits, which are often big enough to fully offset the gross costs of reducing emissions, justifying significant climate action even before considering greenhouse gas externalities [140–145]. Many models omitted bottom-up analysis of market reforms and technology programs, which are important sources of negative net cost (i.e. societally profitable) emissions reduction options [129, 136, 137, 146–151].

Many **benefit–cost models** ignored climate damages for reference cases, assuming economic growth rates in those cases would be unaffected by unrestricted climate change [152]. Many also relied on flawed estimates of climate damages that vastly underestimate the benefits of reducing emissions [153–156]. Standard calculations of discount rates by Nordhaus and others also included a term for business-as-usual economic growth unencumbered by potential climate damages [157], and that assumption led to higher discount rates that make future benefits of climate action appear less valuable than in reality.

Virtually all models ignored the 'long-tail' risks of catastrophic climate change, which lay bare the weaknesses of the benefit–cost framing like no other issue [87, 158]. Benefit–cost models also often ignored the lessons from financial economics on pricing risk and uncertainty [159] and almost always ignored potential cost reductions associated with economies of scale, learning effects, spillovers, network externalities, and irreversibilities [138, 160–167], although sometimes these effects were studied in specific cases [162, 168]. They also almost always ignore people's asymmetrical treatment of losses versus gains (prospect theory), which again biased the results toward delaying mitigation [156].

Some models also omitted benefits from using carbon tax revenues to reduce inefficiencies in the tax system [169], failed to distinguish between endogenous and directed or induced technical change [166, 170, 171], and included only local or regional emissions trading and focused only on carbon dioxide rather than all greenhouse gases [135, 172]. In almost every case, including these factors in the analysis would have shown emissions reductions to be cheaper for society than in the original analysis.

Prematurely excluding cost-competitive mitigation options from analyses like these can raise the apparent cost of reducing emissions. For example, in many places excluding the option of extending the useful life of existing nuclear plants would make achieving emissions goals appear to be more expensive than they would be if this option was kept on the table. How this constraint nets out for a particular technology, policy, or scenario depends on the cost for each option.

Conversely, including options as 'backstop' technologies with optimistic costs and potentials, as has happened in the past, for example, in the case of biomass energy with carbon capture, can make achieving emissions reduction goals appear cheaper than they really would be in practice. On balance, however, the models historically have erred more on the side of making the costs of emissions reductions appear to be more expensive than they really are.

1.6.6 Climate action, equity, and justice

While the costs of climate change fall on all of us, a disproportionate share of those costs will continue fall on the poorest countries, the poorest people, and generations yet to be born [173–176]. The pursuit of equity and justice is central to action on climate, a fact that is uncomfortable for many economists. The economics field has traditionally focused on economic efficiency, not on disparities in income and power relationships, but these issues cannot be ignored in facing the climate challenge.

Climatologist Michael E Mann, writing in his book, *The New Climate War*, wrote:

> …social justice is intrinsic to climate action. Environmental crises, including climate change, disproportionately impact those with the least wealth, the fewest resources, and the least resilience. So simply acting on the climate crisis is acting to alleviate social injustice. It's another compelling reason to institute the systemic changes necessary to avert the further warming of our planet [177].

Some climate policies are more justice-enhancing than others. Unless policies are designed with existing inequities in mind, they risk exacerbating inequality and injustice, so it is not always true that 'acting on the climate crisis is acting to alleviate social injustice' [175, 176, 178]. What is true is that those with the fewest resources will be hit the hardest by climate change [175, 176], climate change is already increasing inequality [176, 179, 180], and society has an obligation to reduce those harms by rapidly reducing emissions and structuring climate policies to address (or at a minimum not exacerbate) existing inequities.

There is some empirical work supporting the differential effects of climate change and other environmental issues on less advantaged groups, for example [176, 181–183], but it's also an intuitively reasonable conclusion. Marginalized people have fewer resources to manage unexpected crises and often live in places susceptible to extreme weather events and environmental pollution. Structural racism exacerbates these inequalities [184, 185].

For example, when extreme temperatures hit, economically disadvantaged communities have less access to well-conditioned private or public spaces. When flooding or hurricanes hit, poorer families have less ability to move to safer ground, either because they are tied to low wage jobs for survival or don't have access to affordable transportation. Communities of color and low-income communities are exposed to more outdoor air pollution than less diverse and more affluent neighborhoods [186], and exposure to air pollution may also be related to increased mortality from COVID-19 and other infections [187]. Many poorer countries are also significantly affected by air and water pollution, and correctly accounting for co-benefits to greenhouse gas emissions reductions should feature prominently in international negotiations on climate targets and commitments [145]. Air pollution from fossil fuels kills millions of people every year [141, 188], and those deaths are borne disproportionately by poorer people.

Climate change is also deeply intertwined with intergenerational justice [69, 189–191]. Those likely to be most affected by a world with a rapidly changing climate are not yet born, while powerful economic interests make their preference for delayed action known loudly in public proceedings and news media around the world.

Another dimension to the question of justice and the climate is the disproportionate effects the wealthy have on emissions [192–194]. The primary beneficiaries of fossil fuel wealth have been rich countries, and rich people everywhere, but they've privatized benefits while socializing costs. That disparity makes it incumbent upon the wealthy to take the lead on climate action. They can afford it, they are historically more to blame for the climate problem than those people who are less well off, and their prominence in society gives them greater ability to influence the opinions of others [195, 196].

As discussed above, humanity's choices for response are threefold: mitigation, adaptation, and suffering. Our decisions on balancing climate mitigation with adaptation and suffering are at their core *moral choices*. How fast and how far we mitigate are not solely questions of economics and are inseparable from questions about equity and justice [173]. The faster and more deeply we reduce emissions and the more we center the correction of existing inequities in our solutions, the less adaptation and suffering we'll do, and the more rapidly justice will be served.

1.6.7 Speed trumps perfection in climate solutions

The seriousness of the climate problem has been obvious to informed observers for at least three decades, but our delay in facing it has virtually guaranteed that little about humanity's response will be optimal. The most important lesson is that we need to get started on rapid emissions reductions as soon as possible. There's no more time to waste.

The IPCC [49], writing in early 2022, emphasized the urgency of responding to the climate problem without delay:

The cumulative scientific evidence is unequivocal: climate change is a threat to human well-being and planetary health. Any further delay in concerted anticipatory global action on adaptation and mitigation will miss a brief and rapidly closing window of opportunity to secure a livable and sustainable future for all.

The IPCC assigned *Very High Confidence* to this statement, meaning there is little doubt that the statement is true, with multiple, independent, and consistent lines of evidence supporting it [197]. The quotation itself is scientist-speak for 'It's an emergency, so get cracking!' and 'It's warming, it's us, we're sure, it's bad, we can fix it (but we'd better hurry)'.

There are lessons for climate solutions from responding to other kinds of emergencies. Dr Michael Ryan, Executive Director of the World Health Organization, in talking about pandemic response in early 2020, said this:

Be fast. Have no regrets. You must be the first mover... In emergency response, if you need to be right before you move, you will never win... Speed trumps perfection...The greatest error is not to move. The greatest error is to be paralyzed by the fear of failure[7].

We conclude from this advice:

- Don't obsess about optimality, just move as quickly as you can.
- Don't obsess about feasibility, just move quickly as you can.
- Don't obsess about obstacles, just move quickly as you can.

For climate, just like for pandemic response, speed trumps perfection, and that's the attitude we need to take in addressing this problem now, because it's a real emergency [198, 199].

1.7 Chapter conclusions

We are in a climate emergency, but most people and institutions aren't acting like it. We need to treat climate like the crisis that it is. That means moving more quickly than in normal times, getting started on rapid emissions reductions immediately. There's no more time to waste.

Keeping global surface temperatures from exceeding 1.5 °C above pre-industrial times will not be easy, nor will drawing down GHGs into climate-positive territory. On the contrary, climate change is the biggest collective challenge modern humanity has ever faced.

We'll need to cut absolute global greenhouse emissions in half by 2030, reaching net zero emissions no later than 2040, then remove emissions with climate-positive processes for many decades thereafter. Industrialized nations, including China, should move even more quickly. Aggressive early action in these nations will drive

---

[7] https://twitter.com/drericding/status/1340997408503853058?s=10

the cost of zero emissions and climate-positive technologies down substantially, benefitting the entire world.

The vast majority of proved fossil fuel reserves will need to be kept in the ground to stabilize the climate. No fossil fuel companies' business plans currently reflect this reality [200], creating what Al Gore called a 'carbon asset bubble' [113]. When this bubble bursts, as it inevitably will, fossil investors will be left holding the bag.

For climate, just like for pandemic response, speed trumps perfection, and that's the attitude we need to take in addressing this problem now. There will always be social and environmental tradeoffs as we scale up climate solutions, but the direct benefits and co-benefits of scaling solutions generally far outweigh the costs, and we already have techniques for minimizing harm while maximizing benefits for nearly everyone. The rest of this book explores tools needed to speed up climate action and truly face the climate challenge.

I'm skeptical that a problem as complex as climate change can be solved by any single branch of science. Technological measures and regulations are important, but equally important is support for education, ecological training and ethics—a consciousness of the commonality of all living beings and an emphasis on shared responsibility.

—Vaclav Havel (*New York Times* 27 September 2007)

Further reading

Burke K D, Williams J W, Chandler M A, Haywood A M, Lunt D J and Otto-Bliesner B L 2018 Pliocene and Eocene provide best analogs for near-future climates *Proc. Natl Acad. Sci.* **115** 13288 https://doi.org/10.1073/pnas.1809600115. This article summarizes current knowledge about past climates going back millions of years.

Dessler A E 2022 *Introduction to Modern Climate Change* (Cambridge: Cambridge University Press). An excellent summary of the latest climate science.

Duane T, Koomey J, Belyeu K and Hausker K 2016 From risk to return: investing in a clean energy economy *Risky Business* 6 December http://riskybusiness.org/fromrisktoreturn/. A business-focused study of reducing emissions in the US energy sector.

Goldstein-Rose S 2020 *The 100% Solution: A Plan for Solving Climate Change* (New York: Melville House). A terrific introduction to the climate problem for non-technical audiences.

Grubler A *et al* 2018 A low energy demand scenario for meeting the 1.5 °C target and sustainable development goals without negative emission technologies *Nat. Energy* **3** 515–27 https://doi.org/10.1038/s41560-018-0172-6. An exemplary aggressive emissions reduction scenario focusing on changes in energy service demands and efficiency.

IPCC 2018 *Global Warming of 1.5°C. An IPCC Special Report on the Impacts of Global Warming of 1.5°C Above Pre-industrial Levels and Related Global Greenhouse Gas Emission Pathways, in the Context of Strengthening the Global Response to the Threat of Climate Change, Sustainable Development, and Efforts to Eradicate Poverty* (Geneva: IPCC) https://www.ipcc.ch/sr15. An interim IPCC report on aggressive emissions scenarios.

IPCC 2021 *Climate Change 2021: The Physical Science Basis. Contribution of Working Group I to the Sixth Assessment Report of the Intergovernmental Panel on Climate Change* ed V Masson-Delmotte *et al* (Cambridge: Cambridge University Press) https://www.ipcc.ch/report/sixth-assessment-report-working-group-i/. The latest IPCC report on the science of climate change.

IPCC 2022 *Climate Change 2022: Impacts, Adaptation and Vulnerability–The Working Group II contribution to the Sixth Assessment Report* (Cambridge: Cambridge University Press) https://www.ipcc.ch/report/sixth-assessment-report-working-group-ii/. The latest IPCC report on impacts of and adaptation to climate change.

IPCC 2022 *Climate Change 2022: Mitigation of Climate Change. Contribution of Working Group III to the Sixth Assessment Report of the Intergovernmental Panel on Climate Change DUE OUT MARCH 2022* (Cambridge: Cambridge University Press) https://www.ipcc.ch/report/sixth-assessment-report-working-group-3/. The latest IPCC report on mitigation of climate change.

Mann M E and Toles T 2016 *The Madhouse Effect: How Climate Change Denial is Threatening our Planet, Destroying our Politics, and Driving Us Crazy* (New York: Columbia University Press). An entertaining introduction to the climate problem from one of the world's top climate scientists and one of the world's top editorial cartoonists.

Mann M E 2021 *The New Climate War: The Fight to Take Back Our Planet* (New York: PublicAffairs) 12 January. The latest summary of the climate issue from a top climate scientist.

Rockstrom J, Gaffney O, Rogelj J, Meinshausen M, Nakicenovic N and Schellenhuber H J 2017 A roadmap for rapid decarbonization *Science* **355** 1269 https://doi.org/10.1126/science.aah3443. A widely cited article defining the 'carbon law' benchmark of halving emissions every decade.

Sovacool B K, Burke M, Baker L, Kumar Kotikalapudi C and Wlokas H 2017 New frontiers and conceptual frameworks for energy justice *Energy Policy* **105** 677–91 https://doi.org/10.1016/j.enpol.2017.03.005

References

[1] Krause F, Bach W and Koomey J 1989 *From Warming Fate to Warming Limit: Benchmarks to a Global Climate Convention* (El Cerrito, CA: International Project for Sustainable Energy Paths) http://mediafire.com/file/pzwrsyo1j89axzd/Warmingfatetowarminglimitbook.pdf

[2] Krause F, Bach W and Koomey J G 1992 *Energy Policy in the Greenhouse* (New York: Wiley)

[3] Stoddard I *et al* 2021 Three decades of climate mitigation: why haven't we bent the global emissions curve? *Annu. Rev. Environ. Resour.* **46** 653–89

[4] Nicholas K 2021 *Under The Sky We Make: How to be Human in a Warming World* (New York: Putnam)

[5] Haustein K, Allen M R, Forster P M, Otto F E L, Mitchell D M, Matthews H D and Frame D J 2017 A real-time global warming index *Sci. Rep.* **7** 15417

[6] Kaufman D, McKay N, Routson C, Erb M, Dätwyler C, Sommer P S, Heiri O and Davis B 2020 Holocene global mean surface temperature, a multi-method reconstruction approach *Sci. Data* **7** 201

[7] Kaufman D *et al* 2020 A global database of holocene paleotemperature records *Sci. Data* **7** 115

[8] Fogt R L, Sleinkofer A M, Raphael M N and Handcock M S 2022 A regime shift in seasonal total Antarctic sea ice extent in the twentieth century *Nat. Clim. Change* **12** 54–62

[9] Zemp M *et al* 2019 Global glacier mass changes and their contributions to sea-level rise from 1961 to 2016 *Nature* **568** 382–6

[10] Slater T, Hogg A E and Mottram R 2020 Ice-sheet losses track high-end sea-level rise projections *Nat. Clim. Change* **10** 879–81

[11] Slater T, Lawrence I R, Otosaka I N, Shepherd A, Gourmelen N, Jakob L, Tepes P, Gilbert L and Nienow P 2021 Review article: Earth's ice imbalance *Cryosphere* **15** 233–46

[12] Dangendorf S, Hay C, Calafat F M, Marcos M, Piecuch C G, Berk K and Jensen J 2019 Persistent acceleration in global sea-level rise since the 1960s *Nat. Clim. Change* **9** 705–10

[13] Sweet W V *et al* 2022 Global and regional sea level rise scenarios for the United States: updated mean projections and extreme water level probabilities along US coastlines *NOAA Technical Report* NOS 01 National Oceanic and Atmospheric Administration, National Ocean Service, Silver Spring, MD https://oceanservice.noaa.gov/hazards/sealevelrise/sea-levelrise-tech-report.html

[14] Lumpkin R *et al* 2020 Global oceans *Bull. Am. Meteorol. Soc.* **101** S129–84

[15] Kramer R J, He H, Soden B J, Oreopoulos L, Myhre G, Forster P M and Smith C J 2021 Observational evidence of increasing global radiative forcing *Geophys. Res. Lett.* **48** e2020GL091585

[16] McSweeney R 2020 Jet stream: is climate change causing more 'blocking' weather events? *Carbon Brief* 12 June https://carbonbrief.org/jet-stream-is-climate-change-causing-more-blocking-weather-events/

[17] Burke K D, Williams J W, Chandler M A, Haywood A M, Lunt D J and Otto-Bliesner B L 2018 Pliocene and eocene provide best analogs for near-future climates *Proc. Natl Acad. Sci.* **115** 13288

[18] Haq B U 2014 Cretaceous eustasy revisited *Global Planet. Change* **113** 44–58

[19] IPCC 2021 *Climate Change 2021: The Physical Science Basis. Contribution of Working Group I to the Sixth Assessment Report of the Intergovernmental Panel on Climate Change* ed V Masson-Delmotte *et al* (Cambridge: Cambridge University Press) https://ipcc.ch/report/sixth-assessment-report-working-group-i/

[20] Friedlingstein P *et al* 2022 Global carbon budget 2021 *Earth Syst. Sci. Data* **14** 1917–2005

[21] Gütschow J, Jeffery M L, Gieseke R, Gebel R, Stevens D, Krapp M and Rocha M 2016 The PRIMAP-hist national historical emissions time series *Earth Syst. Sci. Data* **8** 571–603

[22] Siegenthaler U *et al* 2005 Stable carbon cycle–climate relationship during the late Pleistocene *Science* **310** 1313–7

[23] Petit J R *et al* 1999 Climate and atmospheric history of the past 420,000 years from the Vostok ice core, Antarctica *Nature* **399** 429–36

[24] IPCC 2021 *Summary for Policy Makers. Climate Change 2021: The Physical Science Basis. Contribution of Working Group I to the Sixth Assessment Report of the Intergovernmental Panel on Climate Change* ed V Masson-Delmotte *et al* (Cambridge: Cambridge University Press) https://ipcc.ch/report/sixth-assessment-report-working-group-i/

[25] NAS 2010 *Advancing the Science of Climate Change* (Washington, DC: National Academy of Sciences) http://nap.edu/catalog.php?record_id=12782

[26] Cook J *et al* 2016 Consensus on consensus: a synthesis of consensus estimates on human-caused global warming *Environ. Res. Lett.* **11** 048002

[27] IPCC 2007 *Climate Change 2007: The Physical Science Basis – Contribution of Working Group I to the Fourth Assessment Report of the Intergovernmental Panel on Climate Change* ed S Solomon *et al* (Cambridge: Cambridge University Press) http://ipcc.ch/publications_and_data/publications_and_data_reports.shtml

[28] IPCC 2013 *Climate Change 2013: The Physical Science Basis. Contribution of Working Group I to the Fifth Assessment Report of the Intergovernmental Panel on Climate Change* ed T F Stocker *et al* (Cambridge: Cambridge University Press) http://climatechange2013.org

[29] WMO 2021 Climate change indicators and impacts worsened in 2020 *State of the Global Climate 2020* WMO-No. 1264 World Meteorological Association, Geneva https://public.wmo.int/en/media/press-release/climate-change-indicators-and-impacts-worsened-2020

[30] Basu S, Lehman S J, Miller J B, Andrews A E, Sweeney C, Gurney K R, Xu X, Southon J and Tans P P 2020 Estimating US fossil fuel CO_2 emissions from measurements of 14 °C in atmospheric CO_2 *Proc. Natl Acad. Sci.* **117** 13300

[31] Broecker W S 1975 Climatic change: are we on the brink of a pronounced global warming? *Science* **189** 460–3

[32] Exxon 1982 CO_2 Greenhouse Effect *Technical review* 82EAP 256 Exxon Corporation, Florham Park, NJ https://insideclimatenews.org/wp-content/uploads/2015/09/1982-Exxon-Primer-on-CO2-Greenhouse-Effect.pdf

[33] Jackson R 2020 Eunice Foote, John Tyndall and a question of priority *R. Soc. J. History Sci.* **74** 105–18

[34] Weart S R 2008 *The Discovery of Global Warming* (Cambridge, MA: Harvard University Press)

[35] Rahmstorf S, Cazenave A, Church J A, Hansen J E, Keeling R F, Parker D E and Somerville R C J 2007 Recent climate observations compared to projections *Science* **316** 709

[36] Harries J E, Brindley H E, Sagoo P J and Bantges R J 2001 Increases in greenhouse forcing inferred from the outgoing longwave radiation spectra of the Earth in 1970 and 1997 *Nature* **410** 355–7

[37] Philipona R, Durr B, Marty C, Ohmura A and Wild M 2004 Radiative forcing—measured at Earth's surface—corroborate the increasing greenhouse effect *Geophys. Res. Lett.* **31** L03202

[38] Murphy D M, Solomon S, Portmann R W, Rosenlof K H, Forster P M and Wong T 2009 An observationally based energy balance for the Earth since 1950 *J. Geophys. Res.* **114** D17107

[39] Domingues C M, Church J A, White N J, Gleckler P J, Wijffels S E, Barker P M and Dunn J R 2008 Improved estimates of upper-ocean warming and multi-decadal sea-level rise *Nature* **453** 1090–3

[40] Strong A, Levin K and Tirpak D 2011 Climate science: major new discoveries *Issue Brief* World Resources Institute, Washington, DC http://wri.org/publication/climate-science

[41] Xu Y, Ramanathan V and Victor D G 2018 Global warming will happen faster than we think *Nature* **564** 30–2

[42] Doerr S H and Santín C 2016 Global trends in wildfire and its impacts: perceptions versus realities in a changing world *Phil. Trans. R. Soc.* B **371** 20150345

[43] Burke M, Driscoll A, Heft-Neal S, Xue J, Burney J and Wara M 2021 The changing risk and burden of wildfire in the United States *Proc. Natl Acad. Sci.* **118** e2011048118

[44] Park T, Ganguly S, Tømmervik H, Euskirchen E S, Høgda K-A, Karlsen S R, Brovkin V, Nemani R R and Myneni R B 2016 Changes in growing season duration and productivity of northern vegetation inferred from long-term remote sensing data *Environ. Res. Lett.* **11** 084001

[45] Ortiz-Bobea A, Ault T R, Carrillo C M, Chambers R G and Lobell D B 2021 Anthropogenic climate change has slowed global agricultural productivity growth *Nat. Clim. Change* **11** 306–12

[46] Henson R 2011 *The Rough Guide to Climate Change* 3rd edn (London: Rough Guides)

[47] Xu C, Kohler T A, Lenton T M, Svenning J-C and Scheffer M 2020 Future of the human climate niche *Proc. Natl Acad. Sci.* **117** 11350–5

[48] Zeppetello L R V, Raftery A E and Battisti D S 2022 Probabilistic projections of increased heat stress driven by climate change *Commun. Earth Environ.* **3** 183

[49] IPCC 2022 *Climate Change 2022: Impacts, Adaptation and Vulnerability—The Working Group II Contribution to the Sixth Assessment Report* (Cambridge: Cambridge University Press) https://ipcc.ch/report/sixth-assessment-report-working-group-ii/

[50] Fricko O *et al* 2017 The marker quantification of the shared socioeconomic pathway 2: a middle-of-the-road scenario for the 21st century *Global Environ. Change* **42** 251–67

[51] Hausfather Z and Peters G P 2020 Emissions—the 'business as usual' story is misleading *Nature* **577** 618–20

[52] Hausfather Z and Forster P 2021 Analysis: Do Cop26 Promises Keep Global Warming Below 2 °C? *Carbon Brief* 10 November https://carbonbrief.org/analysis-do-cop26-promises-keep-global-warming-below-2c

[53] Sognnaes I *et al* 2021 A multi-model analysis of long-term emissions and warming implications of current mitigation effortsNat. Clim. Change **11** 1055–62

[54] Moore F C, Lacasse K, Mach K J, Shin Y A, Gross L J and Beckage B 2022 Determinants of emissions pathways in the coupled climate–social system *Nature* **603** 103–11

[55] Koomey J G 2012 *Cold Cash, Cool Climate: Science-Based Advice for Ecological Entrepreneurs* (El Dorado Hills, CA: Analytics)

[56] Ritchie J and Dowlatabadi H 2017 The 1000 GtC coal question: are cases of vastly expanded future coal combustion still plausible? *Energy Econ.* **65** 16–31

[57] Quiggin D, Meyer K D, Hubble-Rose L and Froggatt A 2021 Climate Change Risk Assessment 2021: The Risks Are Compounding, and Without Immediate Action the Impacts Will Be Devastating *Environment and Society Programme Research Paper* Chatham House, London https://chathamhouse.org/sites/default/files/2021-09/2021-09-14-climate-change-risk-assessment-quiggin-et-al.pdf

[58] IPCC 2021 Weather and climate extreme events in a changing climate *Climate Change 2021: The Physical Science Basis. Contribution of Working Group I to the Sixth Assessment Report of the Intergovernmental Panel on Climate Change* ed V Masson-Delmotte *et al* (Cambridge: Cambridge University Press) ch 11 https://ipcc.ch/report/sixth-assessment-report-working-group-i/

[59] Grübler A *et al* 2018 A low energy demand scenario for meeting the 1.5 °C target and sustainable development goals without negative emission technologies *Nat. Energy* **3** 515–27

[60] US Climate Change Science Program 2008 *Weather and Climate Extremes in a Changing Climate—Regions of Focus: North America, Hawaii, Caribbean, and US Pacific Islands* (Washington, DC: US Climate Change Science Program and the Subcommittee on Global Change Research) https://downloads.globalchange.gov/sap/sap3-3/sap3-3-final-all.pdf

[61] Dai A 2011 Drought under global warming: a review *WIREs Climate Change* **2** 45–65

[62] Rahmstorf S and Coumou D 2011 Increase of extreme events in a warming world *Proc. Natl Acad. Sci.* **108** 17905–9

[63] Romm J 2011 The next dust bowl *Nature* **478** 450–1

[64] Singh J, Ashfaq M, Skinner C B, Anderson W B, Mishra V and Singh D 2022 Enhanced risk of concurrent regional droughts with increased ENSO variability and warming *Nat. Clim. Change* **12** 163–70

[65] Hansen J, Sato M and Ruedy R 2012 *The New Climate Dice: Public Perception of Climate Change* (New York: National Aeronautics and Space Administration, Goddard Institute for Space Studies) https://www.giss.nasa.gov/research/briefs/2012_hansen_17/

[66] Ceballos G, Ehrlich P R and Dirzo R 2017 Biological annihilation via the ongoing sixth mass extinction signaled by vertebrate population losses and declines *Proc. Natl Acad. Sci.* **114** E6089

[67] IPBES 2019 *Summary for Policymakers of the Global Assessment Report on Biodiversity and Ecosystem Services, IPBES Plenary at its Seventh Session* (Paris: Intergovernmental Science-Policy Platform on Biodiversity and Ecosystem Services)

[68] Balint M, Domisch S, Engelhardt C H M, Haase P, Lehrian S, Sauer J, Theissinger K, Pauls S U and Nowak C 2011 Cryptic biodiversity loss linked to global climate change *Nature Clim. Change* **1** 313–8

[69] Thompson A 2022 How climate change will hit younger generations *Sci. Am.* 1 February p 76

[70] Anderegg W R L, Abatzoglou J T, Anderegg L D L, Bielory L, Kinney P L and Ziska L 2021 Anthropogenic climate change is worsening North American pollen seasons *Proc. Natl Acad. Sci.* **118** e2013284118

[71] Carlson C J, Albery G F, Merow C, Trisos C H, Zipfel C M, Eskew E A, Olival K J, Ross N and Bansal S 2022 Climate change increases cross-species viral transmission risk *Nature* **607** 555–62

[72] Yong E 2022 We created the 'Pandemicene': by completely rewiring the network of animal viruses, climate change is creating a new age of infectious dangers *The Atlantic* 28 April https://theatlantic.com/science/archive/2022/04/how-climate-change-impacts-pandemics/629699/

[73] Grinsted A and Christensen J H 2021 The transient sensitivity of sea level rise *Ocean Sci.* **17** 181–6

[74] Rahmstorf S 2007 A semi-empirical approach to projecting future sea-level rise *Science* **315** 368–70

[75] Vermeer M and Rahmstorf S 2009 Global sea level linked to global temperature *Proc. Natl Acad. Sci.* **106** 21527–32

[76] Ridgwell A and Schmidt D N 2010 Past constraints on the vulnerability of marine calcifiers to massive carbon dioxide release *Nat. Geosci.* **3** 196–200

[77] Sale P 2011 *Our Dying Planet: An Ecologist's View of the Crisis We Face* (Berkeley, CA: University of California Press)

[78] Dias B B, Hart M B, Smart C W and Hall-Spencer J M 2010 Modern seawater acidification: the response of foraminifera to high-CO_2 conditions in the Mediterranean Sea *J. Geol. Soc.* **167** 843–6

[79] Rantanen M, Karpechko A Y, Lipponen A, Nordling K, Hyvärinen O, Ruosteenoja K, Vihma T and Laaksonen A 2022 The Arctic has warmed nearly four times faster than the globe since 1979 *Commun. Earth Environ.* **3** 168

[80] Lenton T M, Rockström J, Gaffney O, Rahmstorf S, Richardson K, Steffen W and Schellnhuber H J 2019 Climate tipping points—too risky to bet against *Nature* **575** 592–5

[81] Kemp L *et al* 2022 Climate endgame: exploring catastrophic climate change scenarios *Proc. Natl Acad. Sci.* **119** e2108146119

[82] Simpson R D 2022 How do we price an unknowable risk? *Issues in Science and Technology* **38** (Winter) 47–50 https://issues.org/social-cost-of-carbon-economics-simpson/

[83] Orihuela-Pinto B, England M H and Taschetto A S 2022 Interbasin and interhemispheric impacts of a collapsed Atlantic overturning circulation *Nat. Clim. Change* **12** 558–65

[84] Caesar L, McCarthy G D, Thornalley D J R, Cahill N and Rahmstorf S 2021 Current Atlantic meridional overturning circulation weakest in last millennium *Nat. Geosci.* **14** 118–20

[85] Rahmstorf S, Box J E, Feulner G, Mann M E, Robinson A, Rutherford S and Schaffernicht E J 2015 Exceptional twentieth-century slowdown in Atlantic Ocean overturning circulation *Nat. Clim. Change* **5** 475–80

[86] Jackson L C, Biastoch A, Buckley M W, Desbruyères D G, Frajka-Williams E, Moat B and Robson J 2022 The evolution of the North Atlantic meridional overturning circulation since 1980 *Nat. Rev. Earth Environ.* **3** 241–54

[87] Weitzman M L 2009 On modeling and interpreting the economics of catastrophic climate change *Rev. Econ. Stat.* **91** 1–19

[88] Weitzman M L 2010 What is the 'damages function' for global warming—and what difference might it make? *Clim. Change Econ.* **1** 57–69

[89] Weitzman M L 2011 Fat-tailed uncertainty in the economics of catastrophic climate change *Rev. Environ. Econ. Policy* **5** 275–92

[90] Wagner G and Weitzman M W 2016 *Climate Shock: The Economic Consequences of a Hotter Planet* (Princeton, NJ: Princeton University Press)

[91] IPCC 2022 *Climate Change 2022: Mitigation of Climate Change. Contribution of Working Group III to the Sixth Assessment Report of the Intergovernmental Panel on Climate Change* (Cambridge: Cambridge University Press) https://ipcc.ch/report/sixth-assessment-report-working-group-3/

[92] Allen M R, Friedlingstein P, Girardin C A J, Jenkins S, Malhi Y, Mitchell-Larson E, Peters G P and Rajamani L 2022 Net zero: science, origins, and implications *Ann. Rev. Environ. Res.* **47** 849–87

[93] Matthews H D, Gillett N P, Stott P A and Zickfeld K 2009 The proportionality of global warming to cumulative carbon emissions *Nature* **459** 829–32

[94] Zickfeld K, Eby M, Matthews H D and Weaver A J 2009 Setting cumulative emissions targets to reduce the risk of dangerous climate change *Proc. Natl Acad. Sci.* **106** 16129

[95] Ricke K L and Caldeira K 2014 Maximum warming occurs about one decade after a carbon dioxide emission *Environ. Res. Lett.* **9** 124002

[96] Matthews H D and Caldeira K 2008 Stabilizing climate requires near-zero emissions *Geophys. Res. Lett.* **35** L04705

[97] Lahn B 2020 A history of the global carbon budget *WIREs Clim. Change.* **11** e636

[98] Koomey J 2013 Moving beyond benefit-cost analysis of climate change *Environ. Res. Lett.* **8** 041005

[99] Baumol W J 1972 On taxation and the control of externalities *Am. Econ. Rev.* **62** 307–22 http://jstor.org/stable/1803378

[100] Nordhaus W D 1977 Economic growth and climate: the carbon dioxide problem *Am. Econ. Rev.* **67** 341–6 https://www.jstor.org/stable/1815926

[101] Nordhaus W D 1979 *The Efficient Use of Energy Resources* (New Haven, CT: Yale University Press)

[102] Caldeira K, Jain A K and Hoffert M I 2003 Climate sensitivity uncertainty and the need for energy without CO_2 emission *Science* **299** 2052–4

[103] Meinshausen M, Meinshausen N, Hare W, Raper S C B, Frieler K, Knutti R, Frame D J and Allen M R 2009 Greenhouse-gas emission targets for limiting global warming to 2 °C *Nature* **458** 1158–62

[104] IEA 2010 *World Energy Outlook 2010* (Paris: International Energy Agency, Organization for Economic Cooperation and Development) http://worldenergyoutlook.org/

[105] IEA 2011 *World Energy Outlook 2011* (Paris: International Energy Agency, Organization for Economic Cooperation and Development) http://worldenergyoutlook.org/

[106] IEA 2012 *World Energy Outlook 2012* (Paris: International Energy Agency, Organization for Economic Cooperation and Development) http://worldenergyoutlook.org/

[107] IEA 2020 *World Energy Outlook 2020* (Paris: International Energy Agency, Organization for Economic Cooperation and Development) http://worldenergyoutlook.org/

[108] Luderer G, Pietzcker R C, Bertram C, Kriegler E, Meinshausen M and Edenhofer O 2013 Economic mitigation challenges: how further delay closes the door for achieving climate targets *Environ. Res. Lett.* **8** 034033

[109] Rockström J, Gaffney O, Rogelj J, Meinshausen M, Nakicenovic N and Schellnhuber H J 2017 A roadmap for rapid decarbonization *Science* **355** 1269

[110] IPCC 2018 *Global Warming of 1.5 °C. An IPCC Special Report on the Impacts of Global Warming of 1.5 °C above Pre-industrial Levels and Related Global Greenhouse Gas Emission Pathways, in the Context of Strengthening the Global Response to the Threat of Climate Change, Sustainable Development, and Efforts to Eradicate Poverty* (Geneva: IPCC) https://ipcc.ch/sr15

[111] Rogelj J, Meinshausen M, Sedláček J and Knutti R 2014 Implications of potentially lower climate sensitivity on climate projections and policy *Environ. Res. Lett.* **9** 031003

[112] McKibben B 2012 Global warming's terrifying new math *Rolling Stone Mag.* 19 July https://www.rollingstone.com/politics/politics-news/global-warmings-terrifying-new-math-188550/

[113] Gore A and Blood D 2013 The coming carbon asset bubble *The Wall Street Journal* 29 October http://online.wsj.com/news/articles/SB10001424052702304655104579163663464339836?mod=hp_opinion

[114] Matthews H D *et al* 2020 Opportunities and challenges in using remaining carbon budgets to guide climate policy *Nat. Geosci.* **13** 769–79

[115] Rogelj J, Meinshausen M, Schaeffer M, Knutti R and Riahi K 2015 Impact of short-lived non-CO_2 mitigation on carbon budgets for stabilizing global warming *Environ. Res. Lett.* **10** 075001

[116] Rogelj J, Reisinger A, McCollum D L, Knutti R, Riahi K and Meinshausen M 2015 Mitigation choices impact carbon budget size compatible with low temperature goals *Environ. Res. Lett.* **10** 075003

[117] Tokarska K and Matthews D 2021 Refining the remaining 1.5 °C 'carbon budget' *Carbon Brief* 19 January https://carbonbrief.org/guest-post-refining-the-remaining-1-5c-carbon-budget

[118] Matthews H D, Tokarska K B, Rogelj J, Smith C J, MacDougall A H, Haustein K, Mengis N, Sippel S, Forster P M and Knutti R 2021 An integrated approach to quantifying uncertainties in the remaining carbon budget *Commun. Earth Environ.* **2** 7

[119] MacDougall A H, Zickfeld K, Knutti R and Matthews H D 2015 Sensitivity of carbon budgets to permafrost carbon feedbacks and non-CO_2 forcings *Environ. Res. Lett.* **10** 125003

[120] Herman A 2012 *Freedom's Forge: How American Business Produced Victory in World War II* (New York: Random House)

[121] Lovins A B *et al* 2011 *Reinventing Fire: Bold Business Solutions for the New Energy Era* (White River Junction, VT: Chelsea Green) https://rmi.org/insights/reinventing-fire/

[122] Koomey J G, Matthews H S and Williams E 2013 Smart everything: will intelligent systems reduce resource use? *Ann. Rev. Environ. Res.* **38** 311–43

[123] BGR 2020 *BGR Energy Study 2019—Data and Developments Concerning German and Global Energy Supplies* (Hannover: Federal Institute for Geosciences and Natural Resources) https://www.bgr.bund.de/EN/Themen/Energie/Produkte/energy_study_2019_-summary_en.html

[124] Nordhaus W D 1992 An optimal transition path for controlling greenhouse gases *Science* **258** 1315

[125] Wigley T M L, Richels R and Edmonds J A 1996 Economic and environmental choices in the stabilization of atmospheric CO_2 concentrations *Nature* **379** 240–43

[126] Manne A and Richels R 1997 On stabilizing CO_2 concentrations—cost-effective emission reduction strategies *Environ. Model. Assessment* **2** 251–65

[127] IPCC 2014 *Climate Change 2014: Mitigation of Climate Change. Contribution of Working Group III to the Fifth Assessment Report of the Intergovernmental Panel on Climate Change* ed O Edenhofer *et al* (Cambridge: Cambridge University Press) https://www.ipcc.ch/report/ar5/wg3/

[128] Kriegler E *et al* 2014 The role of technology for achieving climate policy objectives: overview of the EMF 27 study on global technology and climate policy strategies *Clim. Change* **123** 353–67

[129] Krause F, Haites E, Howarth R and Koomey J G 1993 *Cutting Carbon Emissions–Burden or Benefit? The Economics of Energy-Tax and Non-Price Policies* (El Cerrito, CA: International Project for Sustainable Energy Paths) http://mediafire.com/file/y4r0n4vcvv4257s/cuttingCemissionsburdenorbenefitbook.pdf

[130] Duane T, Koomey J, Belyeu K and Hausker K 2016 From risk to return: investing in a clean energy economy *Risky Business* http://riskybusiness.org/fromrisktoreturn/

[131] Williams J H, Jones R A, Haley B, Kwok G, Hargreaves J, Farbes J and Torn M S 2021 Carbon-neutral pathways for the United States *AGU Adv.* **2** e2020AV000284

[132] USGCRP 2018 *Impacts, Risks, and Adaptation in the United States: Fourth National Climate Assessment* vol 2 (Washington, DC: US Global Change Research Program) https://nca2018.globalchange.gov

[133] DeCanio S J 2003 *Economic Models of Climate Change: A Critique* (Basingstoke: Palgrave-Macmillan)

[134] Lovins A B, Lovins H, Krause F and Bach W 1981 *Least-Cost Energy: Solving the CO_2 Problem* (Andover, MA: Brick House)

[135] Laitner J A, Decanio S J, Koomey J G and Sanstad A H 2003 Room for improvement: increasing the value of energy modeling for policy analysis *Utilities Policy* **11** 87–94

[136] Krause F, Koomey J and Olivier D 2000 *Cutting Carbon Emissions While Making Money: Climate Saving Energy Strategies for the European Union (Executive Summary for Volume II, Part 2 of Energy Policy in the Greenhouse)* (El Cerrito, CA: International Project for Sustainable Energy Paths) http://mediafire.com/file/bxdgkkb2d5rcjh1/ipsepkyotocosts_eu-report.pdf

[137] Krause F, Baer P and DeCanio S 2001 *Cutting Carbon Emissions at a Profit: Opportunities for the US* (El Cerrito, CA: International Project for Sustainable Energy Paths) http://mediafire.com/file/0aro7bj2d7kqk8w/ipsepcutcarbon_us.pdf

[138] IPCC 2007 *Climate Change 2007: Mitigation of Climate Change—Contribution of Working Group III to the Fourth Assessment Report of the Intergovernmental Panel on Climate Change* ed B O Metz *et al* (Cambridge: Cambridge University Press) http://ipcc.ch/publications_and_data/publications_and_data_reports.shtml

[139] Williams J H, DeBenedictis A, Ghanadan R, Mahone A, Moore J, Morrow W R, Price S and Torn M S 2012 The technology path to deep greenhouse gas emissions cuts by 2050: the pivotal role of electricity *Science* **335** 53–9

[140] Nemet G F, Holloway T and Meier P 2010 Implications of incorporating air-quality co-benefits into climate change policymaking *Environ. Res. Lett.* **5** 014007

[141] Vohra K, Vodonos A, Schwartz J, Marais E A, Sulprizio M P and Mickley L J 2021 Global mortality from outdoor fine particle pollution generated by fossil fuel combustion: results from GEOS-Chem *Environ. Res.* **195** 110754

[142] Shindell D 2020 *Health and Economic Benefits of a 2 °C Climate Policy: Testimony by Professor Drew Shindell* (Washington, DC: US House of Representatives) https://oversight.house.gov/sites/democrats.oversight.house.gov/files/Testimony%20Shindell.pdf

[143] Roberts D 2020 Air pollution is much worse than we thought: ditching fossil fuels would pay for itself through clean air alone *Vox* 12 August https://vox.com/energy-and-environment/2020/8/12/21361498/climate-change-air-pollution-us-india-china-deaths

[144] Cohen A J *et al* 2017 Estimates and 25-year trends of the global burden of disease attributable to ambient air pollution: an analysis of data from the global burden of diseases study 2015 *Lancet* **389** 1907–18

[145] Scovronick N, Anthoff D, Dennig F, Errickson F, Ferranna M, Peng W, Spears D, Wagner F and Budolfson M 2021 The importance of health co-benefits under different climate policy cooperation frameworks *Environ. Res. Lett.* **16** 055027

[146] Brown M A, Levine M D, Short W and Koomey J G 2001 Scenarios for a clean energy future *Energy Policy* **29** 1179–96

[147] Gumerman E, Koomey J G and Brown M A 2001 Strategies for cost-effective carbon reductions: a sensitivity analysis of alternative scenarios *Energy Policy* **29** 1313–23

[148] Sanstad A H, DeCanio S, Boyd G and Koomey J G 2001 Estimating bounds on the economy-wide effects of the CEF policy scenarios *Energy Policy* **29** 1299–312

[149] Lovins A B, Ürge-Vorsatz D, Mundaca L, Kammen D M and Glassman J W 2019 Recalibrating climate prospects *Environ. Res. Lett.* **14** 120201

[150] DeCanio S 1993 Barriers within firms to energy-efficient investments *Energy Policy* **21** 906–14

[151] DeCanio S J 1998 The efficiency paradox: bureaucratic and organizational barriers to profitable energy-saving investments *Energy Policy* **26** 441–54

[152] Bastien-Olvera B A 2019 Business-as-usual redefined: energy systems under climate-damaged economies warrant review of nationally determined contributions *Energy* **170** 862–8

[153] Moore F C and Delavane B D 2015 Temperature impacts on economic growth warrant stringent mitigation policy *Nat. Clim. Change* **5** 127–31

[154] Moore F C, Baldos U, Hertel T and Diaz D 2017 New science of climate change impacts on agriculture implies higher social cost of carbon *Nat. Commun.* **8** 1607

[155] Ackerman F and Stanton E A 2012 Climate risks and carbon prices: revising the social cost of carbon *Economics* **6** 1–25

[156] Kulkarni S, Hof A, van der Wijst K-I and van Vuuren D 2022 Disutility of climate change damages warrants much stricter climate targets *Research Square* https://doi.org/10.21203/rs.3.rs-1788130/v1

[157] Mann G 2022 Check your spillover *London Rev. Books* **44**(10 February) https://lrb.co.uk/the-paper/v44/n03/geoff-mann/check-your-spillover

[158] Cai Y and Lontzek T S 2018 The social cost of carbon with economic and climate risks *J. Political Econ.* **127** 2684–734

[159] Daniel K D, Litterman R B and Wagner G 2019 Declining CO_2 price paths *Proc. Natl Acad. Sci.* **116** 20886

[160] Arthur W B 1990 Positive feedbacks in the economy *Sci. Am.* February pp 92–9

[161] Arthur W B 1994 *Increasing Returns and Path Dependence in the Economy* (Ann Arbor, MI: University of Michigan Press)

[162] Gritsevskyi A and Nakicenovic N 2000 Modeling uncertainty of induced technological change *Energy Policy* **28** 907–21

[163] Rao S, Keppo I and Riahi K 2006 Importance of technological change and spillovers in long-term climate policy *Energy J.* **27** 25–42

[164] Arrow K J 1962 The economic implications of learning by doing *Rev. Econ. Studies* **29** 155–73

[165] Vogt-Schilb A, Meunier G and Hallegatte S 2018 When starting with the most expensive option makes sense: optimal timing, cost and sectoral allocation of abatement investment *J. Environ. Econ. Manag.* **88** 210–33

[166] Fischer C and Newell R G 2008 Environmental and technology policies for climate mitigation *J. Environ. Econ. Manag.* **55** 142–62

[167] Sharpe S and Lenton T M 2021 Upward-scaling tipping cascades to meet climate goals: plausible grounds for hope *Climate Policy* **21** 421–33

[168] Nordhaus W D 2014 The perils of the learning model for modeling endogenous technological change *Energy J.* **35** 1–13

[169] Hamond M J, DeCanio S J, Duxbury P, Sanstad A H and Stinson C H 1997 Tax waste, not work *Challenge* **40** 53–62

[170] Acemoglu D, Aghion P, Bursztyn L and Hemous D 2012 The environment and directed technical change *Am. Econ. Rev.* **102** 131–66

[171] Acemoglu D, Akcigit U, Hanley D and Kerr W 2016 Transition to clean technology *J. Political Econ.* **124** 52–104

[172] de la Chesnaye F C and Weyant J P 2006 EMF 21 multi-greenhouse gas mitigation and climate policy *Energy J. Spec. Issue* **27**

[173] Mohai P, Pellow D and Roberts J T 2009 Environmental justice *Ann. Rev. Environ. Res.* **34** 405–30

[174] Sovacool B K, Burke M, Baker L, Kotikalapudi C K and Wlokas H 2017 New frontiers and conceptual frameworks for energy justice *Energy Policy* **105** 677–91

[175] OECD 2021 The inequalities–environment nexus *OECD Green Growth Papers* No. 2021/01 (Paris: OECD Publishing)

[176] Islam S N and Winkel J 2017 Climate Change and Social Inequality *DESA Working Paper* No. 152 ST/ESA/2017/DWP/152 United Nations, Department of Economic and Social Affairs https://un.org/esa/desa/papers/2017/wp152_2017.pdf

[177] Mann M E 2021 *The New Climate War: The Fight to Take Back Our Planet* (New York: Public Affairs)

[178] Shi L and Moser S 2021 Transformative climate adaptation in the United States: trends and prospects *Science* **372** eabc805

[179] Diffenbaugh N S and Burke M 2019 Global warming has increased global economic inequality *Proc. Natl Acad. Sci.* **116** 9808–13

[180] Zheng Y, Steven J D, Persad G G and Caldeira K 2020 Climate effects of aerosols reduce economic inequality *Nat. Clim. Change* **10** 220–4

[181] Hsiang S *et al* 2017 Estimating economic damage from climate change in the United States *Science* **356** 1362

[182] Tessum C W, Paolella D A, Chambliss S E, Apte J S, Hill J D and Marshall J D 2021 $PM_{2.5}$ polluters disproportionately and systemically affect people of color in the United States *Sci. Adv.* **7** eabf4491

[183] UNEP 2021 *Neglected: Environmental Justice Impacts of Marine Litter and Plastic Pollution* (Nairobi: United Nations Environment Program) https://unep.org/resources/report/neglected-environmental-justice-impacts-marine-litter-and-plastic-pollution

[184] Bullard R D 2008 *Dumping in Dixie: Race, Class, and Environmental Quality* 3rd edn (Boulder, CO: Westview)

[185] Bullard R D (ed) 1999 *Confronting Environmental Racism: Voices from the Grassroots* (Boston, MA: South End)

[186] Zou B, Peng F, Wan N, Mamady K and Wilson G J 2014 Spatial cluster detection of air pollution exposure inequities across the United States *PLoS One* **9** e91917

[187] Petroni M, Hill D, Younes L, Barkman L, Howard S, Howell I B, Mirowsky J and Collins M B 2020 Hazardous air pollutant exposure as a contributing factor to COVID-19 mortality in the United States *Environ. Res. Lett.* **15** 0940a9

[188] Errigo I M *et al* 2020 Human health and economic costs of air pollution in Utah: an expert assessment *Atmosphere* **11** 1238

[189] Howarth R B 2011 Intergenerational justice *The Oxford Handbook of Climate Change and Society* ed J S Dryzek, R B Norgaard and D Schlosberg (Oxford: Oxford University Press) pp 338–52 ch 23

[190] Robinson K S 2009 Time to end the multigenerational Ponzi scheme *What Matters* 22 February 22 https://web.archive.org/web/20120705215631/http://whatmatters.mckinseydigital.com/climate_change/time-to-end-the-multigenerational-ponzi-scheme

[191] Rezai A, Duncan K F and Taylor L 2012 Global warming and economic externalities *Econ. Theory* **49** 329–51

[192] Gore T 2020 *Confronting Carbon Inequality: Putting Climate Justice at the Heart of the COVID-19 Recovery* (Nairobi: Oxfam) https://oxfam.org/en/research/confronting-carbon-inequality

[193] Gore T 2015 *Extreme Carbon Inequality: Why the Paris Climate Deal Must Put the Poorest, Lowest Emitting and Most Vulnerable People First* (Nairobi: Oxfam) https://policy-practice.oxfam.org/resources/extreme-carbon-inequality-why-the-paris-climate-deal-must-put-the-poorest-lowes-582545/

[194] Wiedmann T, Lenzen M, Keyßer L T and Steinberger J K 2020 Scientists' warning on affluence *Nat. Commun.* **11** 3107

[195] Nielsen K S, Nicholas K A, Creutzig F, Dietz T and Stern P C 2021 The role of high-socioeconomic-status people in locking in or rapidly reducing energy-driven greenhouse gas emissions *Nat. Energy* **6** 1011–6

[196] Otto I M, Kim K M, Dubrovsky N and Lucht W 2019 Shift the focus from the super-poor to the super-rich *Nat. Clim. Change* **9** 82–4

[197] Mastrandrea M D *et al* 2010 *Guidance Note for Lead Authors of the IPCC Fifth Assessment Report on Consistent Treatment of Uncertainties, IPCC Cross-Working Group Meeting on Consistent Treatment of Uncertainties Jasper Ridge, CA, USA* (Geneva: Intergovernmental Panel on Climate Change) https://ipcc.ch/site/assets/uploads/2017/08/AR5_Uncertainty_Guidance_Note.pdf

[198] Fischetti M 2021 We are living in a climate emergency, and we're going to say so *Sci. Am.* 12 April https://scientificamerican.com/article/we-are-living-in-a-climate-emergency-and-were-going-to-say-so/

[199] Hickel J 2021 What would it look like if we treated climate change as an actual emergency? *Current Affairs* 15 November https://currentaffairs.org/2021/11/what-would-it-look-like-if-we-treated-climate-change-as-an-actual-emergency

[200] Kühne K, Bartsch N, Tate R D, Higson J and Habet A 2022 'Carbon bombs'—mapping key fossil fuel projects *Energy Policy* **166** 112950

IOP Publishing

Solving Climate Change
A guide for learners and leaders
Jonathan Koomey and Ian Monroe

Chapter 2

Introduction to climate solutions

The best way to predict the future is to invent it.

—Alan Kay

Chapter overview

- Unrestricted climate change threatens human civilization and the global environment. It's an emergency, and society needs to act at a scale and a pace commensurate with such an enormous threat.
- We'll need to treat climate change as the 'adaptive' challenge it is, experimenting and changing course as needed.
- Acting with requisite urgency by deploying technology rapidly will drive down mitigation costs through learning by doing, economies of scale, spillover effects, and network externalities, and open up new opportunities for reducing emissions at even lower costs. Actions now create options later.
- Delaying action is folly, and the longer we delay the sooner we need to get to zero greenhouse gas emissions. That's the inexorable math of temperatures being dependent on cumulative greenhouse gas emissions and it's one reason why we advocate hitting net-zero emissions by 2040 at the latest.
- To stabilize the climate we need to do three things, all as soon as possible: end fossil fuels, minimize emissions of high potency GHGs, and create a climate-positive biosphere.
- Fossil-fuel combustion kills millions every year and injures millions more. The pollution resulting from fossil fuels is a powerful argument for ending combustion virtually everywhere, and it makes reducing greenhouse gas emissions much cheaper than it would be if there were no such negative side effects of fossil combustion.
- Our data, models, and analysis tools all need to improve if we're to truly face the climate challenge, and we need to use appropriate problem framing (abandoning the idea of a unique 'optimal' path for climate stabilization, once and for all).

doi:10.1088/978-0-7503-4032-8ch2

2.1 Chapter introduction

The climate solutions literature is vast and growing all the time. Informed observers have known for many years that greenhouse gas emissions need to get to zero as quickly as possible, but acceptance of this reality is not widespread in the broader society even now.

Most treatments of climate solutions have focused on energy-related CO_2 emissions, treating individual sectors (buildings, transportation, industry, electricity) in isolation. This focus is understandable: CO_2 emissions from energy are the most important driver of historical warming and will be the biggest driver of expected future warming, but that is not the complete picture.

A comprehensive approach to emissions reductions must also tackle emissions reductions for other warming agents (e.g. methane, nitrous oxide, black carbon, and F-gases) and emissions reductions outside the energy sector. The drivers of change, the available levers for change, and the impediments to rapid action are quite different for these other sectors and warming agents than for the energy sector, so they must be treated separately.

This chapter describes a comprehensive framework for thinking about climate solutions that includes all sectors and all warming agents. It also describes at a high level the technical and institutional changes needed to help the world transition to zero emissions. We explore insights available from this framework in subsequent chapters.

2.2 Treat climate like the moral issue it is

Humans are no longer small compared to the Earth. Because of our numbers, our wealth, and most importantly our technology, we are able to affect Earth's life-support systems in irreversible ways [4].

That new-found power raises profound moral and ethical questions [37, 38]. Benefit–cost analysis tries and ultimately fails to boil down these complex issues into a simple calculus of dollars and cents [38–40]. Because the effects of climate change and the costs of eliminating it most often fall on people far apart in time and space [41], because human impacts of climate change are (and will be) unequally distributed [34], because the probability of climate-induced catastrophe is non-zero but unknowable [7–11], and because it's impossible to precisely assess costs and benefits decades hence [4, 42–44], addressing this problem is not solely a question of calculating the most economically efficient path forward [35].

Instead, the focus should be on questions often ignored by the economic discipline: What don't we know? What can't we know? Who benefits? Who loses? How do we choose? What is our responsibility to future generations and to other species? How should we evaluate small but real (and unknowable) risks of catastrophic, civilization-destroying outcomes? Ethical questions like these aren't amenable to economic calculations but addressing them is central to responding responsibly to the climate problem.

2.3 Climate change as an adaptive challenge

Climate change is a messy, open-ended, and ill-defined problem, and is an archetypal example of what some have called an 'adaptive challenge'. To meet such a challenge requires an evolutionary and experimental approach, because not all parts of the solution can be known precisely in advance.

In their book *Moments of Impact* Chris Ertel and Lisa Kay Solomon [45], citing Ronald Heifetz, describe two types of challenges:

- *Technical challenges* are those we can solve using well-known tools and existing institutions. Such problems are well defined and well understood
- *Adaptive challenges*, on the other hand, are 'messy, open-ended, and ill defined'. The tools needed to address them may not yet exist. Such problems require different kinds of leadership and problem-solving skills and cry out for interactive engagement among all the people needed to solve them. They also require institutions to learn, adapt, and evolve.

Climate change is the ultimate adaptive challenge [46], because the rate and scope of the changes needed to solve the problem will stretch us to the limit. In addition, the solutions involve changes in behavior and institutional structure, not just technology [3]. As Koomey argued in *Cold Cash, Cool Climate* back in 2012:

Climate change is probably the biggest challenge modern humanity has ever faced. It's bigger than World War II because it will take decades to vanquish this foe. It's harder than ozone depletion, whose causes were far less intertwined with industrial civilization than fossil fuels and other sources of greenhouse gases. And it's more intractable than the Great Depression (or our current economic malaise) because financial crises eventually pass, assuming we learn from past mistakes and fix the financial system (again!). [4]

We have many options for reducing greenhouse gas emissions, but we'll need new ones, too. Existing options will only get us so far. That's why we must both deploy existing technology and research new ones aggressively. We'll also need to take an evolutionary approach to this problem, one that embraces the adaptive nature of the challenge before us.

That means planning and active management of the transition, integrating planning with markets, using markets to drive costs down, developing a portfolio approach to zero-emissions investments, and guiding markets towards the end goal of a zero-emissions society. That end goal won't happen by itself, but the necessity of planning to help shepherd that process is hard for some to accept.

There are tradeoffs to consider and resources to be allocated. Markets can play a big role, but they also have limits, and someone's got to balance competing factors. That means planning by government, business, and civil society, but that planning needs to be nimble and adaptive to reality as it evolves. It also means the development of new tools that allow for more rapid testing of alternative scenarios. One example is the online analysis tool for the US referenced in

Harvey *et al* [47][1]. Another is the global tool called EN-Roads, developed by Climate Interactive, MIT, and Ventana Systems[2].

2.4 Building new fossil infrastructure makes solving the problem harder

The recognition that constraints on the construction of new fossil infrastructure can contribute to a rapid energy transition is a recent one, but it's gained many converts in the past decade [48–51]. The more high-emission infrastructure we build in the next few years the more we'll have to scrap in the next decade or two, so we need to stop building such infrastructure as soon as we can.

In May 2021, the International Energy Agency released a report analyzing the implications of achieving a net-zero-emissions energy system globally by 2050 [52]. The report states:

> Beyond projects already committed as of 2021, there are no new oil and gas fields approved for development in our [net zero by 2050] pathway, and no new coal mines or mine extensions are required.

The IEA report doesn't give clarity on whether other kinds of new fossil infrastructure will be needed (such as fossil power plants, pipelines, or coal shipping terminals), so that's an area worthy of further analysis. In any case, the faster we move to zero-emission fuels and electricity, the fewer new investments in fossil-fuel infrastructure will be stranded by necessary climate action (of course, that means more existing assets will be stranded sooner, but there's no escaping that outcome at this point).

The IPCC, in the Technical Summary for the latest Working Group III report [12] made this point even more strongly:

> Estimates of future CO_2 emissions from existing fossil-fuel infrastructures already exceed remaining cumulative net CO_2 emissions in pathways limiting warming to 1.5 °C with no or limited overshoot.

If emissions from existing infrastructure will use up the entire remaining carbon budget, there is no scope for additional fossil infrastructure, even if existing fossil capital is retired.

Stopping infrastructure from being built is always easier than shutting profitable facilities down, because owners of those facilities will fight to prevent their facilities from becoming stranded assets. Better to not allow them to operate in the first place.

The fossil-fuel industry has not yet internalized this reality. Recent estimates show plans for almost $1 trillion US for development of new oil and gas resources by 2030

---

[1] https://us.energypolicy.solutions
[2] https://www.climateinteractive.org/en-roads/

for twenty large oil companies who claim to be committed to the goals of the Paris Agreement [53]. The global banking industry allocated more than $700 billion US to the fossil-fuel industry in 2021 alone [54]. Currently planned fossil-fuel developments would lead to a factor of two more emissions than would be allowed under a 1.5 °C limit [36].

2.5 Speeding up the energy transition

One of the most widespread misconceptions in this field is the idea that because energy transitions have taken many decades in the past that they must therefore take just as long in the future [37: section 2.5.2, 38, 39]. This superficially sensible idea is flawed for at least three reasons: (1) the new power of information technology; (2) the lack of policy guiding and accelerating previous energy transitions by promoting early retirement of pre-existing capital; and (3) increasing understanding of the power of learning effects on mass-produced technologies.

Innovation in modern industrial societies has been driven in the past two centuries by a series of what economists call 'general purpose technologies' or GPTs, which have far ranging effects on the way the economy produces value. The most important of these were the steam engine, the telegraph, the electric power grid, the internal combustion engine, and, most recently, computers and related communications technologies [55]. Brynjolfsson and McAfee [56] write:

> GPTs...not only get better themselves over time (and as Moore's law shows, this is certainly true of computers), they also lead to complementary innovations in the processes, companies, and industries that make use of them. They lead, in short, to a cascade of benefits that is both broad and deep...

> Digitization...is not a single project providing one-time benefits. Instead, it's an ongoing process of creative destruction; innovators use both new and established technologies to make deep changes at the level of the task, the job, the process, even the organization itself. And these changes build and feed on each other so that the possibilities offered really are constantly expanding.

The result is that the pace of innovation across virtually all industries is accelerating, which is a direct result of the use of information and communications technology (ICT) to improve production and business processes. And it's not just computers that are improving, it's all businesses that use computers to increase efficiency, improve organizational effectiveness, and reduce costs of manufactured products.

Information and communication technologies have the potential to speed up the energy transition significantly [4, 22, 57, 58]. These technologies allow us to collect data, do real-time analysis and control, replace parts with smarts, move bits instead of moving atoms, 3D print parts from computer designs, and transform institutions and processes more rapidly than we ever could before.

Past energy transitions occurred in an organic way with little policy intervention, and incumbent technologies used their market power and low marginal costs to slow penetration of new technologies. These tactics can be especially effective when industries are natural monopolies and regulators are 'captured' by the industry, but they also apply to technologically sophisticated industries with economies of scale like oil and gas.

To accelerate the energy transition, we can force existing fossil capital to retire rapidly as we accelerate deployment of new technologies. For example, a strong case can be made for the retirement of all coal-fired power plants in advanced nations by 2030, or earlier if possible [59]. The economics of these existing plants (just based on direct costs) have deteriorated sharply in recent years [60], and from society's perspective they haven't been economic for a very long time.

In 2011 three eminent economists [61] declared that coal and oil-fired electricity delivered negative value added to the economy, because the societal costs associated with pollution were so high. Epstein *et al* [62] provide more support for this view for coal plants. The evidence for even higher pollution costs from fossil-fuel and other combustion has accumulated rapidly in the past decade [27, 29, 63–77]. Millions of people die prematurely every year because of fossil combustion, and millions more are injured [27]. Wildfires, which are also big contributors to air pollution [2], will become vastly more common the longer reductions in greenhouse gas emissions are delayed. So will pollen and its associated respiratory effects [6].

There will often be localized transition costs, and we need to help workers and communities affected by the transition to zero emissions [59], but those costs should not stand in the way of action to reduce harm from pollution. The argument against such retirements usually is framed from the perspective of *direct costs and losses to incumbent economic actors*, ignoring pollution costs. When evaluating alternative policies, however, we should instead be focused most intently on the costs *to society* of continuing to burn fossil fuels. When we center that perspective, the imperative to retire existing fossil capital becomes even more clear and compelling.

Another difference between the current energy transition and previous ones is more widespread understanding of the power of learning effects and other forms of increasing returns to scale for mass-produced granular technologies [91: section 2.5.3.3]. The importance of these effects was recognized as early as the 1970s [78, 79] but the experience of the past few decades has made the power of these effects even more clear, for both supply and demand-side technologies [33, 80].

2.6 The false choice between innovation and immediate, rapid emissions reductions

A common belief is that the US should prioritize innovation and R&D in climate-friendly technologies over immediate emission reductions using current technology. For example, an *Atlantic* article about Bill Gates' views on climate solutions [81] concluded:

> This is the Gates Rule: If given a choice of cutting emissions directly or reducing the cost of net-zero technology, the US should choose the latter.

This is a false choice: we can and should do both. Cumulative emissions are what matter to global temperatures and rapid immediate deployment reduces cumulative emissions. Deployment is a key late-stage element of the innovation process. It drives costs down through learning effects, scale, spillovers, and network externalities, and creates new possibilities even without research and development. In addition, *we won't get those cost reductions unless we deploy existing technology*.

Breakthrough technologies that may result from R&D in a decade or two will not replace the need to deploy existing technology immediately. Some or all technologies under development may come to fruition, but they are a supplement to immediate emissions reductions, not a replacement. There is no substitute for deploying existing zero-emissions technology as quickly as we can.

Some potentially promising technologies are at an earlier stage of development. Technologies with minimal societal side effects, high-emissions reduction potential, and reasonable and declining costs are good prospects for research and development followed as quickly as possible by prototype demonstrations and deployment. Speed is of the essence, but some options still need time to mature, and others may never reach the market. We should be supporting the development of promising longer-term options aggressively even as we deploy off-the-shelf technologies at a rapid pace. Not all technologies will make the cut, but we're not lacking for options in most sectors.

2.7 The folly of delay

To solve the climate problem, we must reduce greenhouse gas emissions to zero as quickly as we can [13]. As William Nordhaus said in 2008 [82] 'There is no case for delay' in starting to reduce emissions.

Delay is costly because cumulative emissions are what matter (waiting means we just need to move more quickly later) [14–16, 20, 83]. It is costly because delay means we don't gain the benefits of learning by doing, scale, spillovers, and network externalities from deploying zero-emissions technology [31, 33, 84–86]. It is costly because it adds to the stock of 'stranded assets', fossil-fuel-using capital that will need to be retired before the end of its useful life to meet emissions targets [18, 20, 87–89]. And it is costly because it adds to climate (and other) damages already being incurred by society if we continue to fail to act [20, 90, 91].

These conclusions are not new. Over a decade ago, the International Energy Agency concluded in its 2009 *World Energy Outlook* [92]:

> …each year of delay before moving onto the emissions path consistent with a 2 °C temperature increase would add approximately $500 billion to the global incremental investment cost of $10.5 trillion for the period 2010–2030.

In its 2010 *World Energy Outlook* [17], IEA increased that estimate of losses for each year of delay to $1 trillion.

The White House Council of Economic Advisors under President Obama declared in 2014 [90]:

...a delay that results in warming of 3° Celsius above pre-industrial levels, instead of 2°, could increase economic damages by approximately 0.9 percent of global output. To put this percentage in perspective, 0.9 percent of estimated 2014 US Gross Domestic Product (GDP) is approximately $150 billion [per year].

...net mitigation costs increase, on average, by approximately 40 percent for each decade of delay. These costs are higher for more aggressive climate goals: each year of delay means more CO_2 emissions, so it becomes increasingly difficult, or even infeasible, to hit a climate target that is likely to yield only moderate temperature increases.

The IPCC's special report on 1.5 °C scenarios [20] concluded in 2018:

...every year's delay before initiating emission reductions decreases by approximately two years the remaining time available to reach zero emissions on a pathway still remaining below 1.5 °C...

and

The challenges from delayed actions to reduce greenhouse gas emissions include the risk of cost escalation, lock-in in carbon-emitting infrastructure, stranded assets, and reduced flexibility in future response options in the medium to long term.

Daniel *et al* [30], after applying standard treatments of risk and uncertainty to William Nordhaus's DICE integrated assessment model [23], found:

...delaying implementation by only 1 y costs society approximately $1 trillion. A 5-y delay creates the equivalent loss of approximately $24 trillion, comparable to a severe global depression. A 10-y delay causes an equivalent loss in the order of $10 trillion per year, approximately $100 trillion in total.

This analysis also found that the cost of delay increases quadratically over time, so that a five-year delay costs the world twenty-four times as much as a one-year delay and delaying action by ten years instead of five years increases societal costs four-fold.

Delay is costly, and the faster we move to reduce emissions the easier and cheaper it will be to do so. Conversely, the longer we delay, the more it will cost to fix the problem, and the longer society will incur the huge and avoidable societal costs of pollution from fossil-fuel combustion [27–29, 93]. Delay also means *we need to get to zero emissions earlier to stay under a fixed warming limit.*

As an example, consider fossil energy carbon dioxide emissions in Grubler *et al*'s low energy demand (LED) intervention case [5], which we show in figure 2.1.

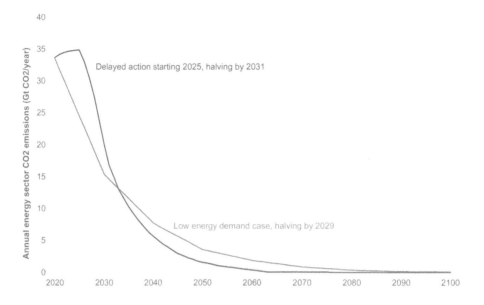

Figure 2.1. Fossil energy CO_2 emissions in the low energy demand scenario compared to a delayed action case. Source: Grubler *et al* [5], calculations from Koomey *et al* [94].

We chose to highlight this scenario because it has no carbon capture from the energy system, simplifying the story related to delaying emissions reductions.

Energy sector carbon dioxide emissions in this scenario decline 5.6% per year (as a % of 2020 emissions) from 2020 to 2030, reaching 56% below 2020 emissions by 2030 (it reaches the 'carbon law' goal of halving emissions a year early, by 2029). The area under that emissions curve equals the cumulative emissions through 2100, which is the 'budget' that we need to meet if we're to keep temperatures from rising more than 1.5 °C.

Figure 2.1 also shows what happens if we don't start reducing emissions until 2025 (the delayed action case). The area under this curve for the delayed action case is the same as for the LED intervention case, so these two cases emit the same amount of carbon from the energy sector through 2100.

Delay makes it harder to solve the problem. Starting later eats up more of the carbon budget, and the needed rate of emissions decline goes up to 8.8% per year from 2025 to 2030 (as a % of 2020) if emissions reductions begin right after emissions peak in 2025. This emissions path halves emissions from 2020 by 2031, representing about a two-year delay compared to the LED intervention case. The delayed action case also reaches zero emissions much earlier, as the inexorable math of cumulative emissions requires.

2.8 Learning by doing only happens if we do!

One of the most important lessons from history is the power of learning by doing for mass-produced products. Consider the last decade or so of technology costs for solar photovoltaics (PVs), onshore wind turbines, and battery packs as shown in figure 2.2. According to Bloomberg New Energy Finance, PV module lithium-ion battery pack

Figure 2.2. Global technology costs over time (2010 to 2020) for solar photovoltaics, onshore wind turbines, and Li-ion battery packs. Source of data: BNEF 2021 (personal communication, Jenny Chase BNEF, 28 January 2021). Used with permission.

prices are down about 90% and onshore wind turbine prices are down more than 50% since 2010, corresponding to learning rates (the percentage change in per unit costs for a doubling of cumulative production experience) from −14% (for wind) to −29% for solar PV modules.

The main reason for these stunning cost reductions is a rapid increase in the deployment of these technologies [85, 95, 96]. Business economists have known for many decades that the cost per unit of mass-produced devices falls between 10% and 30% for each doubling of cumulative production experience [84, 85, 95, 97–102].

This effect is colloquially known as 'learning by doing', but it is often a mixture of economies of scale, learning effects, spillovers, network externalities, induced innovation, and general technological progress in the economy [26, 32, 86]. The total effect on cost is normally treated as a function both of the learning rate per doubling of cumulative production and the total production over time.

Learning by doing creates a powerful argument for a 'deployment-first' strategy to reduce greenhouse gas emissions [33], and it's another reason why 'wait and see' is a terrible approach to solving the climate problem. Of course, we need new technologies and should therefore invest heavily in research and development, but there are vast opportunities for emission reductions using current technologies [24, 25], and cost reductions for these technologies are dependent on implementing them on a large scale (learning by doing only happens if we *do*). The focus in the next few decades should be on aggressive deployment of current low-emissions technologies, bringing new technologies into the mix as they emerge.

We'll also need to focus on improving how fast-changing mass-produced technologies are represented in our analysis tools. For some key technologies (e.g. solar PVs) recent reviews have shown significant lags between recent rapid cost declines and those declines being incorporated into modeling cost data. For example, many global models include costs for solar photovoltaic technologies for 2050 that are much higher even than *current costs* [103] and a strong case can be made that virtually all recent modeling exercises have underestimated the potential contributions of solar PVs to reducing emissions [104–109].

Another implication from acknowledging the importance of learning effects and economics of scale is that until deployment starts in earnest, we won't know precisely how much climate action will cost or how much we can achieve in total. We do know that these learning effects are powerful and rapid, and we know that almost all conventional economic models assessing climate solutions treat these effects cursorily or not at all. That gives us confidence that the costs of reducing emissions will be a lot lower than most people now think.

2.9 How fast should we reduce emissions?

Many studies in the past decade have focused on 2050 as the date for achieving net-zero global emissions [19, 52]. Unfortunately, inadequate emissions reductions to date (and the recent jump in emissions in 2021 from 2020 pandemic lows) have forced even more aggressive action [110, 111].

That's the reality of the climate problem. Warming is driven by cumulative greenhouse gas emissions [110], so the longer we wait to reduce emissions, the faster we must reduce emissions to zero in the future, as we discuss above.

We argue instead for net-zero emissions by 2040. Why are we so aggressive compared to previous analyses?

- Unrestricted climate change is an emergency that threatens the continued orderly development of human civilization, but existing efforts have not reflected that urgency.

- After decades of inadequate global action, we're out of time.
- A closer deadline focuses the mind.
- It's consistent with the latest research on carbon budgets for 1.5 °C [21].
- The nature of path dependent technological systems dominated by increasing returns is that we can't know what's possible until we try, and we haven't tried yet.
- If we shoot for an aggressive goal and fail to meet it, at least it won't be for lack of trying. If we shoot for a less aggressive goal we might miss emissions reductions opportunities that we might otherwise have captured if we were more aggressive.
- There's a fair chance that climate stabilization will be easier than we think.

We are also strong proponents of converting longer-term goals into *annual interim goals*. Many politicians find it easy to embrace goals to be met decades after they'll leave office but converting long-term goals into annual ones allows us to determine if we're on track to meet the temperature goal. If we fall short in one year, we'll need to do more the next year to catch up. That's part of what we mean when we say climate change is an 'adaptive challenge'.

There are those who will say that net-zero emissions by 2040 is not feasible, but they say that about virtually every target, and then proceed to do nothing. If we had acted to reduce emissions three decades ago, as we should have, we could have phased out fossil fuels at a leisurely pace but having dithered for so long we now have no other choice. *We must change political and other constraints so we can do what is necessary.*

2.10 What we must do

The overarching goal of climate policy at this juncture is to stabilize global temperatures, and that means getting to zero net emissions as soon as possible, pulling carbon from the air as technology and policy allow. We summarize this goal as *net zero by 2040, climate positive thereafter*, as shown in figure 2.3.

There are three key elements for achieving this goal, all three to be accomplished as soon as possible:

1. Ending fossil-fuel use.
2. Minimizing emissions of high potency GHGs.
3. Creating a climate-positive biosphere.

These three elements are the essence of climate stabilization.

2.10.1 Ending fossil-fuel use

The most important single driver of greenhouse gas emissions is the production and use of fossil fuels. Many analyses focus on *carbon dioxide emissions* from fossil fuels but solving the climate problem is about more than just cutting carbon dioxide.

It is common for analyses of climate change solutions to talk about carbon dioxide and 'other gases' but we focus instead on splitting greenhouse gases into

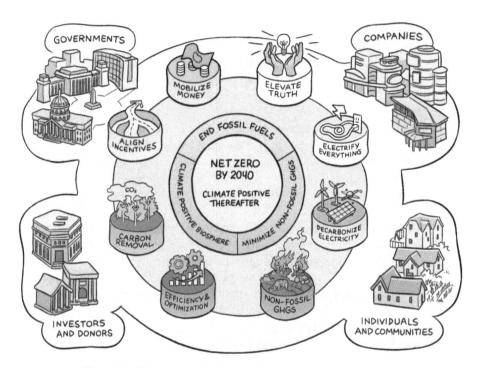

Figure 2.3. What we must do. Source: Copyright 2022 Violet Kitchen.

'fossil' and 'non-fossil' categories. This convention splits emissions of methane and nitrous oxide into those directly associated with fossil fuels (which will be reduced as we reduce and eventually eliminate fossil-fuel use) and those emissions linked to other activities.

Consider figure 2.4, which shows greenhouse gas emissions in 1990 and 2020 for the US [112], split into fossil and non-fossil categories, assuming 100 year global warming potentials (see below).

Carbon dioxide emissions associated with fossil fuels are the biggest single category, but emissions of the other gases are big enough to matter.

Figure 2.5 shows two ways of aggregating emissions, the first bar showing the fossil versus non-fossil split and the second bar showing the more conventional split of fossil plus industrial process carbon dioxide compared to 'other gases'. For the US, land-use carbon dioxide emissions are negative, because forests are growing back after earlier centuries of deforestation, and we exclude the effects of that carbon sink to focus on positive emissions. For the world as a whole, the land-use sector is currently a source of carbon emissions to the atmosphere [113].

Fossil emissions are about 80% of total US emissions, with non-fossil emissions around 20%. Carbon dioxide emissions, including fossil combustion and industrial processes comprise about 80% of the total, with the rest allocated to 'other gases' such as methane, nitrous oxide, and F-gases.

The advantage of the aggregation in the first bar is that it is much clearer and easier to operationalize. Everything in the 'fossil' bin is associated with the

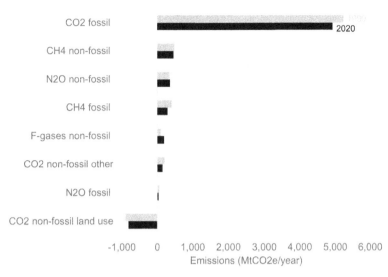

Figure 2.4. US Greenhouse gas emissions in 1990 and 2020 by category (100 year GWPs). Source: US EPA [112]. GWPs from table 7.SM.7 in IPCC [1]. Categories ranked by 2020 emissions using 100 year GWPs.

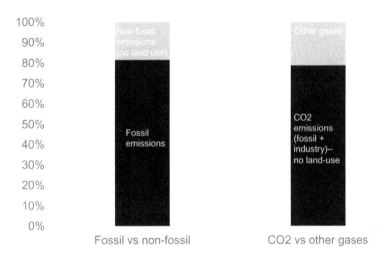

Figure 2.5. US greenhouse gas emissions in 2020 (100 year GWPs). Source: US EPA [112]. GWPs from table 7.SM.7 in IPCC [1].

production and use of fossil fuels, and everything in the 'non-fossil' category represents emissions from industrial processes and other human activities. If we use fewer fossil fuels, everything in the fossil category will go down.

The second aggregation approach (which is still widely used) lumps fossil related 'other gases' (such as methane leakage from coal mines and oil drilling) in with non-fossil emissions of those gases (from industrial processes and agriculture). Carbon dioxide from industrial processes (non-fossil related) is included with fossil CO_2 emissions. We use the aggregation approach embodied in the first bar where data permit to avoid the conceptual confusions embedded in the approach used for the second bar.

2.10.2 Minimizing non-fossil GHGs

Scientists characterize the warming effects of different warming agents in different ways, but it is common to calculate what's called a **global warming potential** (GWP) for each warming agent. The GWP represents the cumulative warming power of each agent as a ratio compared to carbon dioxide over some period. CO_2, by definition, has a GWP of 1.0. Other warming agents, depending on their warming power and residence time in the atmosphere, can have GWPs of many times that of CO_2.

GWP also varies over time. The IPCC calculates GWPs for 20 year, 100 year, and 500 year periods, as shown in table 2.1.

Many greenhouse gases have GWPs much greater than 1.0, which is why we call them **high potency GHGs**. It is particularly important to focus on high potency GHGs with shorter lifetimes, such as methane, black carbon, and HFC-134a because reducing their emissions can have an almost immediate effect on near-term warming trends.

Figure 2.6 shows what happens to the data for the two aggregation methods when 20 year GWPs are used instead of the 100 year values. Because the 20 year GWPs for methane and HFC-134a are a lot higher than the 100 year values, the importance of 'other gases' goes up a lot (this effect is somewhat offset by the lower 20 year GWP values for PFCs, SF_6, and NF_3). Non-fossil emissions also go up as a percentage of the total using these GWPs.

Table 2.1. Characteristics of selected warming agents.

Name	Formula	Mean lifetime years	GWP 20 yrs	GWP 100 yrs	GWP 500 yrs
Carbon dioxide	CO_2	Multiple	1.00	1.00	1.00
Methane	CH_4	12	81	28	8
Nitrous oxide	N_2O	109	273	273	130
Chlorofluorocarbon-11 (CFC-11)	CCl_3F	52	7430	5560	1870
Chlorofluorocarbon-12 (CFC-12)	CCl_2F_2	102	11 400	11 200	5100
Hydrofluorocarbon-23 (HFC-23)	CHF_3	228	12 400	14 600	10 500
Hydrofluorocarbon-32 (HFC-32)	CH_2F_2	5.4	2690	771	220
Hydrofluorocarbon-134a (HFC-134a)	CH_2FCF_3	14	4140	1530	436
Carbon tetrafluoride (PFC-14)	CF_4	50 000	5300	7380	10 600
Sulfur hexafluoride	SF_6	3200	18 300	25 200	34 100
Nitrogen trifluoride	NF_3	569	13 400	17 400	18 200
Black carbon	C	Days to weeks	3200	910	280
Hydrogen	H_2	2	32.2	10.8	NA

1. Black carbon data from Bond *et al* [114].
2. H_2 data from table 11 in Warwick [115]. NA = not available.
3. All other data from IPCC [1] table 7.SM.7.

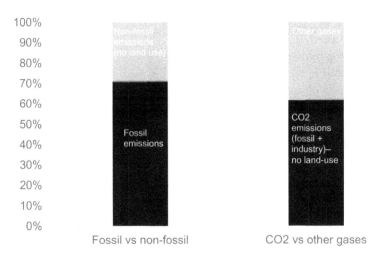

Figure 2.6. US greenhouse gas emissions in 2020 (20 year GWPs). Source: US EPA [112]. GWPs from table 7. SM.7 in IPCC [1].

A strong case can be made for examining results using both 100 year and 20 year GWPs, because the next few decades of climate action are so critical. Examining the 20 year values for scenarios that reach net zero by 2050 can reveal critical information about near-term warming from high potency GHGs.

In the rest of this book, we report greenhouse gas equivalent emissions using 20 year GWPs, referencing 100 year GWPs as appropriate. We also split emissions into 'fossil related' (for which electrification is the main solution) and 'non-fossil related' because this categorization method is conceptually clearer and more useful.

2.10.3 What does 'climate positive' mean?

Because of deforestation and other human activities, the biosphere is currently a source of carbon emissions into the atmosphere. A **climate-positive biosphere** is one that removes carbon from the atmosphere over time, which would make it a **carbon sink**.

There are scientists who are optimistic about engineering solutions to remove carbon from the atmosphere and store it underground that fall under the rubric of 'direct air capture' of CO_2 [116]. Unfortunately, the logistics, economics, and thermodynamics of such schemes remain daunting [116–121], and even if some of these technologies become available it will be decades before they can reach the needed scale.

While we support researching, demonstrating, and deploying such technologies, we are convinced that many well-known techniques for enhancing biospheric sinks are ready for deployment now and should be the focus of initial efforts to turn the biosphere 'climate positive' [122–124]. Such methods include:

- Stopping industrial-scale deforestation everywhere, while creating alternative ways to deliver energy services for those harvesting biomass for subsistence.
- Reforesting and reclaiming denuded land.
- Implementing regenerative farming.

- Creating vast ocean sanctuaries.
- Farming seaweed.
- Weathering of minerals.

The exact mix of techniques for enhancing natural sinks will emerge through rapid deployment and learning by doing, which is why we need to get started immediately.

2.10.4 The eight pillars of climate action

Williams *et al* [25] describe a suite of deep decarbonization scenarios for the United States energy system, and propose four pillars of deep decarbonization:

- Electrify end-uses and fuels
- Decarbonize electricity
- Increase efficiency
- Capture carbon emissions for use or safe storage

To this we add another pillar because it's not just about energy or carbon dioxide:

- Minimize non-fossil GHGs (including land-use, industrial processes, and other gases beyond carbon dioxide).

These five pillars are broadly parallel with those presented in Goldstein-Rose [125] and characterized in great detail in IPCC [12]. Because Williams *et al* focused on the energy sector, they thought of efficiency as *energy* efficiency, but we think of it more broadly as 'optimizing energy and materials flows', so we call it 'efficiency and optimization'. It can also apply to non-fossil emissions, so we move it to be after the 'reduce non-fossil emissions' pillar in our framework. We rename 'capture carbon emissions' to be 'remove carbon' and move it to number five.

We call these first five pillars 'technical pillars' since they organize thinking about ways to reduce emissions based on their technical characteristics. We also add pillars six, seven, and eight to highlight the importance of economics, investments, and framing/communications to accelerating the transition to a zero-emissions world:

- Align incentives
- Mobilize money
- Elevate truth

We call these 'institutional pillars' because they involve institutional change to implement them. By institutional change we mean changes in government, business, and civil society practices, norms, and rules that will be needed to accelerate the transition to zero emissions.

Table 2.2 shows the complete set of pillars, which we rename as 'the eight pillars of climate action'. We organize this book around these eight pillars, discussing under chapters for each pillar how different sectors will be affected by society's shift to net-zero emissions. We summarize each of the eight pillars in turn.

Table 2.2. The eight pillars of climate action.

Technical pillars	Examples
1. Electrify (almost) everything	(a) Energy end-uses. (b) Industrial processes. (c) Electrical fuels (H_2, synthetic methane, ammonia).
2. Decarbonize electricity	(a) Zero-emissions generation sources. (b) Flexible demand/price response. (c) Flexible industrial loads such as the creation of electrical fuels (e.g. H_2, synthetic methane). (d) Storage of electricity and heat. (e) Increased transmission infrastructure. (f) Diversity + 'overbuilding' of generation sources. (g) Improved forecasting of variable generation and loads. (h) Carbon capture from fossil-fuel + biomass combustion.
3. Minimize non-fossil GHGs	(a) F-gases. (b) Methane (manure, food, fossil production/distribution). (c) Soil and forest carbon. (d) N_2O. (e) Black carbon. (f) CO_2 from cement and other industrial processes.
4. Efficiency and optimization	(a) Efficiency/waste reduction. (b) Real-time demand response. (c) Changes in service demand/intensities. (d) Changes in structure of economic activity. (e) Reduce primary energy supply losses. (f) Develop and use new materials to displace old ones.
5. Remove carbon	(a) Stop deforestation and increase afforestation/reforestation. (b) Weathering. (c) Ocean sanctuaries, seaweed farming.
Institutional pillars	**Examples**
6. Align incentives	(a) Make emissions reductions the easy choice. (b) Re-design institutions and reform property rights. (c) Rationalize and tighten regulations. (d) Harness learning effects and social incentives. (e) Make companies liable for climate damages and risks.
7. Mobilize money	(a) End pro-pollution subsidies. (b) Price carbon, other GHGs, and other pollutant damages. (c) Re-deploy fossil spending to build zero-emissions alternatives. (d) Subsidize climate-friendly investments and R&D.

	(e) Use government borrowing/printing money at low interest rates to invest in infrastructure and resilience.
	(f) Get pro-pollution money out of politics.
8. Elevate truth	(a) Debunk misinformation.
	(b) Deplatform disinformation.
	(c) Vote.
	(d) Educate.
	(e) Disrupt propaganda campaigns.
	(f) Ban polluter advertising.
	(g) Hold polluters accountable for lies.

Source: Adapted from Williams et al [25] and Goldstein-Rose [125] with modifications by Koomey and Monroe.

The first pillar in our framework is *Electrify (almost) everything*, basing our choices on the latest research and field experience, and funding RD&D for hard-to-electrify end-uses. In combination with a rapidly decarbonizing electricity grid this effort will reduce emissions more cheaply and quickly than is possible by keeping fuel-fired end-uses in place. It also means expanding access to electricity to the hundreds of millions of people in Africa and Asia who now lack such access.

Pillar number two is *Decarbonize electricity*. To accomplish this goal, we'll need to deploy technologies, policies, and business practices to move electricity generation to zero emissions, while increasing total generation three- to four-fold in coming decades. Key factors include alternative generation technologies, power system controls, increased transmission investments, thermal and electricity storage, carbon sequestration for natural gas and biomass generation, improved forecasting, and 'overbuilding' of key generation investments.

Pillar three is *Minimize non-fossil emissions*. Addressing the short-lived non-fossil pollutants is one way to quickly slow increases in global warming. Agriculture produces non-fossil emissions from fertilizer, deforestation, and changes in land-use patterns. That sector's emissions are also affected by what people chose to eat. Industrial process and some appliances also emit non-fossil greenhouse gases, and these need focused attention.

Pillar four is *Efficiency and optimization*. This chapter focuses on whole system efficiency in the broadest possible sense, that of optimizing energy and material flows. For electric end-uses that means promoting the highest efficiency devices, but it also means enabling information flows and technologies that can better match energy use to times of day when electricity is most plentiful and inexpensive. In addition, it means reducing food waste, improving industrial process controls, and minimizing material flows.

Pillar five is *Remove carbon*. Carbon dioxide stays in the atmosphere for centuries after it's emitted. There are unproven and complicated technical solutions that may help with this effort, but the most likely options involve making the land and seas 'climate positive'. That means stopping industrial-scale deforestation everywhere, aggressively reforesting where it makes sense, creating vast ocean sanctuaries, 'farming' seaweed, and slowing warming in any ways we can (because a warming ocean can hold less carbon dioxide).

Pillar six is *Align incentives*. The fossil-fuel industry has had more than a century to rig the system, privatizing profits and socializing costs. The real costs of pollution are invisible to most market participants, and structural subsidies (like weak regulations, non-existent enforcement, and inadequate bonding requirements for oil and gas wells) fly under the radar. We'll need to re-design markets and re-align incentives throughout the society to reflect the goal of getting to zero emissions as soon as possible.

Pillar seven is *Mobilize money*. Solving the climate problem means replacing high-emissions infrastructure with capital and technology, facilitated by financial incentives. That shift almost always means movement away from periodic expenses for fuel and maintenance to more capital investments. Financial innovation (such as new kinds of standardized loans for homes and small businesses) can help smooth that transition. In addition, significant capital continues to be allocated to fossil-fuel investments, mainly for historical reasons, and that needs to stop immediately, with that money re-allocated to building zero-emissions capital.

Pillar eight is *Elevate truth*. Reducing emissions means confronting the most powerful industry in human history and unwinding the many ways the economy has been tilted towards fossil fuels over more than a century. Those who question the findings of climate science and argue against immediate and rapid emissions reductions are no longer engaged in the scientific process, but in a propaganda exercise. For too long, scientists have taken objections to climate science as good faith arguments from other scientists, but that has enabled the forces of denial and delay. Elevating truth and fighting propaganda are critical to a successful transition to a zero-emissions society.

2.11 Visualizing successful climate action

Our idealized vision for climate action is well characterized by figure 2.7, taken from analysis in [126], which illustrates what we mean by net zero by 2040, climate positive thereafter. By 2035, deforestation and destruction of other undisturbed ecosystems is entirely halted, with afforestation, reforestation, changing agricultural practices, and coastal restoration yielding up to 10 Gt CO_2 uptake per year just after 2040. At the same time, fossil-fuel related emissions decline rapidly and drop to zero just after 2050. Net zero by 2040 is not the end of the journey, it's just an important milestone along the way to a climate-positive world.

2.12 We have to do it all

Under normal circumstances pushing hard on some options would allow us to push less hard on others. For example, shouldn't we be able to worry less about aggressive emissions reductions if we can scale up adaptation, geo-engineering, or direct air capture of carbon? Shouldn't we be able to accept slower reductions in carbon dioxide if we can rapidly scale up reductions in emissions of other warming agents? The answer, unfortunately, is no. We've dithered so long that we now need to push

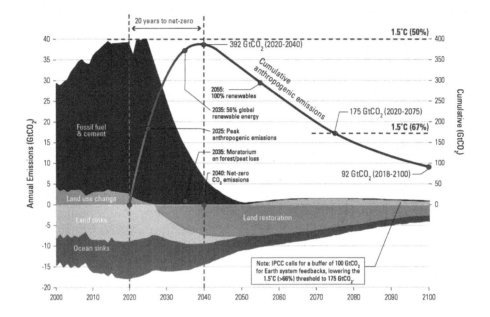

Figure 2.7. Successful climate action means net zero by 2040, climate positive thereafter. Source: Taken from [126: figure 9]. Net zero is achieved in 2040, with carbon removal increasing to 2050, then gradually declining. The biosphere starts offgassing carbon around 2050 when fossil emissions stop, which somewhat diminishes the effects of carbon removal. Reproduced under Creative Commons Attribution 4.0 International License (CC BY 4.0): https://creativecommons.org/licenses/by/4.0/. Design by Karl Burkart.

forward on all options as rapidly as we can. We need to do everything that's feasible and effective, and we need to start immediately.

For example, planting trees and other ways of removing carbon from the atmosphere are treated by some companies as 'offsets', to 'compensate' for emissions elsewhere. This way of thinking conveys the pernicious idea that carbon removal can substitute for emissions reductions. Removing carbon will always be harder, slower, and more expensive than not emitting carbon in the first place, and while we should scale up carbon removal options as fast as we can (as we discuss in chapter 8) they are no substitute for immediate, rapid, and sustained direct emissions reductions.

We must choose alternatives strategically, learning as we go. Focusing on short-lived high potency warming agents (such as methane, HFC-32, HFC-134a, and black carbon) can result in immediate benefits in short run warming [127–129]. That doesn't mean we can slack off on carbon dioxide emissions reductions, we need to reduce ALL emissions quickly. The effects of carbon emission reductions will just take longer to show up in global temperatures, so moving quickly on the other warming agents can pay quicker dividends.

Short-lived warming agents with high GWPs offer the possibility of slowing warming in the near term. We must continue to pursue rapid reductions in carbon dioxide emissions but reducing emissions of other warming agents rapidly must be a strategic priority because of their short-term benefits.

'Doing it all' also doesn't mean throwing money at technologies and policies that are proven not to work. As discussed above, 'adaptive challenges' such as climate change require an experimental and evolutionary approach. The National Research Council [130] calls such a strategy 'iterative risk management', implying that we must try many options, fail fast, and do more of what works and less of what doesn't. It's also an approach consonant with Bayesian updating [131, 132].

Technologies that fail to scale and achieve cost reductions after many attempts must be abandoned, because we can't afford to waste talent, time, money, and effort on things that don't work. It's an emergency and we need to prioritize our efforts for maximal effect. So 'all of the above' for zero-emissions options doesn't mean continuing to invest in failed technologies, it means trying everything and determining what path is quickest and most efficacious, discarding what isn't working.

Some widely cited 'alternatives' also don't address all aspects of the problem. Take 'solar radiation management', which involves increasing the reflectivity of the planet by injecting particles in the atmosphere or other means [133]. While efforts like these might temporarily reduce global temperatures, they do nothing about ocean acidification as increasing CO_2 is absorbed by the oceans.

Of course, there are also other good reasons for avoiding such geo-engineering shenanigans, which could have unfortunate side effects on weather and human health [134] and create what's called 'moral hazard' if policy makers falsely think that these efforts are a substitute for emissions reductions [135]. The picture becomes even more complicated once we accept that our power to change the natural environment far exceeds our ability to predict the effects of those changes, and always will, as David Bella pointed out in the late 1970s [136].

Geo-engineering may ultimately be necessary as an emergency option but they are treated by some as an alternative to immediate, rapid, and sustained emissions reductions, which they most certainly are not. We do not consider them further here.

2.13 Who's responsible?

At http://www.solveclimate.org, we present **Action item checklists** for government, businesses, investors and donors, and individuals, communities, and non-profits. These are concrete actions to jump-start emissions reductions, and many of them will feature prominently in plans to get to climate positive.

Each country, state, or company is different, and not all action items will apply to your situation. That's where your research is needed, to customize recommendations and findings from other studies to the unique context under study.

2.14 Chapter conclusions

We offer a comprehensive vision of solving the climate problem that embraces previous work but expands the focus to include non-fossil emissions, market design, financial innovation, and the fight against propaganda and misinformation. We also identify concrete 'action items' for key actors to start the transition to a zero-emissions society.

Further reading

Davis S J *et al* 2018 Net-zero emissions energy systems *Science* **360** eaas9793 https://doi.org/10.1126/science.aas9793. A widely cited article on achieving net-zero emissions by some of the top researchers in the field.

Dessler A 2022 How greed and politics are slowing the switch to renewable energy *Bull. Atom. Sci.* 17 January https://thebulletin.org/2022/01/how-greed-and-politics-are-slowing-the-switch-to-renewable-energy/. A compact and clear summary of the climate problem and impediments to potential solutions.

Goldstein-Rose S 2020 *The 100% Solution: A Plan for Solving Climate Change* (New York: Melville House). This book is a terrific summary of climate solutions for non-technical audiences. It lists five pillars that have some overlap with our pillars, but is not as comprehensive as our complete list.

Hawken P (ed) 2017 *Drawdown: The Most Comprehensive Plan Ever Proposed to Reverse Global Warming* (New York: Penguin Books). Project Drawdown lays out an extensive list of solutions to the climate problem, and the work is ongoing. Their current list of solutions is here https://drawdown.org/solutions/table-of-solutions.

IPCC 2022 *Climate Change 2022: Mitigation of Climate Change. Contribution of Working Group III to the Sixth Assessment Report of the Intergovernmental Panel on Climate Change* (Cambridge: Cambridge University Press) https://ipcc.ch/report/sixth-assessment-report-working-group-3/. The latest IPCC consensus report on mitigation of climate change.

National Academies of Sciences, Engineering, and Medicine 2021 *Accelerating Decarbonization of the US Energy System* (Washington, DC: The National Academies Press) https://nap.edu/catalog/25932/accelerating-decarbonization-of-the-us-energy-system. High-level view of technology and policy changes needed to speed up decarbonization of the US energy system.

Williams J H, Jones R A, Haley B, Kwok G, Hargreaves J, Farbes J and Torn M S 2021 Carbon-neutral pathways for the United States *AGU Adv.* **2** e2020AV000284 https://doi.org/10.1029/2020AV000284. An exemplary recent study of carbon emissions reductions in the US energy sector.

References

[1] IPCC 2021 *Climate Change 2021: The Physical Science Basis. Contribution of Working Group I to the Sixth Assessment Report of the Intergovernmental Panel on Climate Change* ed V Masson-Delmotte *et al* (Cambridge: Cambridge University Press) https://ipcc.ch/report/sixth-assessment-report-working-group-i/

[2] Marshall B, Driscoll A, Heft-Neal S, Xue J, Burney J and Wara M 2021 The changing risk and burden of wildfire in the United States *Proc. Natl Acad. Sci.* **118** e2011048118

[3] Moore F C, Lacasse K, Mach K J, Yoon A S, Gross L J and Beckage B 2022 Determinants of emissions pathways in the coupled climate–social system *Nature* **603** 103–11

[4] Koomey J G 2012 *Cold Cash, Cool Climate: Science-Based Advice for Ecological Entrepreneurs* (El Dorado Hills, CA: Analytics)

[5] Grübler A *et al* 2018 A low energy demand scenario for meeting the 1.5 °C target and sustainable development goals without negative emission technologies *Nat. Energy* **3** 515–27

[6] Anderegg W R L, Abatzoglou J T, Anderegg L D L, Bielory L, Kinney P L and Ziska L 2021 Anthropogenic climate change is worsening North American pollen seasons *Proc. Natl Acad. Sci.* **118** e2013284118

[7] Kemp L *et al* 2022 Climate endgame: exploring catastrophic climate change scenarios *Proc. Natl Acad. Sci.* **119** e2108146119

[8] Simpson R D 2022 How do we price an unknowable risk? *Issues in Science and Technology* Winter https://issues.org/social-cost-of-carbon-economics-simpson/

[9] Weitzman M L 2009 On modeling and interpreting the economics of catastrophic climate change *Rev. Econ. Stat.* **91** 1–19

[10] Weitzman M L 2010 What is the 'damages function' for global warming—and what difference might it make? *Clim. Change Econ.* **1** 57–69

[11] Weitzman M L 2011 Fat-tailed uncertainty in the economics of catastrophic climate change *Rev. Environ. Econ. Policy* **5** 275–92

[12] IPCC 2022 *Climate Change 2022: Mitigation of Climate Change. Contribution of Working Group III to the Sixth Assessment Report of the Intergovernmental Panel on Climate Change* (Cambridge: Cambridge University Press) https://ipcc.ch/report/sixth-assessment-report-working-group-3/

[13] Allen M R, Friedlingstein P, Girardin C A J, Jenkins S, Malhi Y, Mitchell-Larson E, Peters G P and Rajamani L 2022 Net zero: science, origins, and implications *Ann. Rev. Environ. Resources* **47** 849–87

[14] Matthews H D, Gillett N P, Stott P A and Zickfeld K 2009 The proportionality of global warming to cumulative carbon emissions *Nature* **459** 829–32

[15] Zickfeld K, Eby M, Matthews H D and Weaver A J 2009 Setting cumulative emissions targets to reduce the risk of dangerous climate change *Proc. Natl Acad. Sci.* **106** 16129

[16] Ricke K L and Caldeira K 2014 Maximum warming occurs about one decade after a carbon dioxide emission *Environ. Res. Lett.* **9** 124002

[17] IEA 2010 *World Energy Outlook 2010* (Paris: International Energy Agency, Organization for Economic Cooperation and Development) http://worldenergyoutlook.org/

[18] IEA 2012 *World Energy Outlook 2012* (Paris: International Energy Agency, Organization for Economic Cooperation and Development) http://worldenergyoutlook.org/

[19] Rockström J, Gaffney O, Rogelj J, Meinshausen M, Nakicenovic N and Schellnhuber H J 2017 A roadmap for rapid decarbonization *Science* **355** 1269

[20] IPCC 2018 *Global Warming of 1.5 °C. An IPCC Special Report on the Impacts of Global Warming of 1.5 °C Above Pre-industrial Levels and Related Global Greenhouse Gas Emission Pathways, in the Context of Strengthening the Global Response to the Threat of Climate Change, Sustainable Development, and Efforts to Eradicate Poverty* (Geneva: IPCC) https://ipcc.ch/sr15

[21] Matthews H *et al* 2021 An integrated approach to quantifying uncertainties in the remaining carbon budget *Commun. Earth Environ.* **2** 7

[22] Koomey J G, Matthews H S and Williams E 2013 Smart everything: will intelligent systems reduce resource use? *Annu. Rev. Environ. Res.* **38** 311–43

[23] Nordhaus W D 1992 An optimal transition path for controlling greenhouse gases *Science* **258** 1315

[24] Duane T, Koomey J, Belyeu K and Hausker K 2016 From risk to return: investing in a clean energy economy *Risky Business* http://riskybusiness.org/fromrisktoreturn/

[25] Williams S T *et al* 2021 Carbon-neutral pathways for the United States *AGU Adv.* **2** e2020AV000284

[26] IPCC 2007 *Climate Change 2007: Mitigation of Climate Change—Contribution of Working Group III to the Fourth Assessment Report of the Intergovernmental Panel on Climate Change* ed B Metz *et al* (Cambridge: Cambridge University Press) http://ipcc.ch/publications_and_data/publications_and_data_reports.shtml

[27] Vohra K, Vodonos A, Schwartz J, Marais E A, Sulprizio M P and Mickley L J 2021 Global mortality from outdoor fine particle pollution generated by fossil fuel combustion: results from GEOS-Chem *Environ. Res.* **195** 110754

[28] Roberts D 2020 Air pollution is much worse than we thought: ditching fossil fuels would pay for itself through clean air alone *Vox* 12 August https://vox.com/energy-and-environment/2020/8/12/21361498/climate-change-air-pollution-us-india-china-deaths

[29] Cohen A J *et al* 2017 Estimates and 25-year trends of the global burden of disease attributable to ambient air pollution: an analysis of data from the Global Burden of Diseases Study 2015 *Lancet* **389** 1907–18

[30] Daniel K D, Litterman R B and Wagner G 2019 Declining CO_2 price paths *Proc. Natl Acad. Sci.* **116** 20886

[31] Arthur W B 1990 Positive feedbacks in the economy *Sci. Am.* February pp 92–9

[32] Fischer C and Newell R G 2008 Environmental and technology policies for climate mitigation *J. Env. Econ. Manag.* **55** 142–62

[33] Sharpe S and Lenton T M 2021 Upward-scaling tipping cascades to meet climate goals: plausible grounds for hope *Clim. Policy* **21** 421–33

[34] Mohai P, Pellow D and Roberts J T 2009 Environmental justice *Ann. Rev. Environ. Res.* **34** 405–30

[35] Howarth R B 2011 Intergenerational justice *The Oxford Handbook of Climate Change and Society* ed J S Dryzek *et al* (Oxford: Oxford University Press) ch 23 pp 338–52

[36] Kühne K, Bartsch N, Tate R D, Higson J and Habet A 2022 'Carbon bombs'—mapping key fossil fuel projects *Energy Policy* **166** 112950

[37] Jonas H 1973 Technology and responsibility: reflections on the new task of ethics *Soc. Res.* **40** 31–54

[38] Roe G 2013 Costing the Earth: a numbers game or a moral imperative? *Weather Clim. Soc.* **5** 378–80

[39] Ackerman F and Heinzerling L 2004 *Priceless: On Knowing the Price of Everything and the Value of Nothing* (New York: The New Press)

[40] Köberle A C, Vandyck T, Guivarch C, Macaluso N, Bosetti V, Gambhir A, Tavoni M and Rogelj J 2021 The cost of mitigation revisited *Nat. Clim. Change* **11** 1035–45

[41] Howarth R B 1996 Climate change and overlapping generations *Contemp. Econ. Policy* **14** 100–11

[42] Craig P, Gadgil A and Koomey J 2002 What can history teach us? A retrospective analysis of long-term energy forecasts for the US *Annual Review of Energy and the Environment 2002* ed R H Socolow *et al* (Palo Alto, CA: Annual Reviews) pp 83–118

[43] Koomey J G, Craig P, Gadgil A and Lorenzetti D 2003 Improving long-range energy modeling: a plea for historical retrospectives *Energy J.* **24** 75–92

[44] Scher I and Koomey J G 2011 Is accurate forecasting of economic systems possible? *Clim. Change* **104** 473–9

[45] Ertel C and Solomon L K 2014 *Moments of Impact: How to Design Strategic Conversations that Accelerate Change* (New York: Simon and Schuster)

[46] Koomey J 2014 *Climate Change as an Adaptive Challenge* (Amsterdam: Knovel) http://koomey.com/post/86220629363

[47] Harvey H, Orvis R and Rissman J 2018 *Designing Climate Solutions: A Policy Guide for Low-Carbon Energy* (Washington, DC: Island)

[48] Green F and Denniss R 2018 Cutting with both arms of the scissors: the economic and political case for restrictive supply-side climate policies *Clim. Change* **150** 73–87

[49] Asheim G B, Fæhn T, Nyborg K, Greaker M, Hagem C, Harstad B, Hoel M O, Lund D and Rosendahl K E 2019 The case for a supply-side climate treaty *Science* **365** 325

[50] Newell P and Simms A 2019 Towards a fossil fuel non-proliferation treaty *Clim. Policy* **20** 1043–54

[51] Erickson P, Lazarus M and Piggot G 2018 Limiting fossil fuel production as the next big step in climate policy *Nat. Clim. Change* **8** 1037–43

[52] IEA 2021 *Net Zero by 2050: A Roadmap for the Global Energy Sector* (Paris: International Energy Agency) https://iea.org/reports/net-zero-by-2050

[53] Global Witness 2022 IPCC clarion call puts spotlight on fossil fuel industry's hypocrisy *Global Witness* 12 April https://globalwitness.org/en/campaigns/fossil-gas/ipcc-clarion-call-puts-spotlight-on-fossil-fuel-industrys-hypocrisy/

[54] Oil Change International 2022 *Banking on Climate Chaos 2022: Fossil Fuel Finance Report* 30 March https://priceofoil.org/2022/03/30/banking-on-climate-chaos-2022/

[55] Bresnahan T F and Trajtenberg M 1995 General purpose technologies engines of growth? *J. Econometrics* **65** 83–108

[56] Brynjolfsson E and McAfee A 2011 *Race Against the Machine* (Cambridge, MA: Digital Frontier) https://www.brynjolfsson.com/books

[57] Brynjolfsson E and McAffee A 2014 *The Second Machine Age: Work, Progress, and Prosperity in a Time of Brilliant Technologies* (New York: Norton)

[58] McAfee A and Brynjolfsson E 2017 *Machine, Platform, Cloud: Harnessing Our Digital Future* (New York: Norton)

[59] Bodnar P, Gray M, Grbusic T, Herz S, Lonsdale A, Mardell S, Ott C, Sundaresan S and Varadarajan U 2020 *How to Retire Early: Making Accelerated Coal Phaseout Feasible and Just* (Old Snowmass, CO: Rocky Mountain Institute) https://rmi.org/insight/how-to-retire-early/

[60] Davis R J, Holladay J S and Sims C 2021 Coal-fired power plant retirements in the US *National Bureau of Economic Research Working Paper Series* 28949 National Bureau of Economic Research, Cambridge, MA http://nber.org/papers/w28949

[61] Muller N Z, Mendelsohn R and Nordhaus W 2011 Environmental accounting for pollution in the United States economy *Am. Econ. Rev.* **101** 1649–75

[62] Epstein P R *et al* 2011 Full cost accounting for the life cycle of coal *Ann. NY Acad. Sci.* **1219** 73–98

[63] Greenstone M, Hasenkopf C and Lee K 2021 *Air Quality Life Index: Annual Update* (Chicago, IL: Energy Policy Institute, University of Chicago) https://aqli.epic.uchicago.edu/reports/

[64] Burnett R *et al* 2018 Global estimates of mortality associated with long-term exposure to outdoor fine particulate matter *Proc. Natl Acad. Sci.* **115** 9592

[65] Landrigan P J *et al* 2018 The *Lancet* Commission on pollution and health *Lancet* **391** 462–512

[66] Schraufnagel D E *et al* 2019 Air pollution and noncommunicable diseases: a review by the Forum of International Respiratory Societies 2019; Environmental Committee, part 1: the damaging effects of air pollution *Chest* **155** 409–16

[67] Schraufnagel D E *et al* 2019 Air pollution and noncommunicable diseases: a review by the Forum of International Respiratory Societies 2019; Environmental Committee, part 2: air pollution and organ systems *Chest* **155** 417–26

[68] Gao J *et al* 2018 Public health co-benefits of greenhouse gas emissions reduction: a systematic review *Sci. Total Environ.* **627** 388–402

[69] Choma E F, Evans J S, Hammitt J K, Gómez-Ibáñez J A and Spengler J D 2020 Assessing the health impacts of electric vehicles through air pollution in the United States *Environ. Int.* **144** 106015

[70] Dimanchev E G *et al* 2019 Health co-benefits of sub-national renewable energy policy in the US *Environ. Res. Lett.* **14** 085012

[71] WHO 2015 *Economic Cost of the Health Impact of Air Pollution In Europe: Clean Air, Health, and Wealth* (Copenhagen: World Health Organization, Regional Office for Europe) https://www.euro.who.int/__data/assets/pdf_file/0004/276772/Economic-cost-health-impact-air-pollution-en.pdf

[72] Oudin A, Segersson D, Adolfsson R and Forsberg B 2018 Association between air pollution from residential wood burning and dementia incidence in a longitudinal study in Northern Sweden *PLoS One* **13** e0198283

[73] Ghosh R, Causey K, Burkart K, Wozniak S, Cohen A and Brauer M 2021 Ambient and household $PM_{2.5}$ pollution and adverse perinatal outcomes: a meta-regression and analysis of attributable global burden for 204 countries and territories *PLoS Med.* **18** e1003718

[74] Taylor J H *et al* 2021 A framework for estimating the United States depression burden attributable to indoor fine particulate matter exposure *Sci. Total Environ.* **756** 143858

[75] Wallace-Wells D 2021 Ten million a year *Lond. Rev. Books* 2 December https://lrb.co.uk/the-paper/v43/n23/david-wallace-wells/ten-million-a-year

[76] Lobell D B, Tommaso S D and Burney J A 2022 Globally ubiquitous negative effects of nitrogen dioxide on crop growth *Sci. Adv.* **8** eabm9909

[77] Keswani A, Akselrod H and Anenberg S C 2022 Health and clinical impacts of air pollution and linkages with climate change *NEJM Evidence* **1** https://doi.org/10.1056/EVIDra2200068

[78] Lovins A B 1976 Energy strategy: the road not taken? *Foreign Affairs* **55** 65–92 https://foreignaffairs.com/articles/united-states/1976-10-01/energy-strategy-road-not-taken

[79] Lovins A B 1979 *Soft Energy Paths: Toward a Durable Peace* (New York: Harper Colophon)

[80] Wilson C *et al* 2020 Granular technologies to accelerate decarbonization *Science* **368** 36

[81] Meyer R 2021 *The Weekly Planet*: the big idea from Bill Gates's new climate book—call it the 'Gates Rule' *Atlantic* 16 February https://theatlantic.com/science/archive/2021/02/bill-gatess-new-way-to-understand-climate-action/618043/

[82] Nordhaus W D 2008 *A Question of Balance: Weighing the Options on Global Warming Policies* (New Haven, CT: Yale University Press)

[83] UNEP 2020 *Emissions Gap Report 2020* (Nairobi: United National Environment Programme) https://unep.org/emissions-gap-report-2020

[84] Rubin E S, Azevedo I M L, Jaramillo P and Yeh S 2015 A review of learning rates for electricity supply technologies *Energy Policy* **86** 198–218

[85] Grübler A, Nakicenovic N and Victor D G 1999 Dynamics of energy technologies and global change *Energy Policy* **27** 247–80

[86] Gillingham K and Stock J H 2018 The cost of reducing greenhouse gas emissions *J. Econ. Perspect.* **32** 53–72

[87] Bos K and Gupta J 2019 Stranded assets and stranded resources: implications for climate change mitigation and global sustainable development *Energy Res. Soc. Sci.* **56** 101215

[88] Semieniuk G, Holden P B, Mercure J-F, Salas P, Pollitt H, Jobson K, Vercoulen P, Chewpreecha U, Edwards N R and Viñuales J E 2022 Stranded fossil-fuel assets translate to major losses for investors in advanced economies *Nat. Clim. Change* **12** 532–8

[89] Kemfert C, Präger F, Braunger I, Hoffart F M and Brauers H 2022 The expansion of natural gas infrastructure puts energy transitions at risk *Nat. Energy* **7** 582–7

[90] CEA 2014 *The Cost of Delaying Action to Stem Climate Change* (Washington, DC: White House Council of Economic Advisors) https://obamawhitehouse.archives.gov/sites/default/files/docs/the_cost_of_delaying_action_to_stem_climate_change.pdf

[91] Swiss Re Institute 2021 *The Economics of Climate Change: No Action Not an Option* (Zurich: Swiss Re Institute) https://swissre.com/institute/research/topics-and-risk-dialogues/climate-and-natural-catastrophe-risk/expertise-publication-economics-of-climate-change.html

[92] IEA 2009 *World Energy Outlook 2009* (Paris: International Energy Agency, Organization for Economic Cooperation and Development) http://worldenergyoutlook.org/

[93] McDuffie E E *et al* 2021 Source sector and fuel contributions to ambient $PM_{2.5}$ and attributable mortality across multiple spatial scales *Nat. Commun.* **12** 3594

[94] Koomey J, Schmidt Z, Hausker K and Lashof D 2022 Black boxes revealed: assessing key drivers of 1.5 °C warming scenarios *White Paper* Koomey Analytics, Bay Area, CA

[95] Nemet G 2019 *How Solar Became Cheap: A Model for Low-Carbon Innovation* (New York: Routledge) https://howsolargotcheap.com

[96] Nagy B, Farmer J D, Bui Q M and Trancik J E 2013 Statistical basis for predicting technological progress *PLoS One* **8** e52669

[97] Wright T P 1936 Factors affecting the cost of airplanes *J. Aeronaut. Sci.* **3** 122–8

[98] Alchian A 1963 Reliability of progress curves in airframe production *Econometrica* **31** 679–93

[99] Thompson P 2012 The relationship between unit cost and cumulative quantity and the evidence for organizational learning-by-doing *J. Econ. Perspect.* **26** 203–24

[100] Way R, Lafond F, Lillo F, Panchenko V and Farmer J D 2019 Wright meets Markowitz: how standard portfolio theory changes when assets are technologies following experience curves *J. Econ. Dyn. Control* **101** 211–38

[101] Nemet G F 2006 Beyond the learning curve: factors influencing cost reductions in photovoltaics *Energy Policy* **34** 3218–32

[102] McDonald A and Schrattenholzer L 2001 Learning rates for energy technologies *Energy Policy* **29** 255–61

[103] Krey V *et al* 2019 Looking under the hood: a comparison of techno-economic assumptions across national and global integrated assessment models *Energy* **172** 1254–67

[104] Creutzig F *et al* 2017 The underestimated potential of solar energy to mitigate climate change *Nat. Energy* **2** 17140

[105] Victoria M *et al* 2021 Solar photovoltaics is ready to power a sustainable future *Joule* **5** P1041–56

[106] Jaxa-Rozen M and Trutnevyte E 2021 Sources of uncertainty in long-term global scenarios of solar photovoltaic technology *Nat. Clim. Change* **11** 266–73

[107] Xiao M, Junne T, Haas J and Klein M 2021 Plummeting costs of renewables—are energy scenarios lagging? *Energy Strat. Rev.* **35** 100636

[108] Luderer G *et al* 2022 Impact of declining renewable energy costs on electrification in low-emission scenarios *Nat. Energy* **7** 32–42

[109] Grant N, Hawkes A, Napp T and Gambhir A 2021 Cost reductions in renewables can substantially erode the value of carbon capture and storage in mitigation pathways *One Earth* **4** 1588–601

[110] Hausfather Z 2019 UNEP: 1.5 °C climate target 'slipping out of reach' *Carbon Brief* https://carbonbrief.org/unep-1-5c-climate-target-slipping-out-of-reach

[111] UNEP 2021 *Emissions Gap Report 2021: The Heat Is On—A World of Climate Promises Not Yet Delivered* (Nairobi: United National Environment Programme) https://unep.org/emissions-gap-report-2021

[112] US EPA 2022 *Inventory of US Greenhouse Gas Emissions and Sinks: 1990–2020* (Washington, DC: US Environmental Protection Agency) https://epa.gov/ghgemissions/inventory-us-greenhouse-gas-emissions-and-sinks

[113] Friedlingstein P *et al* 2022 Global carbon budget 2021 *Earth Syst. Sci. Data Discuss.* **14** 1917–2005

[114] Bond T C *et al* 2013 Bounding the role of black carbon in the climate system: a scientific assessment *J. Geophys. Res.: Atmos.* **118** 5380–552

[115] Warwick N, Griffiths P, Keeble J, Archibald A, Pyle J and Shine K 2022 *Atmospheric Implications of Increased Hydrogen Use* (Cambridge: University of Cambridge, NCAS, and University of Reading) https://assets.publishing.service.gov.uk/government/uploads/system/uploads/attachment_data/file/1067144/atmospheric-implications-of-increased-hydrogen-use.pdf

[116] Keith D W, Holmes G, St Angelo D and Heidel K 2018 A process for capturing CO_2 from the atmosphere *Joule* **2** 1573–94

[117] APS 2011 *Direct Air Capture of CO_2 with Chemicals: A Technology Assessment for the APS Panel on Public Affairs* (Washington, DC: American Physical Society) https://aps.org/policy/reports/assessments/upload/dac2011.pdf

[118] Sanz-Pérez E S, Murdock C R, Didas S A and Jones C W 2016 Direct capture of CO_2 from ambient air *Chem. Rev.* **116** 11840–76

[119] Buck H J 2020 Should carbon removal be treated as waste management? Lessons from the cultural history of waste *Interface Focus* **10** 20200010

[120] Grant N, Hawkes A, Mittal S and Gambhir A 2021 The policy implications of an uncertain carbon dioxide removal potential *Joule* **5** 2593–605

[121] Madhu K, Pauliuk S, Dhathri S and Creutzig F 2021 Understanding environmental trade-offs and resource demand of direct air capture technologies through comparative life-cycle assessment *Nat. Energy* **6** 1035–44

[122] National Research Council 2015 *Climate Intervention: Carbon Dioxide Removal and Reliable Sequestration* (Washington, DC: The National Academies Press) https://nap.edu/catalog/18805/climate-intervention-carbon-dioxide-removal-and-reliable-sequestration

[123] National Academies of Sciences, Engineering, and Medicine 2019 *Negative Emissions Technologies and Reliable Sequestration: A Research Agenda* (Washington, DC: The National Academies Press) https://nap.edu/catalog/25259/negative-emissions-technologies-and-reliable-sequestration-a-research-agenda

[124] National Academies of Sciences, Engineering, and Medicine 2021 *A Research Strategy for Ocean-based Carbon Dioxide Removal and Sequestration* (Washington, DC: The National Academies Press) https://nap.edu/catalog/26278/a-research-strategy-for-ocean-based-carbon-dioxide-removal-and-sequestration

[125] Goldstein-Rose S 2020 *The 100% Solution: A Plan for Solving Climate Change* (Brooklyn, NY: Melville House)

[126] Teske S (ed) 2022 Climate sensitivity analysis: all greenhouse gases and aerosols *Achieving the Paris Climate Agreement Goals: Part 2: Science-based Target Setting for the Finance industry—Net-Zero Sectoral 1.5 °C Pathways for Real Economy Sectors* (Cham: Springer International) pp 273–90

[127] Molina M, Zaelke D, Sarma K M, Andersen S O, Ramanathan V and Kaniaru D 2009 Reducing abrupt climate change risk using the Montreal protocol and other regulatory actions to complement cuts in CO_2 emissions *Proc. Natl Acad. Sci.* **106** 20616

[128] Ramanathan V and Yangyang X 2010 The Copenhagen accord for limiting global warming: criteria, constraints, and available avenues *Proc. Natl Acad. Sci.* **107** 8055

[129] Dreyfus G B, Xu Y, Shindell D T, Zaelke D and Ramanathan V 2022 Mitigating climate disruption in time: a self-consistent approach for avoiding both near-term and long-term global warming *Proc. Natl Acad. Sci. USA* **119** e2123536119

[130] NRC 2011 *America's Climate Choices* (Washington, DC: National Research Council of the National Academies, The National Academies Press) https://www.nationalacademies.org/our-work/americas-climate-choices

[131] Howson C and Urbach P 2005 *Scientific Reasoning: The Bayesian Approach* (McLean, VA: Open Court)

[132] Clayton A 2022 I'm a parent and a statistician. There's a smarter way to think about the under-5 vaccine *The New York Times* 1 March https://nytimes.com/2022/03/01/opinion/under-5-vaccine.html

[133] National Academies of Sciences, Engineering, and Medicine 2021 *Reflecting Sunlight: Recommendations for Solar Geoengineering Research and Research Governance* (Washington, DC: The National Academies Press) https://nap.edu/catalog/25762/reflecting-sunlight-recommendations-for-solar-geoengineering-research-and-research-governance

[134] Carlson C J, Colwell R, Sharif Hossain M, Mofizur Rahman M, Robock A, Ryan S J, Shafiul Alam M and Trisos C H 2022 Solar geoengineering could redistribute malaria risk in developing countries *Nat. Commun.* **13** 2150

[135] Piper K 2019 The climate renegade: what happens when someone wants to go it alone on fixing the climate? *Vox* 4 June https://vox.com/the-highlight/2019/5/24/18273198/climate-change-russ-george-unilateral-geoengineering

[136] Bella D A 1979 Technological constraints on technological optimism *Technol. Forecast. Soc. Change* **14** 15–26

Chapter 3

Tools of the trade

Data, Data, Data! I can't make bricks without clay!

—Sherlock Holmes

Chapter overview

- Creating an emissions reduction plan always starts with an emissions inventory.
- For the energy system, build an emissions inventory from an energy balance.
- Next, create a business-as-usual projection for every sector and human activity from the base-year emissions inventory.
- Reviewing previous studies helps in designing the climate-positive scenario.
- Whole-systems integrated design is the way to achieve significant and rapid emissions reductions.
- Following the project template posted at http://www.solveclimate.org gives structure to projects and helps guide students in the right direction.

3.1 Beginning the journey

The first step in assessing how to make society climate positive is always the same: creating an inventory of greenhouse gas emissions associated with the country, community, or company you're analyzing. Such inventories tally emissions from all sources to help decisionmakers decide how best to reduce them.

3.1.1 Emissions inventories

The process of creating an emissions inventory helps organize efforts to reduce emissions, but if you're focused on a well-studied state, province, country, or company, an emissions inventory may already exist [1]. The US [2], California [3], and Europe [4] all have reasonably current official inventories that are periodically updated. The UN reports data from each country, typically at a higher level of aggregation than what the best individual country inventories show[1]. Data availability for companies is dependent on historical attention paid to these issues.

Standards for creating emissions inventories for companies, cities, and individual projects are managed by the **Greenhouse Gas Protocol** (GHG Protocol) group, which is a joint effort of the World Business Council for Sustainable Development and the World Resources Institute[2]. For a recent discussion of data and methodology issues around creating and understanding emissions inventories, see [5].

Boundary choices are critical, particularly for smaller institutions and geographies. The analysis in [6] found that 48 US cities undercounted their emissions on average about 18% mainly because of boundary choices. They also found that self-reporting of emissions without consistent standards limits the reliability of emissions data and calls into question the emissions reduction targets that rely upon them.

3.1.1.1 Energy balances
For fossil fuel combustion, estimating emissions starts with flows of fuel (natural gas, oil, and coal), data for which are typically easily available in what are called 'energy balances', in which all energy flows are tracked from production through to consumption. The canonical source of energy balances is available for virtually all countries from the International Energy Agency [7]. The normal assumption is that carbon in fuels is almost completely (99%) converted to CO_2 through combustion.

There are important special cases where fossil fuels are not completely combusted, with fossil gas venting from oil wells and methane leakage from fossil gas transport and distribution infrastructure being two obvious examples [8–10]. Assessing such leakage is an important part of an emissions inventory. In fact, every molecule of each fuel and their associated emissions need to be tracked and counted.

3.1.1.2 Industrial processes
Industrial processes are another important source of emissions. The manufacture of steel and aluminum, for example, use carbon that results in CO_2 process emissions unrelated to fossil fuel combustion. Redesign of such processes can eliminate the use of carbon and associated emissions but replacing current infrastructure with new plant and equipment can require significant investment. Assessing these options requires detailed technical knowledge of each process and potential alternatives.

3.1.1.3 Agriculture and food systems
Agriculture and food systems are important sources of emissions of methane and nitrous oxide, and tracking these emissions are often more complicated than for fossil fuel combustion and industrial processes. The data are less available and the

---

[1] https://unfccc.int/ghg-inventories-annex-i-parties/2021
[2] https://ghgprotocol.org

chemical and biological processes more complicated than for fuel combustion, but reasonable estimates are still possible in these sectors.

3.1.1.4 Land-use change

Land-use change is another important source of emissions that can be difficult to track precisely, in part because of the dispersed and diverse sources for these emissions. Most often it's CO_2 from deforestation and other land-use changes that are highlighted, but methane and other emissions are associated with such changes as well.

3.1.2 Mapping emissions onto human activities (tasks)

The emissions inventory should map emissions onto human activities (tasks), helping assess the best ways to reduce those emissions. Figure 3.1, shown in the IPCC's Fifth Assessment Report [11], from [12], uses the term 'service' for what we call tasks and 'fuel' for what we call primary energy, but the concept is the same. It shows total anthropogenic emissions in 2010 from all sources mapped onto tasks, sectors, equipment types, device types, final energy, and primary fuel consumption in what is called a Sankey diagram.

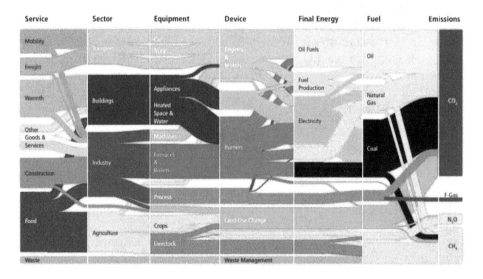

Figure 3.1. Sankey diagram for global emissions circa 2010. Source: Bajželj *et al* [12]. Reproduced under Creative Commons Attribution 4.0 International License (CC BY 4.0): https://creativecommons.org/licenses/by/4.0/. An interactive online version with 2014 data is posted at https://www.refficiency.org/projects/greenhouse-gas-emission-map. Original caption: A Sankey diagram showing the system boundaries of the industry sector and demonstrating how global anthropogenic emissions in 2010 arose from the chain of technologies and systems required to deliver final services triggered by human demand. The width of each line is proportional to GHG emissions released, and the sum of these widths along any vertical slice through the diagram is the same, representing all emissions in 2010.

The detailed data needed to create such an assessment are not always available, and there are often time lags in creating Sankey diagrams (because of data availability issues), but they offer a wonderful high-level map of an emissions inventory.

Because of the nature and urgency of the climate problem, virtually every activity in society will need to be re-engineered to minimize emissions. That's why a comprehensive map like the Sankey diagram in figure 3.1 is so important.

Understanding how emissions result from human activities suggests further inquiries. What activities are the largest sources of emissions? What value does each activity bring? Are there alternative zero-emission ways to generate the same or greater value? What people and institutions are responsible for accomplishing each task? For companies, what is their business model? For individuals, what do they value?

Focusing on tasks makes much bigger improvements possible than would a narrow focus on total emissions. Consider a simple example. If an engineer thinks of her job as burning fossil gas efficiently to reduce CO_2 emissions from operation of a home furnace, physics puts constraints on just how efficient that system can get and how much emissions can be reduced. The first law of thermodynamics says that we can't get more energy out of the furnace than we put in, and there will inevitably be losses.

For a basic new furnace installed in the US nowadays the efficiency as measured by the first law is about 80%. For every unit of energy that flows into the furnace as natural gas, 0.8 units of heat exits the furnace and flows into the heating ducts (some of that heat is lost by the ducts and never makes it to the rooms). The best new gas furnaces are so-called 'condensing' units, and they are around 95% efficient (0.95 units of heat from the furnace for every unit of gas input), but that's about the best we can do. 100% efficiency is an upper limit (according to the first law of thermodynamics) that we can never quite reach.

President Dwight D Eisenhower once famously said 'If you cannot solve a problem, make it bigger', and that dictum applies in this case as well. If the engineer defines the task more broadly as 'keeping rooms warm', more possibilities emerge.

The engineer could tighten the duct work to reduce heat losses so that more hot air gets to the rooms. She could design a super-insulated house that needs little heat to keep it comfortable, including a heat recovery ventilation system to bring in fresh air while recapturing heat from the warmed inside air. She could install active shading to allow sunlight to enter the windows when heat is needed and block the Sun when it's hot outside. She could also replace the gas furnace with a heat pump, which uses one unit of electricity to move three to five units of heat into the house.

Redefining tasks enables us to identify more and better opportunities for emissions reductions. This insight comes from a deep understanding of the second law of thermodynamics [13] and the discipline of whole-systems integrated design [14].

3.2 Rethinking the design process

Whole-systems integrated design encourages engineers to design new systems from scratch to accomplish the task at hand, ignoring constraints imposed by legacy hardware or archaic assumptions, while substantially reducing or eliminating emissions [14, 15]. Amory Lovins is the modern champion of this approach, and he coined the phrase 'factor ten engineering' to describe it [16].

This focus on the task (keeping rooms warm) is essential to achieving emissions reductions and efficiency improvements far beyond what is possible for an engineer working with a much narrower definition of the task (such as burning fossil gas in a furnace to heat a house).

The core insights driving whole-systems integrated design are four-fold:

- New technology only captures existing markets when it is much better than what it replaces. Incremental improvements of ten to twenty percent won't cut it.
- Redesigning the whole system helps new innovations capture multiple benefits, beyond what an incremental analysis would indicate, making it easier to achieve factors of ten or more improvements in the service delivered and associated emissions.
- The focus on dramatic improvements is a way to overcome arbitrary and imagined constraints to achieving such emissions reductions.
- Rapid changes in markets, technologies, and materials open up new possibilities for system design, even months after a product or service is released.

Whole-systems integrated design is a key tool to achieve the rapid emissions reductions we need to stabilize the climate.

3.3 Understanding capital stocks

One of the core concepts needed to understand how energy- and emissions-related systems change is that of capital stocks. An investment in equipment lasts years. In the US, cars typically last about 15 years, furnaces 25 years, building shells 50 to 100 years. Over time, old equipment retires and new equipment takes its place. Retirements can happen because of accidents, breakage, fires, or renovations, and new equipment is typically more efficient than the existing device it replaces.

Capital stock grows to meet increasing service demand. When population grows, the housing stock expands to meet that demand. New homes are constructed to meet that growth as well as to replace existing homes that have retired.

Assessments of emissions reduction potentials need to explicitly represent capital stock turnover. To do so, analysts begin with projections of growth in service demand (such as population, commercial building floor area, or numbers of water heaters) and inventories of existing capital stocks in a base year. They then apply a retirement function to the existing capital stocks. Growth in service demand minus the remaining stock from the base year gives the number of new buildings or devices sold in each year.

Average efficiency or emissions in any year is a function of growth in service demand, retirements, and the difference in efficiency or emissions intensity between the existing stock and new devices or buildings. Appendix B works through a more detailed example of capital stock turnover to help you design simple models for your own use.

3.4 Understanding key drivers of emissions

For understanding the key drivers of changes in greenhouse gas emissions from the energy sector over time, researchers often use the **Kaya identity**. This identity decomposes carbon emissions as a product of aggregate wealth, energy intensity of economic activity, and carbon intensity of the energy supplied. We show the familiar 'four-factor' Kaya identity in equation (3.1):

$$C_{\text{Fossil Fuels}} = P \cdot \frac{GNP}{P} \cdot \frac{PE}{GNP} \cdot \frac{C}{PE} \tag{3.1}$$

where $C_{\text{Fossil Fuels}}$ represents carbon dioxide (CO_2) emissions from fossil fuels combusted in the energy sector;

P is population;

GNP is gross national product, a measure of economic activity;

PE is primary energy, including conversion and energy transmission losses;

C is total net carbon dioxide emitted from the primary energy resource mix;

$\frac{GNP}{P}$ is the average income per person;

$\frac{PE}{GNP}$ is the primary energy intensity of the economy; and

$\frac{C}{PE}$ is the net carbon dioxide intensity of supplying primary energy.

The Kaya identity reflects a more general identity that expresses impact (I) as the product of human population (P), affluence (A), and technology (T) [17, 18], referred to as the **IPAT** identity. Population is the same in both the Kaya and IPAT identities, GNP/person represents affluence, and the other two terms characterize technology.

This formulation implies that a larger number of people with a higher income and more extensive use of certain technologies will have a greater impact on the environment. The role of technology can be ambiguous—technologies that produce and combust fossil fuels are the primary anthropogenic source of carbon dioxide, while technologies for harnessing renewable energy and nuclear power, sequestering carbon, and improving efficiency can reduce net anthropogenic carbon emissions.

Because we care about all emissions that cause warming, decomposing key drivers requires the more comprehensive relationship summarized in equation (3.2), which includes all emissions in terms of carbon dioxide equivalent:

$$C_{\text{Total}}^{\text{eq}} = C_{\text{Fossil Fuels}} + C_{\text{Industry}} + C_{\text{Land-use}} + C_{\text{Non-CO}_2\text{gases}}^{\text{eq}} - CS_{\text{Biomass}} \tag{3.2}$$

where

$C_{\text{Fossil Fuels}}$ is defined in equation (3.1);

C_{Industry} represents carbon dioxide emissions from industrial processes (non-energy uses of fossil fuels that result in emissions, such as cement and aluminum production), some models combine these emissions with fossil fuel combustion emissions, but they should be split out for clarity and internal consistency checks;

$C_{\text{Land-use}}$ represents net carbon dioxide emissions from changes in agriculture and land use that are not associated with emissions reductions from biomass CCS, this term can be negative if there is significant reforestation;

$C_{\text{Non-CO}_2\text{gases}}^{\text{eq}}$ represents emissions of other greenhouse gases converted to CO_2 equivalent using relative factors of global warming potential (GWP); and

CS_{Biomass} represents net negative emissions from sequestering carbon emissions associated with biomass combustion (in effect, such sequestration removes carbon from the biosphere), the emissions reductions from this source must be carefully distinguished from those land-use changes.

If direct air capture of CO_2 is present in future scenarios (as seems likely) an additional term would be needed in equation (3.2).

As many researchers have realized over the years, the Kaya identity as it was originally introduced is incomplete. Appendix D presents an expanded version of that identity to be included in a more comprehensive decomposition. This expanded version was first laid out in detail by Koomey *et al* in 2019 [19] and expanded further by Koomey *et al* in 2022 [20]. In some cases, the original Kaya identity is appropriate, but often the more detailed version is necessary.

Decomposing the key drivers of emissions and emission reductions allows us to make a graph that looks like figure 3.2 (created from the data in van Vuuren *et al* [21], using methods taken from Koomey *et al* [19, 20, 22]). Because of the duration of these projections, this graph uses 100-year GWPs. It shows the effect on emissions over time of all the levers included in the fully expanded decomposition explained in appendix D.

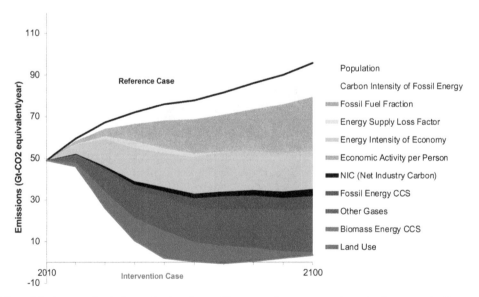

Figure 3.2. Sources of emissions reductions for van Vuuren *et al*'s most aggressive reductions scenario. Source: Scenario results from [21], decomposed using methods from [19, 22]. There are no savings from biomass CCS and a tiny increase in emissions from economic activity per person (not shown) in this scenario.

3.5 Creating structured scenario comparisons

Many comparisons of scenario outputs are done in an ad hoc basis, which in our experience can muddle key findings. We have been strong advocates for more structured comparisons using the decomposition methods highlighted in the previous section and described further in Koomey *et al* [19] and Koomey *et al* [20].

Policy makers should demand such structured comparisons and encourage modeling teams supporting their decision making to produce them. Modeling teams should adopt these methods as a matter of course, to promote sanity checking during scenario creation, improve transparency, speed up the analysis process, and make their results more useful to policy makers. They should also continually improve these methods over time.

We are also strong advocates for focused comparisons of small numbers of well-chosen scenarios, to supplement analyses that analyze dozens or hundreds of scenarios and present the results in aggregate form (such as, for example, in IPCC [23, 24]). Scenario modelers are partial to assessing many models and scenarios as one way of addressing uncertainty [25], but our experience is that key storylines can easily be lost when so many scenarios are presented together. Assessing a handful of scenarios, as we do in Koomey *et al* [20] can yield insights that are just as valuable as those conveyed by analyses covering more scenarios.

3.6 More detailed breakdowns of savings from key options

The analyses of emissions reductions presented in the previous two sections are highly aggregated. More granularity is often required for credible zero emissions scenarios but creating menus of potential savings from different options is situation dependent and requires detailed analysis. Such analysis often draws upon analyses of options from other sources, but these must be adapted and modified for each application.

One high-profile assessment of emissions reductions options for the globe comes from *Project Drawdown*. The original work was presented in 2017 [26] but it has been updated significantly since then. It presents quantitative assessments for dozens of emissions reductions options with documentation for each.

Another great source is the **Deep Decarbonization Pathways Project**, which analyzed business-as-usual and deep decarbonization scenarios (with extensive electrification) for the energy sectors of more than a dozen major countries, including the US, China, India, Indonesia, and Brazil[3].

An interesting recent effort [27] translates the aggregated results of integrated assessment model scenarios into country-specific household level lifestyle changes compatible with those scenarios, which is an area of increasing recent research activity [28]. A focus on service demand and activity levels can help policy makers understand multiple pathways for achieving aggressive emissions reductions.

The latest IPCC Working Group III report on mitigation is another great source, but it is huge, so prepare to search for what you need [23]. A keyword search on

---

[3] https://ddpinitiative.org

single chapters can be quick and will often reveal dozens of relevant references and helpful summary materials.

3.7 A useful way to summarize total emission savings

In analyzing potential climate solutions, it is often convenient to summarize the sources of potential emissions reductions in a simple way. One the most well-known methods is the concept of **climate wedges**, as popularized by Pacala and Socolow [29] and further refined in [30]. It is often referred to as the **stabilization wedges method**.

Most analyses of the climate problem project a business-as-usual (BAU) case that extends current trends into the future, and then compile a list of policies and technologies that would reduce emissions, ranked based on costs and effectiveness (as per the McKinsey cost curve discussed below). Figure 3.3 summarizes what Pacala and Socolow call the stabilization triangle, which would keep emissions constant over 50 years[4]. The triangle is the area between the line representing constant emissions and the business-as-usual case.

Figure 3.4 shows that the stabilization triangle is made up of wedges, each of which avoids 25 billion tons of carbon emissions over 50 years. One wedge could be something like doubling the efficiency of all of the world's light duty vehicles from 30 to 60 miles per gallon, or increasing wind generated electricity by 100 billion kWh per year every year for 50 years. Using the business-as-usual assumptions in their analysis (carbon emissions growth of 1.5% per year), it would take seven or eight wedges to keep emissions constant for 50 years, and more to actually make emissions decline.

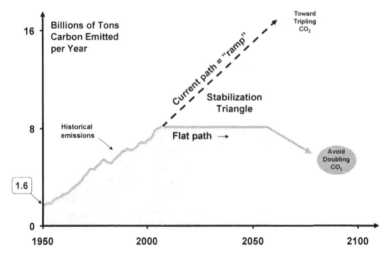

Figure 3.3. The stabilization triangle. Source: Reproduced with permission from Pacala and Socolow [29].

---

[4] http://cmi.princeton.edu/wedges/ has this graph available for download, as well as course materials and other helpful resources.

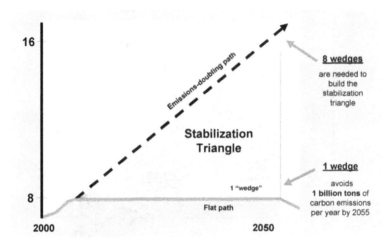

Figure 3.4. The stabilization triangle split into wedges. Source: Reproduced with permission from Pacala and Socolow [29].

The wedges approach has immense heuristic value because it allows people without much technical knowledge to increase their understanding of the problem and potential solutions. It also forces us to confront the enormity of the task ahead in a quantitative way. It was instrumental in convincing many that substantial strides towards a low carbon world were possible with existing technologies.

However, one downside to this and all other conventional approaches is that the number of wedges we'll need to accomplish any specific goal is dependent on the baseline (i.e. the BAU case). If the underlying drivers of emissions are growing more rapidly than we first estimated, we'll need to implement more wedges than we initially expected to achieve any particular level of emissions reductions. The focus on *savings* (or emissions reductions) instead of absolute emissions levels is a weakness of this type of analysis.

The follow-on analysis in [30] discusses another important failing of the way wedges were initially presented. As we know from chapter 1 and appendix A, it is not enough to stabilize greenhouse gas *emissions*. Instead, we need to stabilize greenhouse gas *concentrations*, and that means getting to zero net emissions as quickly as we can, making the stabilization triangle look a bit different, as shown in figure 3.5 (from [30]).

The 'hidden wedges' represent the natural progress in technology and markets compared to a 'frozen technology' baseline. 'Stabilization wedges' are like the ones defined in the original Pacala and Socolow article, needed to stabilize emissions. The 'phase-out wedges' are the ones that actually get us to zero emissions by mid-century. This way of looking at the problem shows that it requires much more effort to truly solve the problem than Pacala and Socolow initially posited.

In any case, this simple way of representing emissions reduction options at a high level can be helpful in explaining scenario results after you've done your analysis. If you understand its limitations, it is a useful heuristic. Given that we're now past

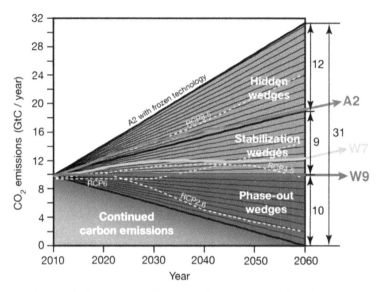

Figure 3.5. Full climate solutions triangle split into wedges. Source: Davis *et al* [30]. Reproduced under Creative Commons Attribution 3.0 Unported License (CC BY 3.0): https://creativecommons.org/licenses/by/3.0/. Original caption: Idealization of future CO_2 emissions under the business-as-usual SRES A2 marker scenario. Future emissions are divided into hidden (sometimes called 'virtual') wedges (brown) of emissions avoided by expected decreases in the carbon intensity of GDP by ~1% per year, stabilization wedges (green) of emissions avoided through mitigation efforts that hold emissions constant at 9.8 GtC y^{-1} beginning in 2010, phase-out wedges (purple) of emissions avoided through complete transition of technologies and practices that emit CO_2 to the atmosphere to ones that do not, and allowed emissions (blue). Wedges expand linearly from 0 to 1 GtC y^{-1} from 2010 to 2060. The total avoided emissions per wedge is 25 GtC, such that altogether the hidden, stabilization and phase-out wedges represent 775 GtC of cumulative emissions.

2020, another revisit of wedges is probably appropriate, with an eye toward shortening their duration to match the time to 2040.

3.8 Understanding technology cost curves

A common way to represent emission reduction potentials is what's called a **supply curve** [31–36]. Figure 3.6 shows an example from the management consulting firm McKinsey for global emission reduction potentials [37, 38]. The *Y*-axis is the cost per saved tonne of greenhouse gas emissions equivalent, with the bars ranked left to right in order of increasing cost per tonne saved. On the *X*-axis is the total savings for each option relative to the reference case in some year, in this case 2030.

Some measures save money, deliver other benefits, and reduce emissions, so those are 'negative cost options'. One example would be light-emitting diode (LED) lamps that last tens of thousands of hours (saving labor) and deliver better light that helps employees be more productive. Another is the recent trend toward installing solar panels on farms where crops are being grown [39]. The panels generate electricity, but also shade the plants and help them grow better during hot weather.

Options above the zero line have positive cost. In principle society should implement the cheapest options first, but that simple idea fails in practice.

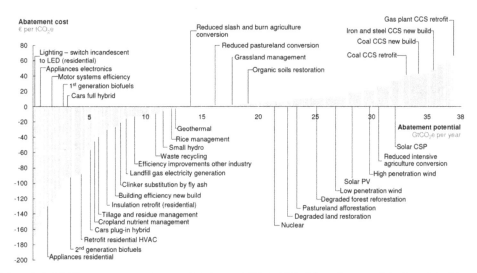

Figure 3.6. McKinsey cost curve. Source: Enkvist *et al* [38]. Exhibit 6 from 'Impact of the financial crisis on carbon economics: Version 2.1 of the global greenhouse gas abatement cost curve', January 2010, McKinsey and Company, www.mckinsey.com. Copyright 2022 McKinsey and Company. Original caption: The curve presents an estimate of the maximum potential of all technical GHG abatement measures below €80 per tCO$_2$e if each lever was pursued aggressively. It is not a forecast of what role different abatement measures and technologies will play.

Some of these measures are dependent on each other and we need to do everything we can think of (and more) if we're to stabilize the climate at safe levels, so we can't afford to be too choosy about the order in which options are implemented.

Graphs like these are constructed by estimating the capital cost of an option, annualizing it using the capital recovery factor[5] (like for a mortgage) and dividing that annual cost by the measure's annual greenhouse gas emissions savings in 2030. For options that save money (like many energy-efficiency measures that save energy AND reduce emissions) the present value of the energy savings is subtracted from the initial cost and the resulting annualized net cost is divided by annual carbon savings.

Gillingham and Stock [31] discuss the McKinsey cost curve and its limitations. Such cost curves are useful for summarizing the broad picture for costs of emissions reductions, but they almost always reflect a static view of current costs rather than an accurate assessment of what costs will prevail decades hence.

One of the key issues raised by Gillingham and Stock is that aggressive implementation of emissions reduction options will drive down costs through learning by doing, economies of scale, network externalities, and other factors that fall under the category of 'increasing returns to scale', as discussed in chapter 1. Such cost curves (as well as virtually all simulation and optimization models of future emissions) usually ignore such factors.

---

[5] https://en.wikipedia.org/wiki/Capital_recovery_factor

Another limitation of such cost curves is that making accurate future cost predictions for path dependent systems dominated by increasing returns to scale is impossible [15, 40–42]. They can still be useful in representing the cost assumptions embodied in modeling exercises but it's important to always keep their limits in mind.

3.9 Scenario simulation tools

Learning about the multidimensional complexity of the climate problem sometimes means experimenting using well-designed scenario creation tools. One of our favorites is from Energy Innovation for the US[6]. This open-source tool allows you to create your own scenario by changing assumptions, save alternative scenarios, and compare scenarios to each other. It includes all sectors and warming agents.

Users can choose assumptions for each sector or end-use and can rapidly see which choices have small or large effects. It runs it through a web browser or natively on a Mac or PC (input data can even be modified on the Mac/PC versions). The tool has also been adapted for some other countries and US states, including China, India, Indonesia, Brazil, Mexico, Canada, Poland, Saudi Arabia, California, Colorado, Virginia, Louisiana, Nevada, and Minnesota.

A similar tool giving the global picture is EN-ROADS, developed by Climate Interactive, MIT, and Ventana Systems[7]. It helps facilitate group learning about climate change solutions for policy makers and business leaders.

A more microsimulation board game is called Energetic[8]. This game teaches about physical and political constraints to reaching the goals embodied in the Paris Climate Agreement in the energy system, focusing on New York City. Players must keep the lights on and win elections to maintain momentum for decarbonization, and it isn't easy to win!

3.10 Life-cycle assessment

Life-cycle assessment (LCA), sometimes also called **life-cycle analysis**, is a way to characterize emissions for products, services, and larger systems, with LCAs typically quantifying climate impacts from mining materials through manufacturing, use, and disposal or recycling [43]. There are two basic approaches, the first is called **process LCA** (and sometimes **product LCA**), which tracks all energy use, materials, and emissions from cradle to grave for specific products. The other common type of LCA is **economic input–output life-cycle assessment** (abbreviated to **EIO-LCA** or **IO LCA**) which uses whole economy or sectoral input–output analysis to track emissions based on categorized expenditures. Many LCA climate footprints also use a hybrid of process LCA and EIO-LCA, since detailed process LCA data

---

[6] https://us.energypolicy.solutions
[7] https://www.climateinteractive.org/en-roads/
[8] https://newyork.thecityatlas.org/energetic/

are often difficult to find for entire product supply chains, and EIO-LCA can be used to fill in data gaps with reasonable approximations.

For analyses related to emissions reductions, process LCAs are most useful for quantifying climate impacts related to a specific product, while EIO-LCA and hybrid LCA approaches are most useful for calculating the climate impacts of larger systems, including households, communities, companies, governments, and investment portfolios. LCA can be a powerful tool to help with climate decision-making, but caution is always needed to ensure that data are complete and comparable. The most important issues with LCAs are choice of **functional unit**, choice of **system boundaries**, and avoiding **time lags** in available data.

A typical LCA starts by defining the goal and scope of the analysis (figure 3.7), including selecting the functional unit that will serve as the basis of comparison. The functional unit can be a physical product (such as a T-shirt or vehicle), or it could be a service (such as a kWh of electricity or unit of transportation distance), or the functional unit can be something much larger and more aggregated (such as the entire annual climate footprint for a company, or the GHGs connected with every $1 invested in a company).

System boundary choices matter because one LCA study may choose different system boundaries than a similar LCA focused on different goals, which can lead to dramatically different conclusions. For example, an LCA with the goal of reducing climate pollution related to T-shirt production may use **cradle-to-gate** systems boundaries that include all LCA impacts until a T-shirt is sold, and the results

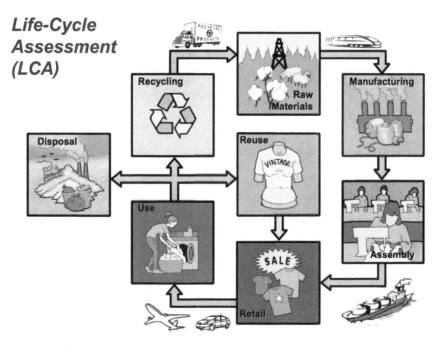

Figure 3.7. Life-cycle assessment scope. Source: Reproduced with permission from http://www.oroeco.org. Copyright 2015 Ian Monroe.

from this analysis may indicate that growing cotton for the T-shirt is the biggest source of climate pollution, but a more complete **cradle-to-grave** LCA analysis that includes climate impacts from the product use and disposal may reveal that the climate pollution from repeatedly washing and drying the T-shirt are even larger than GHGs from the shirt's cotton.

For LCAs related to climate impacts for companies and other complex entities (governments, communities, households, etc), system boundaries are often described in terms of **Scope 1**, **Scope 2**, and **Scope 3 GHG emissions**. Scope 1 emissions are defined as direct climate pollution (e.g. burning natural gas in a company's factories), while Scope 2 emissions come from direct electricity use (e.g. pollution linked to every kWh of power used by a company). Scope 3 emissions encompass all other indirect GHGs linked with operations, including **upstream embodied emissions** linked to every aspect of supply chains for products and services (e.g. raw material extraction, processing, suppliers, and transportation) as well as **downstream emissions** linked to product use and disposal.

While it has become increasingly common for companies and governments to report climate emissions based on LCA data, GHG reporting often only includes Scope 1 and Scope 2 emissions, leaving out Scope 3 GHGs even though these upstream and downstream emissions are also often significant (figure 3.8). Analysis by Etho Capital (of which Ian was a co-founder) has shown that Scope 3 emissions represent over 80% of

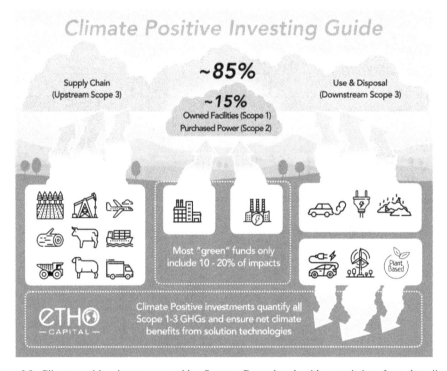

Figure 3.8. Climate-positive investment guide. Source: Reproduced with permission from http://www.ethocapital.com. Copyright 2022 Etho Capital.

the climate footprint for most companies, and Scope 3 GHGs can be nearly 99% of the climate footprint for a technology company like Apple, where the vast majority of emissions are linked to third party suppliers in Asia as well as downstream electricity consumption when Apple's products are used. Scope 3 lCA analysis can also help quantify the climate benefits for companies that produce climate solutions, revealing which companies are net climate positive when the emissions from producing their products and services are weighed against their downstream climate benefits.

Another example of how important system boundaries can be comes from the oil and gas sector. Often analyses of emissions for oil and gas have focused on combustion emissions of fuel products, with the upstream emissions for exploration/extraction and the midstream emissions from refining being tracked in the industrial sector, disconnected from final combustion by consumers. When oil and gas companies report their emissions, they generally only include Scope 1 and Scope 2 emissions, leaving out the Scope 3 emissions connected with burning the fuels they produce after they're sold. This choice of boundaries for oil and gas masks differences in total life-cycle emissions that are big enough to matter [8–10, 44], undercounting the full climate impacts from fossil fuels and hiding opportunities for emissions reductions in the oil and gas sector that are only now becoming well understood [9, 10, 45]. When Scope 3 upstream and downstream impacts are included (including methane leakage and fuel product combustion), ExxonMobil's company-wide climate footprint is over 6 times higher than an LCA that only include Scope 1–2 climate system boundaries.

Time lags matter because many LCAs require detailed product-specific data and by the time the data become available, the products on which the analysis were conducted can be five to ten years old. This isn't as important for slow moving industries, but for rapidly changing sectors like computing and electric vehicles, using old data can result in outdated calculations and conclusions.

For example, figure 2.2, in chapter 2, showed that costs for lithium-ion battery packs fell about 90% from 2010 to 2020. Those costs include all the energy and materials costs for making that battery, so if total costs fell 90%, you can bet that energy and materials use (and associated emissions) also fell substantially (although probably not by exactly 90%). Process LCAs can be a useful supplement to the other analytical tools described above, but like all tools, they need to be used with caution and full awareness of their benefits and limitations.

3.11 Understanding energy systems

For background on energy systems concepts and terminology, there's nothing better than IIASA's Energy Primer [46]. It's a foundational work that provides true conceptual understanding, enabling students to dig further. It has three goals:

- To present a systems perspective on energy, emphasizing interconnections and relationships between different parts of the system,
- To explore the importance of energy service demand and how it affects energy systems, and
- To present long-run historic context on energy systems and how fast they can change.

The focus on energy service demand (what we call **tasks** in chapter 2) is critically important and provides the foundation for analysis of electrification.

3.12 Following good analytical practice

Early in our class we discuss good data analysis practices for student projects, focusing on avoiding pitfalls in data acquisition and handling, digging into the numbers, creating consistent comparisons, doing back-of-the-envelope calculations, building simple models, telling good stories, making clear tables and graphs, and creating adequate documentation, based on the material in Koomey [47]. We find that explaining these expectations up front helps students do better projects and saves time for the instructors. We encapsulate some of those expectations in the project template posted at http://www.solveclimate.org.

3.13 Chapter conclusions

Most students who take our class have taken previous classes on energy systems, at a minimum. Familiarity with the sources described in this chapter provides the additional core knowledge needed to assess the potential for reducing emissions to zero.

It is fashionable today to assume that any figures about the future are better than none. To produce figures about the unknown, the current method is to make a guess about something or other—called an 'assumption'—and to derive an estimate from it by subtle calculation. The estimate is then presented as the result of scientific reasoning, something far superior to mere guesswork. This is a pernicious practice that can only lead to the most colossal planning errors, because it offers a bogus answer where, in fact, an entrepreneurial judgment is required.

—E F Schumacher

Further reading

CA ARB 2021 *California Greenhouse Gas Emissions for 2000 to 2019: Trends of Emissions and Other Indicators* (Sacramento, CA: California Air Resources Board) https://ww2.arb.ca.gov/ghg-inventory-data. An emissions inventory for California, the US state that has been among the most aggressive at reducing emissions.

EEA 2020 *Trends and Drivers of EU Greenhouse Gas Emissions* (Luxembourg: European Environment Agency) https://eea.europa.eu/themes/climate/eu-greenhouse-gas-inventory/eu-greenhouse-gas-inventory. A high level source for emissions in Europe, a region that has led on reducing emissions.

Enkvist P-A, Nauclér T and Rosander J 2007 A cost curve for greenhouse gas reduction *McKinsey Quarterly* 1 February https://mckinsey.com/business-functions/sustainability/our-insights/a-cost-curve-for-greenhouse-gas-reduction. The first iteration of the McKinsey cost curve.

Enkvist P-A, Dinkel J and Lin C 2010 Impact of the financial crisis on carbon economics: version 2.1 of the global greenhouse gas abatement cost curve *McKinsey Sustainability Report* 1 January https://mckinsey.com/business-functions/sustainability/our-insights/impact-of-the-financial-crisis-on-carbon-economics-version-21. The second iteration of the McKinsey cost curve.

Gordon D 2022 *No Standard Oil: Managing Abundant Petroleum in a Warming World* (New York: Oxford University Press) https://nostandardoil.com. Life-cycle emissions for oil and gas are not well understood, but this book summarizes the current state of knowledge, giving history and context for this complex area.

Gordon D, Koomey J, Brandt A and Bergerson J 2022 *Know Your Oil and Gas: Generating Climate Intelligence to Cut Petroleum Industry Emissions* (Boulder, CO: Rocky Mountain Institute) https://rmi.org/insight/kyog/. This report presents the latest research and practice on total life-cycle emissions for oil.

Grübler A N, Nakicenovic S, Pachauri H-H and Smith K R 2015 *Energy Primer: Based on Chapter 1 of the Global Energy Assessment* (Laxenburg: International Institute for Applied Systems Analysis) https://iiasa.ac.at/web/home/research/Flagship-Projects/Global-Energy-Assessment/Chapter1.en.html. A foundational text on energy systems.

IEA 2020 *World Energy Balances: 2020 Edition* (Paris: International Energy Agency) http://wds.iea.org/wds/pdf/WORLDBAL_Documentation.pdf. Energy balances for every country in the world.

IPCC 2022 *Climate Change 2022: Mitigation of Climate Change. Contribution of Working Group III to the Sixth Assessment Report of the Intergovernmental Panel on Climate Change* (Cambridge: Cambridge University Press) https://ipcc.ch/report/sixth-assessment-report-working-group-3/. This report summarizes consensus knowledge on climate solutions, but lags recent developments by several years because of the time needed for the IPCC process.

Koomey J 2017 *Turning Numbers into Knowledge: Mastering the Art of Problem Solving* 3rd edn (El Dorado Hills, CA: Analytics). A sourcebook for teaching good analytical practice.

Krishnan M 2022 *The Net-Zero Transition: What It Would Cost, What It Could Bring* (New York: McKinsey Global Institute) https://mckinsey.com/business-functions/sustainability/our-insights/the-economic-transformation-what-would-change-in-the-net-zero-transition. This report grew out of the McKinsey cost curve work from 2007 and 2010.

Lovins A, Bendewald M, Kinsley M, Bony L, Hutchinson H A, Sheikh I and Acher Z 2010 *Factor Ten Engineering Design Principles* X10–10 (Old Snowmass, CO: Rocky Mountain Institute) https://rmi.org/our-work/areas-of-innovation/office-chief-scientist/10xe-factor-ten-engineering/. This report gives guidance on integrated whole-systems design, which is a discipline that can be taught, but often isn't.

Mulvaney D 2020 *Sustainable Energy Transitions: Socio-Ecological Dimensions of Decarbonization* (New York: Palgrave Macmillan) http://dustinmulvaney.com/ set. This exemplary interdisciplinary book discusses technical, social, economic, and ecological issues related to energy transitions.

Rocky Mountain Institute *Factor Ten Engineering* http://rmi.org/rmi/10xE. More guidance on integrated whole-systems design.

US EPA 2021 *Inventory of US Greenhouse Gas Emissions and Sinks: 1990—2019* (Washington, DC: US Environmental Protection Agency) https://epa.gov/ghgemissions/inventory-us-greenhouse-gas-emissions-and-sinks-1990-2019. The official emissions inventory for the US.

References

[1] Roten D, Marland G, Bun R, Crippa M, Gilfillan D, Jones M W, Janssens-Maenhout G, Marland E and Andrew R 2022 CO_2 emissions from energy systems and industrial processes: inventories from data- and proxy-driven approaches *Balancing Greenhouse Gas Budgets* (Amsterdam: Elsevier) https://elsevier.com/books/balancing-greenhouse-gas-budgets/poulter/978-0-12-814952-2

[2] US EPA 2022 *Inventory of US Greenhouse Gas Emissions and Sinks: 1990—2020* (Washington, DC: US Environmental Protection Agency) https://epa.gov/ghgemissions/inventory-us-greenhouse-gas-emissions-and-sinks

[3] CA ARB 2021 *California Greenhouse Gas Emissions for 2000 to 2019: Trends of Emissions and Other Indicators* (Sacramento, CA: California Air Resources Board) https://ww2.arb.ca.gov/ghg-inventory-data

[4] EEA 2020 *Trends and Drivers of EU Greenhouse Gas Emissions* (Luxembourg: European Environment Agency) https://eea.europa.eu/themes/climate/eu-greenhouse-gas-inventory/eu-greenhouse-gas-inventory

[5] Poulter B, Canadell J, Hayes D and Thompson R (ed) 2022 *Balancing Greenhouse Gas Budgets: Accounting for Natural and Anthropogenic Flows of CO_2 and Other Trace Gases* (Amsterdam: Elsevier) https://elsevier.com/books/balancing-greenhouse-gas-budgets/poulter/978-0-12-814952-2

[6] Gurney K R, Liang J, Roest G, Song Y, Mueller K and Lauvaux T 2021 Under-reporting of greenhouse gas emissions in US cities *Nat. Commun.* **12** 553

[7] IEA 2020 *World Energy Balances: 2020 Edition* (Paris: International Energy Agency) http://wds.iea.org/wds/pdf/WORLDBAL_Documentation.pdf

[8] Koomey J, Gordon D, Brandt A and Bergeson J 2016 *Getting Smart about Oil in a Warming World* (Washington, DC: Carnegie Endowment for International Peace) http://carnegieendowment.org/2016/10/04/getting-smart-about-oil-in-warming-world-pub-64784

[9] Gordon D 2022 *No Standard Oil: Managing Abundant Petroleum in a Warming World* (New York: Oxford University Press) https://nostandardoil.com

[10] Gordon D, Koomey J, Brandt A and Bergerson J 2022 *Know Your Oil and Gas: Generating Climate Intelligence to Cut Petroleum Industry Emissions* (Boulder, CO: Rocky Mountain Institute) https://rmi.org/insight/generating-climate-intelligence-to-cut-petroleum-industry-emissions/

[11] IPCC 2014 *Climate Change 2014: Mitigation of Climate Change. Contribution of Working Group III to the Fifth Assessment Report of the Intergovernmental Panel on Climate Change*

ed O Edenhofer *et al* (Cambridge: Cambridge University Press) https://www.ipcc.ch/report/ar5/wg3/

[12] Bajželj B, Julian M A and Cullen J M 2013 Designing climate change mitigation plans that add up *Environ. Sci. Technol.* **47** 8062–9

[13] Carnahan W, Ford K W, Prosperetti A, Rochlin G I, Rosenfeld A H, Ross M, Rothberg J, Seidel G and Socolow R H 1975 Efficient use of energy: part 1—a physics perspective, a report of the Research Opportunities Group of a summer study held in Princeton, New Jersey in July 1974 *AIP Conf. Proc.* **25** PB-242773 https://www.osti.gov/biblio/7134080

[14] Stansinoupolos P, Smith M H, Hargroves K and Desha C 2008 *Whole System Design: An Integrated Approach to Sustainable Engineering* (New York: Routledge)

[15] Koomey J G 2012 *Cold Cash, Cool Climate: Science-Based Advice for Ecological Entrepreneurs* (El Dorado Hills, CA: Analytics)

[16] Lovins A, Bendewald M, Kinsley M, Bony L, Hutchinson H, Pradhan A, Sheikh I and Acher Z 2010 *Factor Ten Engineering Design Principles X10-10* (Old Snowmass, CO: Rocky Mountain Institute (RMI)) https://rmi.org/our-work/areas-of-innovation/office-chief-scientist/10xe-factor-ten-engineering/

[17] Ehrlich P R and Holdren J P 1971 Impact of population growth *Science* **171** 1212–7

[18] Ehrlich P R and Holdren J P 1972 One-dimensional ecology *Bull. Atomic Sci.* **28** 18–27

[19] Koomey J, Schmidt Z, Hummel H and Weyant J 2019 Inside the black box: understanding key drivers of global emission scenarios *Environ. Model. Softw.* **111** 268–81

[20] Koomey J, Schmidt Z, Hausker K and Lashof D 2022 Black boxes revealed: assessing key drivers of 1.5 °C warming scenarios *White Paper* Koomey Analytics, Bay Area, CA

[21] van Vuuren D P *et al* 2018 Alternative pathways to the 1.5 °C target reduce the need for negative emission technologies *Nat. Clim. Change* **8** 391–7

[22] Koomey J, Schmidt Z, Hausker K and Lashof D 2022 Exploring the black box: applying macro decomposition tools for scenario comparisons *Environ. Model. Softw.* **155** 105426

[23] IPCC 2022 *Climate Change 2022: Mitigation of Climate Change. Contribution of Working Group III to the Sixth Assessment Report of the Intergovernmental Panel on Climate Change* (Cambridge: Cambridge University Press) https://ipcc.ch/report/sixth-assessment-report-working-group-3/

[24] IPCC 2018 *Global Warming of 1.5°C. An IPCC Special Report on the Impacts of Global Warming of 1.5°C Above Pre-industrial Levels and Related Global Greenhouse Gas Emission Pathways, in the Context of Strengthening the Global Response to the Threat of Climate Change, Sustainable Development, and Efforts to Eradicate Poverty* (Geneva: IPCC) https://ipcc.ch/sr15

[25] Guivarch C *et al* 2022 Using large ensembles of climate change mitigation scenarios for robust insights *Nat. Clim. Change* **12** 428–35

[26] Hawken P (ed) 2017 *Drawdown: The Most Comprehensive Plan Ever Proposed to Reverse Global Warming* (New York: Penguin)

[27] Hanmer C, Wilson C, Edelenbosch O Y and van Vuuren D P 2022 Translating global integrated assessment model output into lifestyle change pathways at the country and household level *Energies* **15** 1650

[28] van den Berg N J, Hof A F, Timmer V J and van Vuuren D P 2022 Current lifestyles in the context of future climate targets: analysis of long-term scenarios and consumer segments for residential and transport *Environ. Res. Commun.* **4** 095003

[29] Pacala S and Socolow R 2004 Stabilization wedges: solving the climate problem for the next 50 years with current technologies *Science* **305** 968–72

[30] Davis S J, Cao L, Caldeira K and Hoffert M I 2013 Rethinking wedges *Environ. Res. Lett.* **8** 011001

[31] Gillingham K and Stock J H 2018 The cost of reducing greenhouse gas emissions *J. Econ. Perspect.* **32** 53–72

[32] Grubb M, Edmonds J, ten Brink P and Morrison M 1993 The costs of limiting fossil-fuel CO_2 emissions: a survey and analysis *Annu. Rev. Energy Env.* **18** 397–478

[33] Jackson T 1991 Least-cost greenhouse planning: supply curves for global warming abatement *Energy Policy* **19** 35–46

[34] Meier A 1982 Supply curves of conserved energy *PhD Thesis* Energy and Resources Group, University of California, Berkeley http://escholarship.org/uc/item/20b1j10d

[35] Rosenfeld A, Atkinson C, Koomey J G, Meier A, Mowris R and Price L 1993 Conserved energy supply curves *Contemp. Policy Issues* **11** 45–68

[36] Krause F and Koomey J G 1989 Unit costs of carbon savings from urban trees, rural trees, and electricity conservation: a utility cost perspective *Conf. on Urban Heat Islands (Berkeley, CA, 23–24 February)*

[37] Enkvist P-A, Nauclér T and Rosander J 2007 A cost curve for greenhouse gas reduction *McKinsey Quart.* 1 February https://mckinsey.com/business-functions/sustainability/our-insights/a-cost-curve-for-greenhouse-gas-reduction

[38] Enkvist P-A, Dinkel J and Lin C 2010 Impact of the financial crisis on carbon economics: version 2.1 of the global greenhouse gas abatement cost curve *McKinsey Sustainability Report* 1 January https://mckinsey.com/business-functions/sustainability/our-insights/impact-of-the-financial-crisis-on-carbon-economics-version-21

[39] Rosen E 2022 Can dual-use solar panels provide power and share space with crops? *New York Times* 28 June https://nytimes.com/2022/06/28/business/dual-use-solar-panels-agrivoltaics-blue-wave-power.html?referringSource=articleShare

[40] Craig P, Gadgil A and Koomey J 2002 What can history teach us? A retrospective analysis of long-term energy forecasts for the US *Annual Review of Energy and the Environment 2002* ed R H Socolow, D Anderson and J Harte (Palo Alto, CA: Annual Reviews) pp 83–118

[41] Koomey J G, Craig P, Gadgil A and Lorenzetti D 2003 Improving long-range energy modeling: a plea for historical retrospectives *Energy J.* **24** 75–92

[42] Scher I and Koomey J G 2011 Is accurate forecasting of economic systems possible? *Clim. Change* **104** 473–79

[43] Matthews H S, Hendrickson C T and Matthews D 2018 *Life Cycle Assessment: Quantitative Approaches for Decisions that Matter* (Pittsburgh, PA: LCAtextbook) https://lcatextbook.com

[44] Gordon D, Brandt A, Bergeson J and Koomey J 2015 *Know Your Oil: Creating a Global Oil-Climate Index* (Washington, DC: Carnegie Endowment for International Peace) http://goo.gl/Jly9Op

[45] Brandt A R, Mohammad S M, Englander J G, Koomey J and Gordon D 2018 Climate-wise choices in a world of oil abundance *Environ. Res. Lett.* **13** 044027

[46] Grübler A, Nakicenovic N, Pachauri S, Rogner H-H and Smith K R 2015 *Energy Primer: Based on Chapter 1 of the Global Energy Assessment* (Laxenburg: International Institute for Applied Systems Analysis) https://iiasa.ac.at/web/home/research/Flagship-Projects/Global-Energy-Assessment/Chapter1.en.html

[47] Koomey J 2017 *Turning Numbers into Knowledge: Mastering the Art of Problem Solving* 3rd edn (El Dorado Hills, CA: Analytics)

IOP Publishing

Solving Climate Change
A guide for learners and leaders
Jonathan Koomey and Ian Monroe

Chapter 4

Electrify (almost) everything

We don't just need to change our fuels; we need to change our machines.
—Saul Griffith

Chapter overview

- Electrifying everything is a cornerstone of achieving zero emissions quickly.
- Local air pollution and waste from combustion will become a thing of the past.
- Over time, emissions from electrified equipment will decline as the electricity system decarbonizes.
- End-uses that are harder to electrify (such as long-haul aviation and shipping) in the near term may require the use of fuels derived from electricity ('electric fuels') such as hydrogen and synthetic methane.
- Electrifying everything also means bringing clean electricity to the hundreds of millions of people in Africa and Asia who do not now have it.

4.1 Introduction

Solving the climate problem ultimately means the end of fossil fuels virtually everywhere. The easiest and most effective way to accomplish this goal with almost all energy end-uses is to electrify them at breakneck speed [1–7].

Electrifying everything is a daunting task, but will result in a cleaner, safer, and more prosperous world than we have now. We need to choose electric technologies at every opportunity and, in many cases, scrap existing fossil capital early.

Electrical storage technologies have progressed rapidly in the past decade [8] and advances in computing, electric motors, and power controls have been dramatic. The world is on the cusp of a rapid shift towards electrification [1].

doi:10.1088/978-0-7503-4032-8ch4   4-1

For most heating applications, electrifying means switching to electric heat pumps, for others (mainly high temperature heat) that means electric resistance or induction heat. For light vehicles, trucks, small planes, snowmobiles, tractors, small boats, and landscaping equipment, that means electric motors powered by batteries.

Some end-uses will be harder to electrify until new technologies become available, such as long-haul aviation and big ships, but for virtually all the rest, there are electrified options now that offer many advantages:

1. No local pollutant emissions of any kind.
2. Less noise than internal combustion engines.
3. Fewer moving parts and higher reliability.
4. Lower maintenance costs.
5. Lower operating costs.
6. More accurate control of processes.
7. (Increasingly) lower capital costs, especially after scaling up production.
8. Higher efficiency, both at the point of use and for the larger system.
9. Avoided energy use associated with discovering, refining, and transporting fuels.

We listed avoided pollutant emissions first for a reason. Air pollution is one of the largest causes of premature deaths globally [9–16]. In fact, outdoor air pollution alone kills about nine million people every year, more than tobacco, obesity, alcohol, and diseases such as TB/AIDS/malaria, and indoor air pollution kills another three million people every single year [17]. Many of these deaths are avoidable if we electrify everything and switch electricity generation to non-combustion sources such as renewables and nuclear power. Correctly counting the benefits of reducing such pollution makes climate action even more cost effective than just considering the benefits of stabilizing the climate.

The many benefits of electrification are already driving adoption for some important uses, such as electric vehicles. The International Energy Agency reports that global market share for battery electric automobiles doubled from 2020 to 2021 to almost 9% of all cars sold after increasing more than 50% from 2019 to 2020. The electric car market share in China (the world's largest automobile market) hit 20% in December 2021 [18].

Even with all the demonstrable benefits of electrifying, it won't be easy. All replacement equipment needs to be electric from now on. Displacing existing capital stocks at the needed speed requires aggressive policies, changes in business models, and new financing tools.

As Griffith [1, 19] explains, the climate problem in the energy sector, writ large, is the problem of rapidly displacing fossil fuel use with electrified capital investments. That means helping consumers and businesses finance those investments and (in many cases) compensating them for the costs of early retirement (we discuss these issues in the chapters titled 'Align incentives' and 'Mobilize money').

They key to enabling electrification is **capital stock turnover**. The **natural retire-ment rate** for capital goods is often characterized by an average lifetime (see appendix B). When equipment reaches the end of its useful life, there's an opportunity to replace it with electric options, but this opportunity doesn't come up very often. Gas furnaces in homes typically last 25 years, gas water heaters 10–12 years, gas cooktops 15–20 years, residential buildings 50 to 100 years, and commercial buildings and many industrial plants 30 to 50 years (but are often renovated).

The urgency of the climate problem combined with long lifetimes for capital equipment imply that we'll need to encourage many consumers and businesses to scrap their equipment even earlier than natural replacement cycles would suggest, if we're to stabilize the climate near 1.5 °C.

4.2 Creating or adopting a business-as-usual (BAU) scenario

Let's begin by defining boundaries. Electrifying (almost) everything applies to almost all current uses of fossil fuels, with the possible exceptions (at least in the near term) of bulk shipping, long-haul aviation, and petrochemicals. At the end of this chapter, we describe options for reducing emissions from those activities, but for the rest, electrification is the best way to reduce emissions from fossil fuels.

Table 4.1 categorizes fossil emissions from various parts of the fuel cycle. Combustion/use emissions are almost always dominant, but other parts of the life cycle can contribute significant emissions. As fossil fuel-fired end-uses are electrified, virtually all emissions identified in table 4.1 will decline, so we treat them as linked. Fossil methane emissions from coal mines or oil and gas drilling, for example, are part of 'fossil emissions' in our framework.

4.2.1 Creating a base-year emissions inventory

Developing a strategy for electrification starts with an 'energy balance' for a base year, which is available for almost every country from the International Energy Agency [20]. It is analogous to an emissions inventory, but for energy flows. Just as for emissions, a Sankey diagram is the key to understanding the full system of energy production through to energy end-use and services delivered (tasks performed). Figure 4.1 shows a Sankey diagram for the US in 2021, from Lawrence Livermore National Laboratory (https://flowcharts.llnl.gov). A much more detailed interactive online version for the US from Saul Griffith's Otherlab can be accessed at http://departmentof.energy, and summary Sankey diagrams by sector are contained in Griffith's recent book [1].

Another way to show electricity and fuel consumption is with a snapshot of primary or final energy, as showing in figure 4.2 (taken from Jadun et al [21]). Almost three quarters of primary energy in buildings in 2015 was supplied by electricity, about one third in industry, and almost none in transportation.

From energy flows we can calculate fossil emissions. Figure 4.3 shows the same data as figure 2.4 above but adjusted to 20 year GWPs for methane, nitrous oxides, and F-gases. Fossil CO_2 emissions are by far the biggest category, but fossil methane and nitrous oxide emissions are big enough to matter. For the US, all categories but F-gases and non-fossil nitrous oxides show emissions declining to 2020, but for many countries in the developing world, emissions increased strongly over that

Table 4.1. Anthropogenic fossil pollution by type and life-cycle stage.

	Production	Processing	Transport	Combustion/use	Disposal	Petrochemicals
GHGs/warming agents						
CO_2	X	X	X	X	X	X
CH_4	X	X	X	X	X	X
N_2O	X	X	X	X	X	X
CO	X	X	X	X	X	X
Volatile organic compounds	X	X	X	X	X	X
Black carbon	X	X	X	X	X	X
Cooling agents						
Particulates	X	X	X	X	X	X
NO_x	X	X	X	X	X	X
SO_2	X	X	X	X	X	X
Air Pollution						
Particulates NO_x SO_2	XXX	XXX	XXX	XXX	XXX	XXX
Ozone	X	X	X	X	X	X
Benzene	X	X	X	X	X	X
Trace metals	X	X	X	X	X	X
Volatile organic compounds	X	X	X	X	X	X
Other hydrocarbons	X	X	X	X	X	X
Water Pollution						
Fracking water	X					
Coal ash					X	
Crude oil	X		X		X	
Trace metals	X	X	X	X	X	X
Land Pollution						
Coal ash					X	
Gasoline leaks			X	X	X	
Crude oil					X	
Trace metals	X	X	X	X	X	X
Other pollution/effects						
Noise	X	X	X	X	X	
Thermal		X		X	X	X
Explosions	X	X	X	X	X	

X implies it's an issue for that pollutant and life-cycle stage.

period. In addition, the declines in US emissions are modest compared to what is required for the world to achieve climate stabilization.

Figure 4.4 shows the same data as for figure 4.3 for 2020 but sliced a bit differently, showing fossil, non-fossil, and non-fossil land use by gas. In these data, non-fossil emissions are about 29% of positive emissions (excluding land use CO_2)

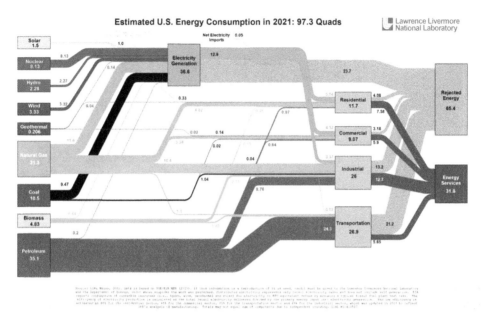

Figure 4.1. Sankey diagram for the US energy system in 2021. Quads = quadrillion(1015) Btus of energy. Source: https://flowcharts.llnl.gov. Lawrence Livermore National Laboratory and the Department of Energy.

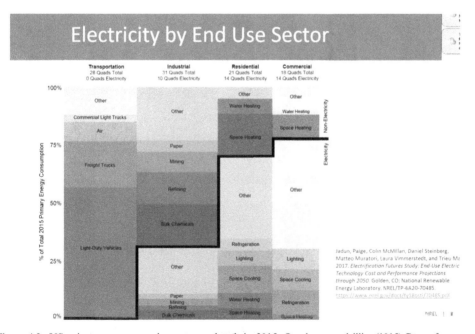

Figure 4.2. US primary energy use by sector and task in 2015. Quads = quadrillion(1015) Btus of energy. Source: Jadun *et al* [21]. National Renewable Energy Laboratory.

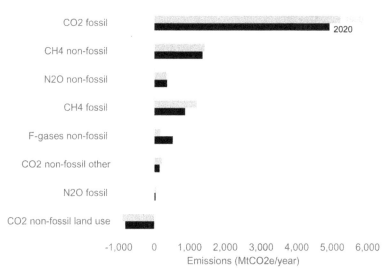

Figure 4.3. US greenhouse gas emissions in 1990 and 2020 by category (20 year GWPs). Source: US EPA [22]. GWPs from table 7.SM.7 in IPCC [23]. Order of categories is the same as figure 2.4, which are ranked by 2020 emissions using 100 year GWPs.

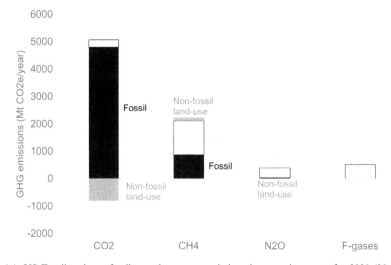

Figure 4.4. US Fossil and non-fossil greenhouse gas emissions by warming agent for 2020 (20 year GWPs). Source: US EPA [22]. GWPs from table 7.SM.7 in IPCC [23].

but recent work shows an even larger fraction of total emissions associated with non-fossil sources [24] integrated over the next few decades.

CO_2 emissions are dominated by fossil combustion while methane emissions are about two thirds non-fossil and one third fossil related. N_2O and F-gases are dominated by non-fossil sources.

Figure 4.5 shows carbon dioxide emissions from fossil fuel combustion broken out by sector and fuel. Petroleum for transportation is the largest source of these

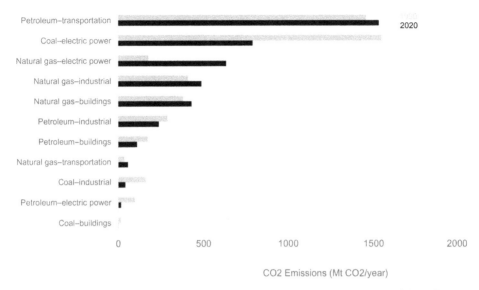

Figure 4.5. US fossil combustion related carbon dioxide emissions by category for 1990 and 2020. Source: US EPA [22]. GWPs from table 7.SM.7 in IPCC [23]. The GWP for carbon dioxide is the same for all analysis periods by definition.

emissions, followed by coal combustion in the electric power sector. Coal use for electricity generation declined by about half from 1990 to 2020 but ticked up again in 2021 as the economy recovered after the pandemic and fossil gas prices spiked. Natural gas use for electricity generation grew rapidly to 2020, reflecting the importance of this resource for displacing coal (other major contributors to reducing coal use included rapid growth in renewable power generation and declines in total electricity use).

In the electric power sector, replacing combustion generation with non-combustion zero-emissions generation is one pillar of climate stabilization. For direct fuel use in industrial, buildings, and transportation, the solution is electrification, with very few exceptions.

It's also important to treat fossil methane and nitrous oxide emissions, shown for 1990 and 2020 in figures 4.6 and 4.7, respectively. Methane leakage from fossil fuel extraction and use is responsible for virtually all fossil methane emissions. Almost all fossil nitrous oxide emissions come from combustion in a nitrogen–oxygen atmosphere, but the total warming effect of these emissions is tiny compared to carbon dioxide and methane.

Data availability will affect the level of detail possible for your BAU projection. In most cases, reaching the disaggregation represented in the EPA data for the US above is difficult, so you'll need to determine what your data allow.

Information technology and some other industries move so quickly that you may need to tweak your BAU case to reflect current developments [25]. A salient recent example is that of cryptocurrency, which has created significant new electricity

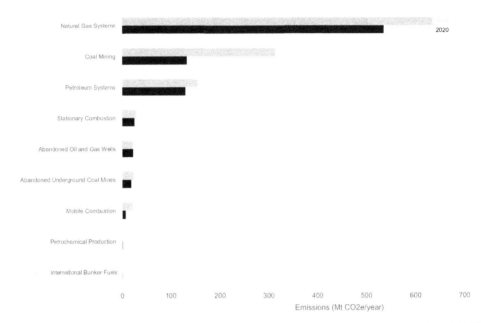

Figure 4.6. US fossil methane emissions by category for 1990 and 2020 (20 year GWPs). Source: US EPA [22]. GWPs from table 7.SM.7 in IPCC [23].

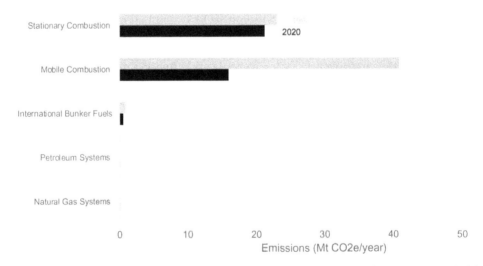

Figure 4.7. US fossil nitrous oxide emissions by category for 1990 and 2020 (20 year GWPs). Source: US EPA [22]. GWPs from table 7.SM.7 in IPCC [23].

demands in the span of just a few years [26–28]. There is significant debate on the value of that new technology in particular, but the local effects of new mining facilities (or even new conventional data centers) can be big enough to matter and are unlikely to be fully reflected in reference cases generated even a few years ago.

4.2.2 Projecting base-year emissions to mid-century

Once you've found or created a base-year energy balance and emissions inventory, you'll need to create a projection of energy use by end-use and task into the future, assuming current trends in technologies and policies continue. These projections include current expectations about how growth in service demand (i.e. tasks) will evolve over time.

Such projections are inherently uncertain, but you will need a baseline against which to measure progress and assess options. Some common sources of baseline projections include the International Energy Agency for the world and major countries and regions [29], British Petroleum for the world and major regions [30], and the US Energy Information Administration for the US [31]. Many countries, non-governmental organizations, states, and local utilities create such business-as-usual cases. You'll need to find the one best suited to your chosen geography.

As an example, table 4.2 shows selected key indicators of projected US service demand from the Energy Information Administration's Annual Energy Outlook 2022, which is the official business-as-usual projection for the US government. It is updated annually. Note that cooling and heating degree days reflect a warming climate, which changes heating and cooling demand.

Figure 4.8 shows an example of a business-as-usual (BAU) projection for natural use in the US commercial sector from the Energy Information Administration's Annual Energy Outlook 2020. This level of detail is the minimum required to do credible analysis of electrification in the commercial sector. The 'Other uses' category is often large enough that more analysis is needed to understand trends in key end-uses and prospects for electrifying them.

4.3 Analyzing electrification for a climate-positive scenario

The next step is to characterize prospects for electrification, focusing on the largest and fastest growing fossil fuel end-uses first. For each task we'll need alternatives to electrify fuel end-uses and to increase efficiency in electricity end-uses. We'll also need to specify how to accelerate stock turnover so that old equipment can be replaced at a more rapid pace than standard retirement cycles would suggest.

The climate-positive scenario uses the same drivers by task used in the business-as-usual case but sometimes policies and technologies affect them. An example would be projections of vehicle miles traveled (VMT), which generally assume that the structure of settlements will continue unchanged. We can change our settlement patterns and use technology (such as telecommuting) to reduce VMT, so each key driver should be assessed for ways to modify it in the future. We don't need to assume that the future needs to be the past writ large.

4.3.1 Understanding key technologies

As always, we begin with tasks. The three critical tasks for electrification, cutting across sectors and activities, are delivering heat, moving people and goods, and

Table 4.2. Selected indicators/drivers of projected US service demand, AEO 2022.

	Units	2021	2030	2040	2050
Residential					
Cooling degree days	CDD (base 65F)	1480	1639	1767	1898
Heating degree days	HDD (base 65F)	4072	3920	3757	3594
Average house area	Square feet	1795	1846	1901	1953
Mobile homes	Millions	7	7	7	6
Multifamily homes	Millions	32	35	37	39
Single-family homes	Millions	86	94	101	108
Total homes	Millions	125	135	144	153
Commercial					
Cooling degree days	CDD (base 65F)	1480	1639	1767	1898
Heating degree days	HDD (base 65F)	4072	3920	3757	3594
Total floorspace, new addition	B square feet	2	2	2	3
Total floorspace, surviving	B square feet	92	102	113	125
Total floorspace	B square feet	94	104	116	127
Industrial					
Combined heat and power capacity	GW	29	31	33	36
Combined heat and power generation	TWh/year	154	168	180	197
Value added, agriculture mining and construction	B2012$/year	2683	2844	3128	3498
Value added, manufacturing	B2012$/year	6592	7548	8608	9937
Value added, total	B2012$/year	9275	10,392	11736	13435
Aircraft					
Seat-miles demanded, narrow body aircraft	B miles	662	1050	1349	1712
Seat-miles demanded, regional jets	B miles	95	118	127	140
Seat-miles demanded, wide body aircraft	B miles	128	328	402	492
Seat-miles demanded, total	B miles	885	1497	1878	2344
Travel demand, revenue passenger miles, domestic	B miles	536	864	1072	1321
Travel demand, revenue passenger miles, international	B miles	94	434	557	707
Travel demand, revenue passenger miles, total	B miles	630	1299	1629	2028
Travel demand, revenue-ton miles, freight	B miles	47	60	75	93

Source: US DOE [31].

real-time controls (with associated data collection). The most important advantage of electricity is that it has high 'exergy' (the ability to do work), as described in [33, 34].

4.3.1.1 Heat

The three most important means to deliver heat using electricity are heat pumps (such as an air conditioner running in reverse), electric resistance (such as your toaster oven), and induction (which heats a pan directly by inducing currents in ferrous metal). Each of these is applicable in different situations, but the most widely used technology will be heat pumps, because they excel at moving heat across relatively low temperature gradients. Most heat needed in modern societies is low

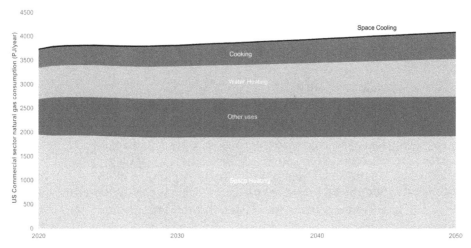

Figure 4.8. Projected US natural gas consumption, current trends continued. Source: US DOE [32]. PJ = 10^{15} joules of final energy.

temperature heat (think heating and cooling of buildings and water heating). Sensible application of building shell design, Passive House, and other low-energy design principles can make application of heat pumps even easier [35].

Heat pumps are getting better at delivering medium temperature heat. There are even heat pumps that use carbon dioxide as the refrigerant and can reach higher temperatures than typical heat pumps, while avoiding the high warming impacts of fluorine-based refrigerants.

Electric resistance and induction heating will also have uses, particularly when very high temperatures are needed [36]. Efforts to improve the efficiency of heat delivery and storage become increasingly important when electrifying, particularly for high temperature heat [8].

4.3.1.2 Mobility

Electric motors are a key enabling technology for electrification. They have higher power density, greater torque, instant response, no direct emissions, and no idling losses compared to combustion engines. They are also available in a range of sizes, from nanometer scale [37] to giant motors with output measured in tens of megawatts[1]. Combined with energy storage (usually batteries) electric mobility is cheaper, cleaner, and more performant than the internal combustion engines it replaces.

4.3.1.3 Real-time controls and data collection

Another benefit of electrified heat and mobility is the ease of integrating those technologies with precise electronic controls and sophisticated data collection. We'll

---

[1] https://www.gepowerconversion.com/news/ge-successfully-completed-no-load-testing-one-worlds-largest-80-megawatt-induction

talk more about these benefits in chapter 7, but for now suffice it to say that electrified end-uses are far easier to control in a precise way, which creates new possibilities for operating and optimizing power systems.

4.3.2 Understanding key policies

We list selected policies and other institutional actions needed to accelerate electrification in the checklists of action items below at http://www.solveclimate. org. Anything that accelerates adoption of electric technologies and accelerates retirement of existing fossil capital is fair game, including banning sales of combustion equipment by some date, banning expansion of the fossil gas grid, removing subsidies for combustion equipment, giving rebates for electrification, and setting purchase requirements for large institutions to only buy electrified equipment.

They key is that the system is now rigged to favor fossil fuels, and we need to unrig it and then push as hard as we can to get electrified equipment installed. Every single rule and regulation needs to be evaluated with a view towards speeding up electrification [1].

4.4 Data sources

The easiest way to create a climate-positive scenario is to find a pre-existing study with summary results that you can apply to your project. Fortunately, many such studies have been conducted around the world in recent years. The Deep Decarbonization Project created decarbonization analyses of exemplary technical specificity, including extensive electrification for more than a dozen major countries, including the US, China, India, Indonesia, and Brazil[2]. Those analyses also include business-as-usual scenarios that you can use, but these only cover the energy sector.

The International Energy Agency recently conducted a state-of-the-art study of how the world's energy sector could achieve net zero emissions by 2050 [38] that also included extensive electrification. As with all such studies, there are limitations, but it's a well-constructed authoritative study that uses the working towards a goal framing and gives significant detail.

Another important source for the US is the National Renewable Energy Laboratory's *Electrification Futures Study*[3]. That group has completed six reports to date, covering demand-side and supply-side technology data as well as power system design and operations.

Lawrence Berkeley National Laboratory, which is one of the world's leading centers for the study of energy and the environment, recently studied electrification for overall electrification strategies [39], long-haul trucking [40–42], US industrial boilers [43], and China's building sector [44].

Other research groups have studied electrification for medium and heavy duty shipping [45], electrification of European heat [46, 47], industrial heat [36], electricity

---

[2] https://ddpinitiative.org
[3] https://www.nrel.gov/analysis/electrification-futures.html

based plastics [48], energy-intensive industries [49, 50], and industrial policy to achieve net zero emissions in the industrial sector [51].

The recently completed IPCC Working Group III report on climate mitigation [52] is another terrific source, but it will take some digging to find what you need. It is thousands of pages long but the full report is a gold mine of references on different aspects of decarbonization, including electrification of buildings, industry, and transportation.

If you can't find an already-completed study that applies to your country, state, or company, you'll need to create one yourself. Once energy uses are associated with tasks, you'll need to find 'technology data' describing the technologies embedded in each system [21]. Such data characterize things like efficiency, operating hours per year, equipment lifetime, and equipment size/capacity/service delivered.

Efficiency of energy-using devices is measured using *test procedures*. For most equipment in buildings and industry in the US, those test procedures are defined by the National Institutes of Standards and Technology[4] in consultation with other federal agencies. For light vehicles, those procedures are defined by the US Department of Energy and the US Environmental Protection Agency[5]. Test procedures describe the operating conditions under which equipment will be measured and they are essential for ensuring that tests are conducted in a consistent way.

Efficiency changes over time, even in the BAU case, and those changes need to be explicitly characterized, for both the BAU case and the case where everything is electrified. Efficiency over time can be driven by changes in fuel prices, efficiency standards, labeling, and incentive programs.

Operating hours are usually calculated using consumer surveys or direct measurements. Operating hours matter most for products such as lighting, where user behavior has a huge influence on energy use. For appliances such as refrigerators, which aren't much affected by user behavior, these aren't typically estimated at the device level (although operating hours for individual components inside these devices are sometimes collected). For some information technology equipment, such as game consoles, manufacturers can collect usage data over the network for virtually all existing devices.

Equipment lifetimes are important for understanding stock turnover. The average lifetime embodies a distribution of equipment lifetimes. Some appliances retire early, because of renovations or fires, while others last much longer than the average. When the average lifetime is combined with what's called a 'retirement function' it's possible to estimate how much of the stock existing in some base year will still exist in some future year assuming 'natural' retirements (see appendix B for a specific example of a retirement function and how to use it).

Equipment size, capacity, and service delivered are all measures related to the task being performed. For heating and cooling equipment, capacity is measured

---

[4] https://www.nist.gov
[5] https://fueleconomy.gov/feg/how_tested.shtml

based on how much heat or cool air a device can produce per hour. Freezer volume, refrigerator through-the-door icemakers, and lighting quality are other examples.

Technology data are often available for energy used in buildings and related equipment, but a handful of experts around the world also study industrial equipment and processes. Once you identify these experts, case studies for industries and technologies can help in your assessment of electrification and emissions reduction options more generally. Tracking down the authors of recent studies, such as Zuberi et al [43] and Silvia et al [53] often leads to other related work.

One of the most important sources of technology data for appliances and equipment in the US is the Department of Energy's page summarizing the state of standards for dozens of different devices[6]. Implementing a standard in the US follows a set process that involves creating what's called a 'technical support document' or TSD, which contains exhaustive analysis on the technology and economics of improving efficiency in the device under study (after consultation with industry and other stakeholders). TSDs can be a technology data gold mine. To locate TSDs on the DOE's site, click on the product of interest then click on the link for 'Ongoing Rulemaking for Standards'.

Another source of such data is ENERGY STAR, a set of voluntary labeling programs run jointly by the US Environmental Protection Agency and the US Department of Energy[7]. These programs, which are designed based on extensive technical and economic analysis as well as industry consultations, typically apply the ENERGY STAR label to products that are in the top quartile of efficiency for a given product. The data and reports created by ENERGY STAR give a clear picture of what's happening to the most efficient products on the market.

Because the market for many products is global, the TSDs and data from ENERGY STAR can often be useful outside the US. Product regulators in many countries collect technology data applicable to those countries for designing efficiency standards and labels.

Looking afresh at industrial processes can yield innovations that substantially reduce or eliminate emissions while improving service and reducing costs [54]. For example, process CO_2 emissions for steel can be eliminated by using renewables to create hydrogen that is used to directly reduce iron ore, skipping the process that normally uses carbon and emits carbon dioxide [55, 56]. Another new process was created by Apple and Alcoa to eliminate process emissions for aluminum, which they expect ultimately to be cheaper than current processes [57]. Industries using high emissions processes can almost always redesign them if given the right incentives.

4.5 Assessing increases in electricity demand

Switching natural gas-fired energy uses to electricity will reduce natural gas consumption and increase electricity consumption. One purpose of an end-use level

---

[6] https://www.energy.gov/eere/buildings/standards-and-test-procedures
[7] http://www.energystar.gov

analysis of electrification is to get an accurate picture of just how much electricity demand will go up, so we can plan to meet that demand.

In his book *Electrify*, Saul Griffith [1] estimates the US will consume three to four times as much electricity as it now does if we electrify everything, and that means building a lot more generating capacity. It also means changing the way we design, expand, and operate the utility system, which we explore more in chapter 5. Each country, state, or region varies in its resource endowments and energy demand, which is why detailed analysis is needed to fully understand what steps are needed.

Both energy and capacity requirements need to be assessed for each end-use. There's recent work on energy and demand requirements for long-haul trucking electrification [40], among other end-uses. There's also work on increased electricity demand for electrifying basic materials industries in Europe [58].

Because of the local and system efficiency benefits from electrification, total primary energy use can be about half of current levels while delivering three times the electricity and the same energy services (accomplishing the same tasks as we would have otherwise). We can save about a quarter of current US primary energy consumption by switching generation to non-combustion sources such as solar, wind, hydro, and nuclear power. Electrifying transportation will save another 15% of primary energy, and eliminating the extraction, processing, and transportation of fossil fuels saves another 11% [1].

4.6 What activities can't be easily electrified now?

The toughest activities to electrify, such as long-haul aircraft and large ships, will require new technology and innovative thinking [54, 59, 60]. One possibility being researched is biofuels, another is what you could call an electrical fuel, hydrogen derived from non-combustion electricity generation, and still another is ammonia.

Even more exotic is the idea of 'synthetic methane', derived from carbon extracted from the atmosphere combined with hydrogen. In principle, such synthetic methane could allow for the creation of net zero methane, as long as the carbon really is extracted from the atmosphere. It wouldn't avoid other problems of such a fuel, like methane leakage and other pollutant emissions from combustion.

All such fuels are problematic because of the large losses involved in creating them. It is likely that only hydrogen will be a significant contributor to a fully electrified system, to be used in creation of plastics, steel, fertilizer, and other industrial processes. It can also store energy for the electric grid and may (eventually) power ships and planes in a liquified state, although we'll need to track hydrogen emissions, because they have a warming impact more than ten times greater than carbon dioxide per unit mass, according to the latest analysis [61–63] .

Global shipping is an interesting case, because roughly 40% of that shipping by weight is fossil fuels [64], so in a zero-emissions world, all that shipping goes away. This example shows why thinking about the whole system is critical for understanding options for electrifying and reducing emissions.

There's also the petrochemical industry. Because fossil fuels emit greenhouse gases throughout the supply chain, even fossil fuels used to create plastics, pesticides, herbicides, and other materials have a climate impact, even if those products don't create greenhouse gases when used or reach end-of-life (and even that last assumption isn't always true, because sometimes plastics are burned). That means that our high-end goal of 'ending fossil fuels' really does mean ending fossil fuels in all their applications.

The first step, of course, is to track sources of plastics by task, just as for other sources of pollution [65–68], and to project uses for plastic into the future to assess business-as-usual service demands [69]. We'll ultimately need to create chemical feedstocks from biological sources or directly from electrolyzed hydrogen and recaptured carbon dioxide [48, 70], but that transition will take years. In the meantime, the mantra of 'reduce, re-use, recycle' (plus shifting to more sustainable materials) must be our first steps toward a more sustainable path for petrochemicals.

4.7 A different type of electrification

The word 'electrification', as we discuss it above, means fully electrifying an existing energy system. It is also used to describe the process of bringing electricity to those who don't yet have easy or inexpensive access, and this effort must proceed in parallel with electrifying existing energy systems. There is a great deal of historical experience with rural electrification around the world [71–75], but this process hasn't proceeded nearly quickly enough for the 800 million people still without access to electricity[8].

The IPCC [52: Technical Summary] indicates with 'high confidence' that 'providing access to modern energy services universally would increase global GHG emissions by a few percent at most'. It's a huge societal benefit for modest cost, and it's the right thing to do.

The most important lesson from our discussion of electrifying existing energy systems is that building fossil infrastructure at this late date is a colossal error. Bringing energy access those without it means bringing *zero-emissions electricity* to them, in sufficient amounts at reasonable cost. There is no case for building additional natural gas or oil pipelines to address energy access issues, and in fact such infrastructure is counterproductive. The long lifetime of pipelines will make them stranded assets in short order [76], and no one should fall for fossil fuel companies disingenuous use of 'energy access' as an excuse to build more fossil infrastructure.

For many end-uses, such as fuel-based lighting, electrification using distributed solar photovoltaics, batteries, and LED lamps brings huge and immediate health, safety, and economic benefits at low cost [77, 78]. The one short-term exception to this rule may be the use of Liquified Petroleum Gas (LPG) for cooking in some rural areas of Africa and Asia.

Indoor air pollution from cooking with traditional fuels takes a horrific toll on the health of people in these regions, and LPG cooking is a fast way to eliminate traditional fuels in cooking [79]. It also doesn't require pipeline capacity, since it is delivered in

---

[8] https://ourworldindata.org/energy-access, calculated using World Bank data: http://data.worldbank.org/data-catalog/world-development-indicators

trucks that are used to refill canisters, so the amount of likely stranded assets would be low (repurposing a supply chain is a lot easier than repurposing a pipeline).

If this idea is implemented, it needs to be clearly understood that the use of LPG will be temporary until electricity arrives in sufficient amount and quality to allow for electric cooking. Electric induction hobs are also relatively cheap nowadays, so the advent of electricity will likely displace traditional fuels for cooking more quickly than many now think.

Promoting rural electrification is a complicated policy problem, but it has been extensively studied in the modern context [80–82]. Because distributed renewables have become so much cheaper in recent years, and because technology for smaller-scale grids (micro-grids, mini-grids) has developed quickly, the options for bringing electricity to every person on Earth have never been more attractive (the same statement applies to clean water). As with many such issues, understanding local context and mobilizing private capital are key to success.

4.8 Chapter conclusions

A cornerstone of reaching zero emissions in the energy sector is to electrify almost everything. We need to accelerate turnover of capital stock, and every time we replace equipment from now on, we need to make it electric. Over time, emissions from electrified equipment will decline as the electricity sector decarbonizes, costs for electrifying will decline as we scale up production of electric equipment, and the wasteful thermal losses associated with combustion will become a thing of the past.

We also need to bring clean electricity (and clean water, for that matter) to every human on the planet. Electrification will result in a better world for all who live in it, and it is long past time to make it a priority.

Further reading

Griffith S 2021 *Electrify: An Optimist's Playbook for Our Clean Energy Future* (Cambridge, MA: MIT Press). This book describes the case for electrification and gives high level insight into shifting our economy from spending on fuels to more spending on zero-emissions capital. This site, created by Saul Griffith and his colleagues, gives practical advice for homeowners wanting to electrify: https://www.rewiringamerica.org.

IEA 2020 *World Energy Balances: 2020 Edition* (Paris: International Energy Agency) http://wds.iea.org/wds/pdf/WORLDBAL_Documentation.pdf. The definitive source on energy balances by country.

Jacobson M Z 2020 *100% Clean, Renewable Energy and Storage for Everything* (Cambridge: Cambridge University Press) https://www.cambridge.org/highere-ducation/books/100-clean-renewable-energy-andstorage-for-everything/26F962411A4A4E1402479C5AEE680B08#overview. A summary of Jacobson's work on 100% renewable energy systems, in which electrification plays a critical role, updated from his earlier work.

Jadun P, McMillan C, Steinberg D, Muratori M, Vimmerstedt L and Mai T 2017 *Electrification Futures Study: End-Use Technology Cost and Performance Projections through 2050* (Golden, CO: National Renewable Energy Laboratory) NREL/TP-6A20-70485 https://www.nrel.gov/analysis/electrification-futures.html. This report is an authoritative source of technology data on end-use technologies for the US.

Lovins A B 2021 Decarbonizing our toughest sectors—profitably *MIT Sloan Manag. Rev* Fall, 4 August https://sloanreview.mit.edu/article/decarbonizing-our-toughest-sectors-profitably/. Lovins has been ahead of the curve on 'hard to decarbonize' sectors for years, and this article describes his latest work for a business audience.

Lovins A B 2021 *Profitably Decarbonizing Heavy Transport and Industrial Heat: Transforming These 'Harder-to-Abate' Sectors Is Not Uniquely Hard and Can Be Lucrative* (Old Snowmass, CO: Rocky Mountain Institute) July https://www.rmi.org/profitable-decarb/. This report gives more technical detail on decarbonizing heavy transport and industrial heat.

Mai T, Jadun P, Logan J, McMillan C, Muratori M, Steinberg D, Vimmerstedt L, Jones R, Haley B and Nelson B 2018 *Electrification Futures Study: Scenarios of Electric Technology Adoption and Power Consumption for the United States* (Golden, CO: National Renewable Energy Laboratory) NREL/TP-6A20-71500 https://www.nrel.gov/analysis/electrification-futures.html. This study explores electrification scenarios in great detail.

Project Drawdown (https://www.drawdown.org/solutions) gives significant detail on fossil and emissions reduction options, focusing on end-use efficiency and renewable electricity generation.

References

[1] Griffith S 2021 *Electrify: An Optimist's Playbook For Our Clean Energy Future* (Cambridge, MA: MIT Press)
[2] Dennis K 2015 Environmentally beneficial electrification: electricity as the end-use option *Electr. J.* **28** 100–12
[3] Dennis K, Colburn K and Lazar J 2016 Environmentally beneficial electrification: the dawn of 'emissions efficiency *Electr. J.* **29** 52–8
[4] Binsted M 2022 An electrified road to climate goals *Nat. Energy* **7** 9–10
[5] Aalto P (ed) 2021 *Electrification: Accelerating the Energy Transition* (Amsterdam: Academic)
[6] Aalto P, Haukkala T, Kilpeläinen S and Kojo M 2021 Introduction: electrification and the energy transition *Electrification* ed P Aalto (New York: Academic) ch 1 pp 3–24
[7] Victoria M, Zeyen E and Brown T 2022 Speed of technological transformations required in Europe to achieve different climate goals *Joule* **6** 1066–86
[8] Blair N *et al* 2022 Storage futures study: key learnings for the coming decades *Report* NREL/TP-7A40-81779 National Renewable Energy Laboratory, Golden, CO https://nrel.gov/analysis/storage-futures.html

[9] Vohra K, Vodonos A, Schwartz J, Marais E A, Sulprizio M P and Mickley L J 2021 Global mortality from outdoor fine particle pollution generated by fossil fuel combustion: results from GEOS-chem *Environ. Res.* **195** 110754

[10] Roberts D 2020 Air pollution is much worse than we thought: ditching fossil fuels would pay for itself through clean air alone *Vox* 12 August https://vox.com/energy-and-environment/2020/8/12/21361498/climate-change-air-pollution-us-india-china-deaths

[11] Scovronick N, Anthoff D, Dennig F, Errickson F, Ferranna M, Peng W, Spears D, Wagner F and Budolfson M 2021 The importance of health co-benefits under different climate policy cooperation frameworks *Environ. Res. Lett.* **16** 055027

[12] Schraufnagel D E *et al* 2019 Air pollution and noncommunicable diseases: a review by the Forum of International Respiratory Societies 2019; Environmental Committee, part 1: the damaging effects of air pollution *Chest* **155** 409–16

[13] Schraufnagel D E *et al* 2019 Air pollution and noncommunicable diseases: a review by the Forum of International Respiratory Societies 2019; Environmental Committee, part 2: air pollution and organ systems *Chest* **155** 417–26

[14] Choma E F, Evans J S, Hammitt J K, Gómez-Ibáñez J A and Spengler J D 2020 Assessing the health impacts of electric vehicles through air pollution in the United States *Environ. Int.* **144** 106015

[15] Copat C, Cristaldi A, Fiore M, Grasso A, Zuccarello P, Signorelli S S, Conti G O and Ferrante M 2020 The role of air pollution (PM and NO_2) in COVID-19 spread and lethality: a systematic review *Environ. Res.* **191** 110129

[16] Castells-Quintana D, Dienesch E and Krause M 2021 Air pollution in an urban world: a global view on density, cities and emissions *Ecol. Econ.* **189** 107153

[17] Errigo I M *et al* 2020 Human health and economic costs of air pollution in Utah: an expert assessment *Atmosphere* **11** 1238

[18] Paoli L and Gül T 2022 Electric cars fend off supply challenges to more than double global sales *International Energy Agency* 30 January https://www.iea.org/commentaries/electric-cars-fend-off-supply-challenges-to-more-than-double-global-sales

[19] Griffith S 2020 *Solving Climate Change with a Loan* (San Francisco, CA: Otherlab) https://medium.com/otherlab-news/solving-climate-change-with-a-loan-d1aac4b8259a

[20] IEA 2020 *World Energy Balances: 2020 Edition* (Paris: International Energy Agency) http://wds.iea.org/wds/pdf/WORLDBAL_Documentation.pdf

[21] Jadun P, McMillan C, Steinberg D, Muratori M, Vimmerstedt L and Mai T 2017 Electrification futures study: end-use technology cost and performance projections through 2050 *Report* NREL/TP-6A20-70485 National Renewable Energy Laboratory, Golden, CO https://www.nrel.gov/analysis/electrification-futures.html

[22] US EPA 2022 *Inventory of US Greenhouse Gas Emissions and Sinks: 1990–2020* (Washington, DC: US Environmental Protection Agency) https://epa.gov/ghgemissions/inventory-us-greenhouse-gas-emissions-and-sinks

[23] IPCC 2021 *Climate Change 2021: The Physical Science Basis. Contribution of Working Group I to the Sixth Assessment Report of the Intergovernmental Panel on Climate Change* ed V Masson-Delmotte *et al* (Cambridge: Cambridge University Press) https://ipcc.ch/report/sixth-assessment-report-working-group-i/

[24] Dreyfus G B, Xu Y, Shindell D T, Zaelke D and Ramanathan V 2022 Mitigating climate disruption in time: a self-consistent approach for avoiding both near-term and long-term global warming *Proc. Natl Acad. Sci. USA* **119** e2123536119

[25] Masanet E, Shehabi A, Lei N, Smith S and Koomey J 2020 Recalibrating global data center energy-use estimates *Science* **367** 984

[26] Koomey J 2019 *Estimating Bitcoin Electricity Use: A Beginner's Guide* (Washington, DC: Coin Center) https://coincenter.org/entry/bitcoin-electricity

[27] Koomey J and Masanet E 2021 Does not compute: avoiding pitfalls assessing the internet's energy and carbon impacts *Joule* **5** 1625–8

[28] Lei N, Masanet E and Koomey J 2021 Best practices for analyzing the direct energy use of blockchain technology systems: review and policy recommendations *Energy Policy* **156** 112422

[29] IEA 2021 *World Energy Outlook 2021* (Paris: International Energy Agency, Organization for Economic Cooperation and Development (OECD)) http://worldenergyoutlook.org/

[30] British Petroleum 2020 *BP Energy Outlook: 2020 Edition* (London: British Petroleum) https://bp.com/en/global/corporate/energy-economics/energy-outlook.html

[31] US DOE 2022 *Annual Energy Outlook 2022, with Projections to 2050* (Washington, DC: Energy Information Administration, US Department of Energy) https://eia.gov/aeo

[32] US DOE 2020 *Annual Energy Outlook 2020, with Projections to 2050* (Washington, DC: Energy Information Administration, US Department of Energy) https://eia.gov/aeo

[33] Eyre N 2021 From using heat to using work: reconceptualising the zero carbon energy transition *Energy Efficiency* **14** 77

[34] Nakicenovic N, Ishitani H, Johansson T, Marland G, Moreira J R and Rogner H-H 1996 Energy primer *Climate Change 1995: Impacts, Adaptations and Mitigation of Climate Change* ed R T Watson, M C Zinyowera and R H Moss (Cambridge: Cambridge University Press) pp 77–92 https://www.ipcc.ch/report/ar2/wg2/

[35] Harvey L D 2015 *A Handbook on Low-Energy Buildings and District-Energy Systems: Fundamentals, Techniques and Examples* (New York: Routledge)

[36] Sandalow D, Friedmann J, Aines R, McCormick C, McCoy S and Stolaroff J 2019 ICEF industrial heat decarbonization roadmap *ICEF* December https://icef.go.jp/pdf/summary/roadmap/icef2019_roadmap.pdf

[37] Tierney H L, Murphy C J, Jewell A D, Baber A E, Iski E V, Khodaverdian H Y, McGuire A F, Klebanov N and Sykes E C H 2011 Experimental demonstration of a single-molecule electric motor *Nat. Nanotechnol.* **6** 625–9

[38] IEA 2021 *Net Zero by 2050: A Roadmap for the Global Energy Sector* (Paris: International Energy Agency) https://iea.org/reports/net-zero-by-2050

[39] Deason J, Wei M, Leventis G, Smith S and Schwartz L 2018 *Electrification of Buildings and Industry in the United States: Drivers, Barriers, Prospects, and Policy Approaches* (Berkeley, CA: Lawrence Berkeley National Laboratory) https://eta-publications.lbl.gov/sites/default/files/electrification_of_buildings_and_industry_final_0.pdf

[40] Tong F, Wolfson D, Jenn A, Scown C D and Auffhammer M 2021 Energy consumption and charging load profiles from long-haul truck electrification in the United States *Environ. Res.: Infrastruct. Sustain.* **1** 025007

[41] Tong F, Jenn A, Wolfson D, Scown C D and Auffhammer M 2021 Health and climate impacts from long-haul truck electrification *Environ. Sci. Technol.* **55** 8514–23

[42] Phadke A, Khandekar A, Abhyankar N, Wooley D and Rajagopal D 2021 *Why Regional and Long-Haul Trucks Are Primed for Electrification Now* (Berkeley, CA: Lawrence Berkeley National Laboratory) https://energyanalysis.lbl.gov/publications/why-regional-and-long-haul-trucks-are

[43] Zuberi M, Jibran S, Hasanbeigi A and Morrow W R 2021 *Electrification of Boilers in US Manufacturing* (Berkeley, CA: Lawrence Berkeley National Laboratory) LBNL-2001436 https://energyanalysis.lbl.gov/publications/electrification-boilers-us

[44] Feng W, Zhou N, Wang W, Khanna N, Liu X and Hou J 2021 *Pathways for Accelerating Maximum Electrification of Direct Fuel Use in China's Building Sector Drivers, Impacts, Barriers, Prospects, and Policy Recommendations* (Berkeley, CA: Lawrence Berkeley National Laboratory) https://energyanalysis.lbl.gov/publications/pathways-accelerating-maximum

[45] Forrest K, MacKinnon M, Tarroja B and Samuelsen S 2020 Estimating the technical feasibility of fuel cell and battery electric vehicles for the medium and heavy duty sectors in California *Appl. Energy* **276** 115439

[46] Thomaßen G, Kavvadias K and Navarro J P J 2021 The decarbonisation of the EU heating sector through electrification: a parametric analysis *Energy Policy* **148** 111929

[47] Papapetrou M, Kosmadakis G, Cipollina A, Commare U L and Micale G 2018 Industrial waste heat: estimation of the technically available resource in the EU per industrial sector, temperature level and country *Appl. Therm. Eng.* **138** 207–16

[48] Palm E, Nilsson L J and Åhman M 2016 Electricity-based plastics and their potential demand for electricity and carbon dioxide *J. Cleaner Prod.* **129** 548–55

[49] Bataille C *et al* 2018 A review of technology and policy deep decarbonization pathway options for making energy-intensive industry production consistent with the Paris Agreement *J. Clean. Prod.* **187** 960–73

[50] Bataille C G F 2020 Physical and policy pathways to net-zero emissions industry *WIREs Clim. Change* **11** e633

[51] Nilsson L J *et al* 2021 An industrial policy framework for transforming energy and emissions intensive industries towards zero emissions *Clim. Policy* **21** 1053–65

[52] IPCC 2022 *Climate Change 2022: Mitigation of Climate Change. Contribution of Working Group III to the Sixth Assessment Report of the Intergovernmental Panel on Climate Change.* (Cambridge: Cambridge University Press) https://ipcc.ch/report/sixth-assessment-report-working-group-3/

[53] Madeddu S *et al* 2020 The CO_2 reduction potential for the European industry via direct electrification of heat supply (power-to-heat)*Environ. Res. Lett.* **15** 124004

[54] Lovins A B 2021 *Profitably Decarbonizing Heavy Transport and Industrial Heat: Transforming These 'Harder-to-Abate' Sectors Is Not Uniquely Hard and Can Be Lucrative* (Old Snowmass, CO: Rocky Mountain Institute)

[55] Valentin V, Åhman M and Nilsson L J 2018 Assessment of hydrogen direct reduction for fossil-free steelmaking *J. Cleaner Prod.* **203** 736–45

[56] Steve G, Sovacool B K, Kim J, Bazilian M and Uratani J M 2021 Industrial decarbonization via hydrogen: a critical and systematic review of developments, socio-technical systems and policy options *Energy Res. Soc. Sci.* **80** 102208

[57] Nellis S 2019 Apple buys first-ever carbon-free aluminum from Alcoa-Rio Tinto venture *Reuters* 5 December https://reuters.com/article/us-apple-aluminum-idUSKBN1Y91RQ

[58] Stefan L, Nilsson L J, Åhman M and Schneider C 2016 Decarbonising the energy intensive basic materials industry through electrification—implications for future EU electricity demand *Energy* **115** 1623–31

[59] Gnadt A R, Speth R L, Sabnis J S and Barrett S R H 2019 Technical and environmental assessment of all-electric 180-passenger commercial aircraft *Prog. Aerosp. Sci.* **105** 1–30

[60] Schäfer A W, Barrett S R H, Doyme K, Dray L M, Gnadt A R, Self R, O'Sullivan A, Synodinos A P and Torija A J 2019 Technological, economic and environmental prospects of all-electric aircraft *Nat. Energy* **4** 160–6

[61] Warwick N, Griffiths P, Keeble J, Archibald A, Pyle J and Shine K 2022 *Atmospheric Implications of Increased Hydrogen Use* (Cambridge: University of Cambridge, NCAS, and University of Reading) https://assets.publishing.service.gov.uk/government/uploads/system/uploads/attachment_data/file/1067144/atmospheric-implications-of-increased-hydrogen-use.pdf

[62] Richard D, Simmonds P, O'Doherty S, Manning A, Collins W and Stevenson D 2006 Global environmental impacts of the hydrogen economy *Int. J. Nucl. Hydrogen Prod. Appl.* **1** 57–67

[63] Derwent R 2018 Hydrogen for heating: atmospheric impacts—a literature review *BEIS Research paper* Number 21 Department for Business, Energy and Industrial Strategy https://assets.publishing.service.gov.uk/government/uploads/system/uploads/attachment_data/file/760538/Hydrogen_atmospheric_impact_report.pdf

[64] UNCTAD 2021 *Review of Maritime Transport 2021* (Geneva: United Nations Conference on Trade and Development) https://unctad.org/webflyer/review-maritime-transport-2021

[65] Roland G, Jambeck J R and Law K L 2017 Production, use, and fate of all plastics ever made *Sci. Adv.* **3** e1700782

[66] Brooks A L, Wang S and Jambeck J R 2018 The Chinese import ban and its impact on global plastic waste trade *Sci. Adv.* **4** eaat0131

[67] Jambeck J R, Geyer R, Wilcox C, Siegler T R, Perryman M, Andrady A, Narayan R and Law K L 2015 Plastic waste inputs from land into the ocean *Science* **347** 768–71

[68] Law K L, Starr N, Siegler T R, Jambeck J R, Mallos N J and Leonard G H 2020 The United States' contribution of plastic waste to land and ocean *Sci. Adv.* **6** eabd0288

[69] Borrelle S B *et al* 2020 Predicted growth in plastic waste exceeds efforts to mitigate plastic pollution *Science* **369** 1515–8

[70] Meys R, Kätelhön A, Bachmann M, Winter B, Zibunas C, Suh S and Bardow A 2021 Achieving net-zero greenhouse gas emission plastics by a circular carbon economy *Science* **374** 71–6

[71] Altawell N 2021 Rural electrification projects *Rural Electrification* (New York: Academic) ch 8 pp 149–56 https://sciencedirect.com/science/article/pii/B9780128224038000084

[72] Altawell N, Milne J, Seowou P and Sykes L (ed) 2020 *Rural Electrification: Optimizing Economics, Planning and Policy in an Era of Climate Change and Energy Transition* (Amsterdam: Academic)

[73] Hasenöhrl U 2018 Rural electrification in the British Empire *Hist. Retail. Consump.* **4** 10–27

[74] Hirsh R 2022 *Powering American Farms: The Overlooked Origins of Rural Electrification* (Baltimore, MD: Johns Hopkins University Press) https://press.jhu.edu/books/title/12690/powering-american-farms

[75] Torero M 2015 The impact of rural electrification: challenges and ways forward *Rev. Écon. Dév.* **23** 49–75

[76] Kemfert C, Präger F, Braunger I, Hoffart F M and Brauers H 2022 The expansion of natural gas infrastructure puts energy transitions at risk *Nat. Energy* **7** 582–7

[77] Mills E 2016 Identifying and reducing the health and safety impacts of fuel-based lighting *Energy Sust. Dev.* **30** 39–50

[78] Apple J, Vicente R, Yarberry A, Lohse N, Mills E, Jacobson A and Poppendieck D 2010 Characterization of particulate matter size distributions and indoor concentrations from kerosene and diesel lamps *Indoor Air* **20** 399–411

[79] Smith K R and Pillarisetti A 2017 *Household Air Pollution from Solid Cookfuels and Its Effects on Health* (Washington, DC: The International Bank for Reconstruction and Development/The World Bank) http://europepmc.org/abstract/MED/30212117 http://europepmc.org/books/NBK525225 https://ncbi.nlm.nih.gov/books/NBK525225

[80] Bhattacharyya S C and Palit D 2016 Mini-grid based off-grid electrification to enhance electricity access in developing countries: what policies may be required? *Energy Policy* **94** 166–78

[81] Williams N J, Jaramillo P, Taneja J and Selim Ustun T 2015 Enabling private sector investment in microgrid-based rural electrification in developing countries: a review *Renew. Sustain. Energy Rev.* **52** 1268–81

[82] Graber S, Adesua O, Agbaegbu C, Malo I and Sherwood J 2019 *Electrifying the Underserved: Collaborative Business Models for Developing Minigrids under the Grid* (Old Snowmass, CO: Rocky Mountain Institute) https://rmi.org/insight/undergrid-business-models

Chapter 5

Decarbonize electricity

Ever-cheaper renewable energy technologies give electricity the edge in the race to zero.
—International Energy Agency, *Net Zero by 2050*, May 2021

Chapter overview

- Decarbonizing the electricity system reduces emissions from electrified end-uses over time.
- We'll need to increase total electricity generation several-fold if we are to electrify everything, and the whole system needs to be nearly zero emissions.
- Most current scenarios show that east-cost power systems will incorporate 80% or more variable renewable generation.
- We can create high-renewables systems now using existing gas generation as an emergency backup, but we'll need to change how those generators are paid to minimize their emissions and only allow them to operate during emergencies.
- The reliability of a power system is a characteristic of the whole system and is not just dependent on the details of individual generators.
- Power markets and operational heuristics will need to change to accommodate the shift to variable renewables. These changes also increase the flexibility of the power system and will make integration of all new resources easier over time, while preserving reliability.

5.1 Introduction

Researchers have known for a long time that the potential for cheaply and rapidly reducing emissions in the electricity sector has been greater than for other sectors. In the past decade or so, researchers combined that knowledge with the realization that electrifying currently non-electric end-uses can result in rapid efficiency improvements and emissions reductions for those otherwise hard-to-abate sectors.

Electrifying all energy end-uses will require significantly more electricity generation than currently installed, by some estimates three to four times more [1, 2]. That means building new zero-emissions generation well beyond replacing current power plants while minimizing costs and ensuring system reliability. It also means changing the design and operation of the power system (and associated institutions) to make integrating lots of variable renewable generation easier and cheaper than it has been in the past.

5.2 Bringing the future into focus

Designing a future power system is complicated. The key uncertainty for the next couple of decades is how a zero-emissions system dominated by variable renewables can meet system demands under extreme conditions.

Fortunately, we don't need to have it all figured out now. The simplest way to make a high-variable-renewables system work is to keep a lot of existing fossil gas capacity around [3], maintain it well, and design incentive structures to that it only runs in emergencies. You could call it a 'strategic reserve' of gas capacity that is available to meet big swings in demand when solar and wind are less available.

As we develop more zero-emission ways to balance the utility system in extreme situations, the strategic reserve can be phased out, but in the interim, it's a simple and proven way to have a 95+% renewables system that would respond well to periodic lulls in wind, for example. As long as the gas plants only run a few percent of the year on average, the battle for zero-emissions electricity is essentially won.

A key finding from Williams *et al* for the US (and corroborated by Victoria *et al* [4] for Europe) is that although 'carbon neutral pathways diverge in energy strategy, resource use, and cost primarily after 2035', a key set of actions are common across all pathways over the next decade:

- Rapid construction (>500 GW by 2030) of new wind and solar generation capacity.
- Phase out of coal generation by 2030.
- Maintaining current nuclear and fossil gas generation capacity.
- Electrification of light vehicles (>50% market share for electric vehicles by 2030).
- Electrification of residential heating (>50% market share for heat pumps by 2030).

The study concludes with an acknowledgment of the deep uncertainty about future energy systems combined with the necessary first steps toward rapid emissions reductions:

...taking decisive near-term action in the areas that are well understood, combined with laying the necessary groundwork in the areas of uncertainty, puts the United States on a carbon-neutral pathway right away while allowing the most difficult decisions and tradeoffs to be made with better information in the future.

Knowing that a low-emissions future will involve high penetration of variable renewable generation brings the future into focus [5]. The things that will enable more variable renewables to operate successfully on the grid will create new possibilities for whatever new sources of generation emerge in coming decades.

A high-renewables grid will likely rely on some or all of the following resources [6]:

- 'Clean firm' generation sources, such as nuclear power, fossil gas or biomass with carbon capture and storage (CCS), hydropower, and geothermal [7, 8].
- Diversity of renewable generation sources.
- 'Overbuilding'[1] of renewable capacity to increase flexibility, reduce risk, and ensure adequate supplies during seasonal periods of low generation.
- More transmission capacity to allow broader geographic aggregation of variable renewables.
- Battery storage for individual buildings and at utility scale [9].
- Long-term electricity storage other than batteries, such as pumped hydro [10].
- Seasonal, weekly, and daily thermal storage for buildings and industry.
- Flexible/interruptible industrial loads.
- Creation of electrical fuels for energy storage, such as electrolysis of hydrogen or creation of synthetic methane.
- Better load and supply forecasting.
- Real-time prices for large numbers of customers.
- Load aggregation and better price responsiveness of demand [11].

All these improvements will make it easier to operate and optimize a high-renewables electricity grid, regardless of the mix of resources that end up supplying power.

There will also be changes in more conventional technologies and practices so that they will integrate better into a high-renewables grid. One example is changing the design and operation of power markets [12, 13]. The economics of electricity systems has for many decades been strongly influenced by the short run operating costs (marginal costs) of fossil fuel generators [14]. Rate design and utility business models have been structured around the idea of marginal costs being one of the primary drivers of utility system dispatch and investments.

---

[1] By overbuilding we mean building far more renewables capacity then a strict least-cost rule would indicate in order to gain diversity benefits of spreading renewables out geographically and by resource type. This technique has become commonplace in analyses over the past decade or so, and renewables are now cheap enough that even if some of their generation is 'spilled' (i.e. not used) the economics of those investments are still acceptable, and worth it to gain those diversity benefits.

The shift to a system dominated by capital investment means that market designs for the electricity system need to evolve [6, 12, 13, 15–28]. The end state of having a large proportion of electricity generation being low or nearly zero marginal costs with significant storage in the system [9, 10] will force system operators to develop alternative heuristics, improved contracts, and new market structures for dispatching power plants, paying for the construction of new plants, and determining how demand-size aggregators and storage resources will participate in these markets. This will also mean developing new analytical methods for assessing the *value* of different operating rules and power system investments, not just assessing costs [29–31].

Another important aspect of shifting to non-combustion generation is that the primary energy associated with combustion losses simply disappears. As Griffith [32] points out, we don't need to replace primary energy that was simply thrown away all along. For fossil power plants, that's half to two-thirds of the fuel input that heats the surroundings without performing useful work.

5.3 What about system reliability?

A key concern about power systems is maintaining reliability. Outages are expensive for customers, for the utility, and for society [33–37][2]. Fortunately, the innovations described above, properly implemented, can lead to higher-reliability power systems.

Reliability can only be assessed by examining the grid as a whole system, not just focusing on the characteristics of individual power plants. That is why the common attention paid to 'baseload' resources is misguided (and is an artifact of the way power systems used to operate). In fact, high-capacity-factor generation sources can affect system reliability negatively in poorly designed systems, and their large size creates other challenges for operating the system reliably [12, 28].

5.4 What about 100% renewables?

The question of whether a grid comprised 100% of renewable power is feasible or economic has preoccupied many in recent years [5, 38–48, 51]. This acrimonious debate has had the beneficial effect of pushing the discussion to better reflect recent technological developments in a more accurate way. Renewables have come a long way in the past decade.

Worrying too much about this question is a distraction, however. Getting to 80% renewables is pretty clearly feasible almost everywhere [5, 47, 48], and we don't have to decide about the last 20% for years. We should unquestionably research, develop, and deploy many options for integrating high-renewables generation fractions, but we won't know exactly how the system will evolve for years, and that's fine.

---

[2] The Lawrence Berkeley National Laboratory Interruption Cost Estimate calculator summarizes these data in a convenient way: https://icecalculator.com/home.

5.5 Creating or adopting a business-as-usual scenario

Just as for electrification, the first step is to understand base year emissions. Figure 5.1 shows carbon dioxide emissions for the US power sector in 1990 and 2020, revealing dramatic changes over this period. Power-sector coal emissions fall by about half while natural gas generation emissions go up by almost a factor of four.

Natural gas power plants displaced coal plants over this period, but declining electricity demand and rapid growth in renewable electricity generation (wind and solar in particular) also had measurable effects.

The rate of decline in coal generation is about 1.7% per year expressed as a percentage of 1990 emissions, but much of that decline occurred in the past decade. From 2010 to 2020, US coal generation declined almost 6% per year as a percentage of 2010 generation. In the UK, from 2012 to 2020, coal generation declined 12% per year as a percentage of 2012 generation (which was the UK peak)[3]. Eliminating 2020 coal generation by 2030 (as climate stabilization requires) means declines of 10% per year expressed as a percentage of 2020 emissions.

For every emissions segment we need a baseline projection that reflects current trends or business-as-usual (BAU). For most countries, states, or utility systems, the local authorities have already created such a scenario, but it may not have as much detail as needed for your purposes. Just as for energy systems and emissions inventories more generally, we want electricity use (and fuel use to be electrified)

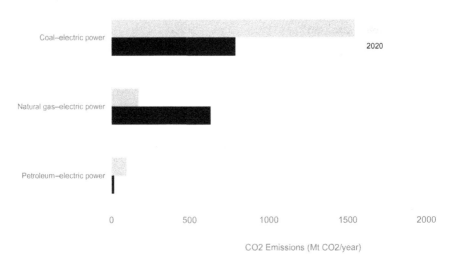

Figure 5.1. Power-sector fossil combustion carbon dioxide emissions by fuel for 1990 and 2020. Source: US EPA [49]. GWPs from table 7.SM.7 in IPCC [50]. The GWP for carbon dioxide is the same for all analysis periods by definition.

---

[3] https://www.gov.uk/government/statistics/solid-fuels-and-derived-gases-chapter-2-digest-of-united-kingdom-energy-statistics-dukes

to be mapped to human activities (tasks) and then projected to 2040 for each task, based on expectations for changes in service demand. This yields a projection of future demand and electricity generation if current trends continue (see chapter 4).

The fuel and generation mix for the BAU case allows us to calculate baseline emissions for the electricity sector. Even the BAU case electricity generation will be getting cleaner, because of the rapid drops in the costs of renewable generation technologies in recent years (even subsidized fossil fuels can't compete against solar and wind at current prices in many cases). In the intervention case, case electricity generation will get cleaner even more quickly.

5.6 Creating the climate-positive scenario

The climate-positive scenario from chapter 4 supplies future electricity demand with nearly complete electrification by 2040. The projected future demand tells us how much zero-emissions electricity generation we'll need to build to meet the demand, which helps us design the electrification scenario on the supply side.

5.6.1 Understanding key technologies

The key technologies for zero-emissions generation all involve non-combustion technologies such as solar, wind, nuclear, and geothermal, combined with rapid improvements in the efficiency of electrified end-uses. While these technologies involve some emissions to build them and (sometimes) minor emissions to operate them, they all have nearly zero emissions in operation. Key technologies involve energy storage in many forms, load aggregation, price responsive demand, more transmission, better forecasting, and creation of electrical fuels for storage and manufacturing (in particular hydrogen).

5.6.2 Understanding key policies

As with electrification, we list key policies in the Checklists of Action Items at http://www.solveclimate.org. The name of the game is rapid displacement of fossil fuel generation—we need to do everything we can to speed up this process. These policies include rapid retirement schedules for fossil plants, carbon pricing, emissions targets, non-fossil generation targets market design, and streamlining of regulatory approvals for power plants and transmission projects.

5.7 Data sources

As with all work of this type, researchers need to ensure that the data sources they use are suitable and representative for their chosen geography. The data sources below are widely cited and used but will almost always need to be checked and tweaked to suit your needs.

The most important data sources on power systems tend to be local. Because of the long-lived nature of their capital assets and because they are often regulated, utilities usually collect data for resource planning and publish it in reports that are publicly available. Their regulators often require certain data to be made public because utility resource plans affect electricity prices. Look for the utility's business-as-usual projection as a starting point.

More forward-looking utilities create alternative scenarios that reduce emissions, although these are usually not as aggressive as climate stabilization requires. That means you'll need to alter assumptions in those scenarios and estimate potential emissions reductions from more aggressive action.

US utilities are regulated at the state level, so look to the state's public utility commission or public service commission for data. The US Energy Information Administration[4] also compiles data at the utility, state, and national levels. Other countries regulate utilities at either the state or national levels so you'll need to find available data sources that suit your chosen geography.

The International Energy Agency[5] compiles electricity consumption and production numbers for most countries around the world. Unfortunately, much of these data are behind a paywall (at least for now).

EMBER[6] is a terrific source for power system data in many countries in the world. They compile and combine EIA, IEA, and local data to generate historical data on power generation and emissions and make those data freely available.

The Deep Decarbonization Pathways Project[7] has created climate stabilization pathways for many countries around the world, and their modeling of the electricity sector is generally detailed and well documented. Their work can inform both BAU and climate stabilization scenario design.

The US National Renewable Energy Laboratory (NREL) has worked for many years on electrification futures, with focus on both the supply and demand sides[8]. The work is a treasure trove for researchers creating scenarios for a zero-emissions power grid.

Another important data source from NREL is their life-cycle assessment (LCA) harmonization project for electricity generation technologies[9]. The indirect emissions for these technologies vary a great deal and can be big enough to matter, and harmonized LCA results can be important when creating total emissions estimates for electricity generation.

The state-by-state US scenario outputs from Christopher Clack at Vibrant Clean Energy[10] are another great data source, exploring different clean energy and clean electricity scenarios to 2050.

Professor Mark Z Jacobson at Stanford has created detailed 100% renewable electricity plans for US states[11] and 145 countries around the world[12].

---

[4] http://www.eia.gov

[5] http://www.iea.org

[6] https://ember-climate.org/data/

[7] https://ddpinitiative.org

[8] https://www.nrel.gov/analysis/electrification-futures.html

[9] https://www.nrel.gov/analysis/life-cycle-assessment.html

[10] https://zero2050usa.com

[11] https://web.stanford.edu/group/efmh/jacobson/Articles/I/WWS-USA.html

[12] https://web.stanford.edu/group/efmh/jacobson/Articles/I/WWS-145-Countries.html

5.8 Chapter conclusions

Decarbonizing electricity requires massive investments in non-combustion generation technology. Total generation must increase several-fold if we're to electrify everything. To make the decarbonized power system a reality, we'll need to modify how we build and operate the utility system, enhancing flexibility, encouraging demand response, overbuilding renewables, creating more transmission, doing better forecasting, increasing diversity of renewable generation, installing storage of different durations, and building dispatchable zero-emissions generation when we can.

Further reading

Blair N, Augustine C, Cole W, Denholm P, Frazier W, Geocaris M, Jorgenson J, McCabe K, Podkaminer K, Prasanna A and Sigrin B 2022 *Storage Futures Study: Key Learnings for the Coming Decades* (Golden, CO: National Renewable Energy Laboratory) NREL/TP-7A40-81779, April https://www.nrel.gov/analysis/storage-futures.html. Energy storage is a critical enabling technology for the energy transition. This comprehensive study reviews key insights on this technology.

Denholm P, Brown P, Cole W, Brown M, Jadun P, Ho J, Mayernik J, McMillan C and Sreenath R 2022 *Examining Supply-Side Options to Achieve 100% Clean Electricity by 2035* (Golden, CO: National Renewable Energy Laboratory) NREL/TP-6A40–81644 https://www.nrel.gov/docs/fy22osti/81644.pdf. This report describes the latest scenarios from NREL, including an electrification scenario on the demand side combined with rapid changes in the electricity generation system.

Faruqui A 2020 The coming transformation of the electricity sector: a conversation with Amory Lovins *Electr. J.* **33** 106827 https://doi.org/10.1016/j.tej.2020.106827. Lovins gives a high-level view of upcoming developments in the electricity sector.

IPCC 2022 *Climate Change 2022: Mitigation of Climate Change. Contribution of Working Group III to the Sixth Assessment Report of the Intergovernmental Panel on Climate Change.* Section 6.2.2 (Cambridge: Cambridge University Press) https://www.ipcc.ch/report/sixth-assessment-report-working-group-3/. The IPCC process is often slow, but this IPCC report is a treasure trove, giving summary findings and references on building zero-emissions power systems.

Jacobson M Z, von Krauland A-K, Coughlin S J, Palmer F C and Smith M M 2022 Zero air pollution and zero carbon from all energy at low cost and without blackouts in variable weather throughout the US with 100% wind–water–solar and storage *Renew. Energy* **184** 430–42 https://doi.org/10.1016/j.renene.2021.11.067. This article summarizes Jacobson's latest work on 100% renewable power grids.

Jacobson M Z 2020 *100% Clean, Renewable Energy and Storage for Everything* (Cambridge: Cambridge University Press) https://doi.org/10.1017/9781108786713. This book summarizes historical and recent research on 100% renewables power grids.

Jenkins J D, Luke M and Thernstrom S 2018 Getting to zero carbon emissions in the electric power sector *Joule* **2** 2498–510 https://doi.org/10.1016/j.joule.2018.11.013. This article summarizes the authors' review of 40 electricity-sector studies focusing on implications for zero-emissions power grids.

Lovins A B 2017 Reliably integrating variable renewables: moving grid flexibility resources from models to results *Electr. J.* **30** 58–63 https://doi.org/10.1016/j.tej.2017.11.006. Lovins summarizes the pitfalls and promise of high renewables power grids.

Lovins A B 2017 Do coal and nuclear generation deserve above-market prices? *Electr. J.* **30** 22–30 https://doi.org/10.1016/j.tej.2017.06.002. There has been much confusion in the public policy debates around the role of large-scale 'baseload' plants in high-renewables power grids, and this article critically assess some of the most common claims in that literature.

Luderer G *et al* 2022 Impact of declining renewable energy costs on electrification in low-emission scenarios *Nat. Energy* **7** 32–42 https://doi.org/10.1038/s41560-021-00937-z. An analysis of how rapidly declining renewables costs affect electrification.

National Academies of Sciences, Engineering, and Medicine 2021 *Accelerating Decarbonization of the US Energy System* (Washington, DC: The National Academies Press) https://doi.org/10.17226/25932. A consensus report on how to accelerate decarbonization, reviewing both technical and institutional changes needed to make zero-emissions energy systems a reality.

Olson A, Hull S, Ming Z, Schlag N and Duff C 2021 *Scalable Markets for the Energy Transition: A Blueprint for Wholesale Electricity Market Reform* (San Francisco, CA: Energy and Environmental Economics) https://www.ethree.com/scalable-markets-for-the-energy-transition-a-new-e3-report/. Market design is critical for getting incentives right in the utility sector, and this report delves deeply into the relevant issues.

Sepulveda N A, Jenkins J D, de Sisternes F J and Lester R K 2018 The role of firm low-carbon electricity resources in deep decarbonization of power generation *Joule* **2** 2403–20 https://doi.org/10.1016/j.joule.2018.08.006. One of the key sources exploring the need for and issues around firm low-emissions electricity generation.

Sepulveda N A, Jenkins J D, Edington A, Mallapragada D S and Lester R K 2021 The design space for long-duration energy storage in decarbonized power systems *Nat. Energy* **6** 506–16 https://doi.org/10.1038/s41560-021-00796-8. Long duration storage would make operating high renewables grids a lot easier, and this study explores the tradeoffs, constraints, and opportunities for known technologies.

Williams J H, Jones R A, Haley B, Kwok G, Hargreaves J, Farbes J and Torn M S 2021 Carbon-neutral pathways for the United States *AGU Adv.* **2** e2020AV000284 https://doi.org/10.1029/2020AV000284. Detailed analysis of many carbon neutral pathways for the US.

Williams J H, Jones R A and Torn M S 2021 Observations on the transition to a net-zero energy system in the United States *Energy Clim. Change* **2** 100050 https://doi.org/10.1016/j.egycc.2021.100050. Useful summary of lessons for net-zero-emissions energy systems.

Zhou E and Mai T 2021 *Electrification Futures Study: Operational Analysis of US Power Systems with Increased Electrification and Demand-Side Flexibility* (Golden, CO: National Renewable Energy Laboratory) NREL/TP-6A20-79094 https://www.nrel.gov/analysis/electrification-futures.html. Useful analysis of how to operate an increasingly electrified energy system.

References

[1] IEA 2021 *Net Zero by 2050: A Roadmap for the Global Energy Sector* (Paris: International Energy Agency) https://iea.org/reports/net-zero-by-2050

[2] Griffith S 2021 *Electrify: An Optimist's Playbook For Our Clean Energy Future* (Cambridge, MA: MIT Press)

[3] Williams J H, Jones R A and Torn M S 2021 Observations on the transition to a net-zero energy system in the United States *Energy Clim. Change* **2** 100050

[4] Victoria M, Zeyen E and Brown T 2022 Speed of technological transformations required in Europe to achieve different climate goals *Joule* **6** 1066–86

[5] Denholm P, Brown P, Cole W, Brown M, Jadun P, Ho J, Mayernik J, McMillan C and Sreenath R 2022 *Examining Supply-Side Options to Achieve 100% Clean Electricity by 2035* (Golden, CO: National Renewable Energy Laboratory) NREL/TP-6A40-81644 https://nrel.gov/docs/fy22osti/81644.pdf

[6] Lovins A B 2017 Reliably integrating variable renewables: moving grid flexibility resources from models to results *Electr. J.* **30** 58–63

[7] Jenkins J D, Luke M and Thernstrom S 2018 Getting to zero carbon emissions in the electric power sector *Joule* **2** 2498–510

[8] Sepulveda N A, Jenkins J D, de Sisternes F J and Lester R K 2018 The role of firm low-carbon electricity resources in deep decarbonization of power generation *Joule* **2** 2403–20

[9] Blair N *et al* 2022 *Storage Futures Study: Key Learnings for the Coming Decades* (Golden, CO: National Renewable Energy Laboratory) NREL/TP-7A40-81779 https://nrel.gov/analysis/storage-futures.html

[10] Sepulveda N A, Jenkins J D, Edington A, Mallapragada D S and Lester R K 2021 The design space for long-duration energy storage in decarbonized power systems *Nat. Energy* **6** 506–16

[11] Zhou E and Mai T 2021 *Electrification Futures Study: Operational Analysis of US Power Systems with Increased Electrification and Demand-Side Flexibility* (Golden, CO: National

Renewable Energy Laboratory) NREL/TP-6A20-79094 https://nrel.gov/analysis/electrifica-tion-futures.html

[12] Robbie O and Aggarwal S 2018 Refining competitive electricity market rules to unlock flexibility *Electr. J.* **31** 31–7

[13] Olson A, Hull S, Ming Z, Schlag N and Duff C 2021 *Scalable Markets for the Energy Transition: A Blueprint for Wholesale Electricity Market Reform* (San Francisco, CA: Energy and Environmental Economics) https://ethree.com/scalable-markets-for-the-energy-transi-tion-a-new-e3-report/

[14] Kahn E 1988 *Electric Utility Planning and Regulation* (Washington, DC: American Council for an Energy-Efficient Economy)

[15] Lo H, Blumsack S, Hines P and Meyn S 2019 Electricity rates for the zero marginal cost grid *Electr. J.* **32** 39–43

[16] MIT 2016 *Utility of the Future: An MIT Energy Initiative Response to an Industry in Transition* (Cambridge, MA: Massachusetts Institute of Technology) http://energy.mit.edu/research/utility-future-study/

[17] Frew B A, Milligan M, Brinkman G, Bloom A, Clark K and Denholm P 2016 *Revenue Sufficiency and Reliability in a Zero Marginal Cost Future* (Golden, CO: National Renewable Energy Laboratory) NREL/CP-6A20–66935 https://nrel.gov/docs/fy17osti/66935.pdf

[18] Bielen D, Burtraw D, Palmer K and Steinberg D 2017 The future of power markets in a low marginal cost world *Working paper* RFF WP 17-26 Resources for the Future, Washington, DC https://rff.org/publications/working-papers/the-future-of-power-markets-in-a-low-mar-ginal-cost-world/

[19] Gramlich R and Lacey F 2020 *Who's the Buyer?: Retail Electric Market Structure Reforms in Support of Resource Adequacy and Clean Energy Deployment* (Washington, DC: Wind Solar Alliance) https://windsolaralliance.org/wp-content/uploads/2020/03/WSA-Retail-Structure-Contracting-FINAL.pdf

[20] Aggarwal S, Corneli S, Gimon E, Gramlich R, Hogan M, Orvis R and Pierpont B 2019 *Wholesale Electricity Market Design for Rapid Decarbonization* (San Francisco, CA: Energy Innovation) https://energyinnovation.org/wp-content/uploads/2019/07/Wholesale-Electricity-Market-Design-For-Rapid-Decarbonization.pdf

[21] Corneli S 2020 *A Prism-Based Configuration Market for Rapid, Low Cost and Reliable Electric Sector Decarbonization* (Washington, DC: World Resources Institute workshop on Market Design for the Clean Energy Transition: Advancing Long-Term Approaches) https://wri.org/events/2020/12/market-design-clean-energy-transition-advancing-long-term

[22] Gimon E 2020 *Let's Get Organized! Long-Term Market Design for a High Penetration Grid* (Washington, DC: World Resources Institute workshop on Market Design for the Clean Energy Transition: Advancing Long-Term Approaches) https://wri.org/events/2020/12/mar-ket-design-clean-energy-transition-advancing-long-term

[23] Tierney S F 2020 *Wholesale Power Market Design in a Future Low-Carbon Electric System: A Proposal for Consideration* (Boston, MA: Analysis Group) https://wri.org/events/2020/12/market-design-clean-energy-transition-advancing-long-term

[24] Taylor J A, Sairaj V D and Duncan S C 2016 Power systems without fuel *Renew. Sustain. Energy Rev.* **57** 1322–36

[25] Pierpont B 2020 *A Market Mechanism for Long-Term Energy Contracts to Support Electricity System Decarbonization* (Washington, DC: World Resources Institute workshop

on Market Design for the Clean Energy Transition: Advancing Long-Term Approaches) https://wri.org/events/2020/12/market-design-clean-energy-transition-advancing-long-term

[26] WRI and RFF 2020 *Proc. Conf. on Market Design for the Clean Energy Transition: Advancing Long-Term Approaches (16–17 December)* (Washington, DC: World Resources Institute and Resources for the Future) https://wri.org/events/2020/12/market-design-clean-energy-transition-advancing-long-term

[27] Brown T and Reichenberg L 2021 Decreasing market value of variable renewables can be avoided by policy action *Energy Econ.* **100** 105354

[28] Lovins A B 2017 Do coal and nuclear generation deserve above-market prices? *Electr. J.* **30** 22–30

[29] Mallapragada D S, Sepulveda N A S and Jenkins J D 2020 Long-run system value of battery energy storage in future grids with increasing wind and solar generation *Appl. Energy* **275** 115390

[30] Das S, Hittinger E and Williams E 2020 Learning is not enough: diminishing marginal revenues and increasing abatement costs of wind and solar *Renew. Energy* **156** 634–44

[31] Johansson V, Thorson L, Goop J, Göransson L, Odenberger M, Reichenberg L, Taljegard M and Johnsson F 2017 Value of wind power—implications from specific power *Energy* **126** 352–60

[32] Griffith S 2020 *Solving Climate Change with a Loan* (San Francisco, CA: Otherlab) https://medium.com/otherlab-news/solving-climate-change-with-a-loan-d1aac4b8259a

[33] Koomey J, Calwell C, Laitner S, Thornton J, Brown R E, Eto J, Webber C and Cullicott C 2002 Sorry, wrong number: the use and misuse of numerical facts in analysis and media reporting of energy issues *Annual Review of Energy and the Environment 2002* ed R H Socolow, D Anderson and J Harte (Palo Alto, CA: Annual Reviews) pp 119–58

[34] LaCommare K, Hamachi J H, Eto L N, Dunn and Sohn M D 2018 Improving the estimated cost of sustained power interruptions to electricity customers *Energy* **153** 1038–47

[35] Sullivan M J, Myles T C, Schellenberg J A and Larsen P H 2018 *Estimating Power System Interruption Costs: A Guidebook for Electric Utilities* (Berkeley, CA: Lawrence Berkeley National Laboratory) LBNL-2001164 https://emp.lbl.gov/publications/estimating-power-system-interruption

[36] Sanstad A H, Zhu Q, Leibowicz B, Larsen P H and Eto J H 2020 *Case Studies of the Economic Impacts of Power Interruptions and Damage to Electricity System Infrastructure from Extreme Events* (Berkeley, CA: Lawrence Berkeley National Laboratory) https://eta-publications.lbl.gov/publications/case-studies-economic-impacts-power

[37] Baik S, Hanus N L, Sanstad A H, Eto J H and Larsen P H 2021 *A Hybrid Approach to Estimating the Economic Value of Enhanced Power System Resilience* (Berkeley, CA: Lawrence Berkeley National Laboratory) https://eta-publications.lbl.gov/sites/default/files/hybrid_paper_final_22feb2021.pdf

[38] Denholm P *et al* 2021 The challenges of achieving a 100% renewable electricity system in the United States *Joule* **5** 1331–52

[39] Zozmann E, Göke L, Kendziorski M, Angel C R D, Hirschhausen C V and Winkler J 2021 100% renewable energy scenarios for North America—spatial distribution and network constraints *Energies* **14** 658

[40] Clack C T M *et al* 2017 Evaluation of a proposal for reliable low-cost grid power with 100% wind, water, and solar *Proc. Natl Acad. Sci.* **114** 6722

[41] Jacobson M Z, Delucchi M A, Cameron M A and Frew B A 2015 Low-cost solution to the grid reliability problem with 100% penetration of intermittent wind, water, and solar for all purposes *Proc. Natl Acad. Sci.* **112** 15060–5

[42] Jacobson M Z *et al* 2017 100% clean and renewable wind, water, and sunlight all-sector energy roadmaps for 139 countries of the world *Joule* **1** 108–21

[43] Jacobson M Z, Delucchi M A, Cameron M A and Mathiesen B V 2018 Matching demand with supply at low cost in 139 countries among 20 world regions with 100% intermittent wind, water, and sunlight (WWS) for all purposes *Renew. Energy* **123** 236–48

[44] Deason W 2018 Comparison of 100% renewable energy system scenarios with a focus on flexibility and cost *Renew. Sustain. Energy Rev.* **82** 3168–78

[45] Bistline J E 2017 Economic and technical challenges of flexible operations under large-scale variable renewable deployment *Energy Econ.* **64** 363–72

[46] Bistline J E T and Blanford G J 2020 Value of technology in the US electric power sector: impacts of full portfolios and technological change on the costs of meeting decarbonization goals *Energy Econ.* **86** 104694

[47] Breyer C *et al* 2022 On the history and future of 100% renewable energy systems research *IEEE Access.* **10** 78176–218

[48] Hand M M, Baldwin S, DeMeo E, Reilly J M, Mai T, Arent D, Porro G, Meshek M and Sandor D 2012 *Renewable Electricity Futures Study* (Golden, CO: National Renewable Energy Laboratory) NREL/TP-6A20-52409 https://nrel.gov/analysis/re-futures.html

[49] US EPA 2022 *Inventory of US Greenhouse Gas Emissions And Sinks: 1990—2020* (Washington, DC: US Environmental Protection Agency) https://epa.gov/ghgemissions/inventory-us-greenhouse-gas-emissions-and-sinks

[50] IPCC 2021 *Climate Change 2021: The Physical Science Basis. Contribution of Working Group I to the Sixth Assessment Report of the Intergovernmental Panel on Climate Change* ed V Masson-Delmotte (Cambridge: Cambridge University Press) https://ipcc.ch/report/sixth-assessment-report-working-group-i/

[51] IPCC 2022 *Climate Change 2022: Mitigation of Climate Change. Contribution of Working Group III to the Sixth Assessment Report of the Intergovernmental Panel on Climate Change.* (Cambridge: Cambridge University Press) section 6.6.2.2 https://www.ipcc.ch/report/sixth-assessment-report-working-group-3/

IOP Publishing

Solving Climate Change
A guide for learners and leaders
Jonathan Koomey and Ian Monroe

Chapter 6

Minimize non-fossil warming agents

You never change things by fighting existing reality. To change something, build a new model that makes the existing model obsolete.

—R Buckminster Fuller

Chapter overview

- The climate solutions literature has in general paid far less attention to non-fossil emissions options than to energy-related carbon dioxide emissions.
- Methane emissions are the largest source of warming after fossil carbon dioxide emissions, with more than half associated with the agricultural and food sectors, about a third related to waste management (landfills and wastewater treatment), and the rest to land-use change.
- Non-fossil carbon dioxide emissions from industrial production are significant and require innovation in processes, which in many cases can reduce emissions to zero or close to it.
- Non-fossil carbon dioxide and methane emissions from land-use change are significant but vary greatly depending on location. Eliminating those emissions requires halting industrial-scale deforestation immediately, changing land-use practices, and accelerating reforestation projects.
- Non-fossil nitrous oxide emissions are mostly related to soil management and agricultural practices, with small amounts associated with industrial processes. Changing agricultural practices and capturing emissions from wastewater treatment can substantially reduce those emissions, while industrial production-related emissions will require innovation to fix.
- Non-fossil emissions of F-gases are all related to industrial production and will require innovation to reduce them, just like when CFCs were phased under the Montreal Protocol.

doi:10.1088/978-0-7503-4032-8ch6

6.1 Introduction

The climate problem can't be solved by focusing solely on carbon dioxide emissions from fossil fuels. There are many emissions that contribute to warming, and while CO_2 from fossil fuels is the largest single factor, all sources of such warming agents will need to be reduced or eliminated on the path to net-zero emissions [8]. This chapter explores ways to reduce non-fossil emissions.

6.2 Sources of non-fossil emissions

Table 6.1 characterizes what we mean when we say 'non-fossil emissions' that affect Earth's temperature [1–3, 13, 14]. The dominant ones from a greenhouse gas perspective are methane (CH_4) and nitrous oxide (N_2O) from agriculture, carbon dioxide from cement production for construction, and carbon dioxide and F-gases from manufacturing processes. Other environmental impacts can also be important (and won't necessarily decline as the energy transition proceeds) but in general, a world without fossil fuels will also have greatly reduced environmental impacts compared to the current system.

X implies it's an issue for that pollutant and activity.

Table 6.1. Anthropogenic non-fossil greenhouse gas pollution by type and major activity group.

	Manufacturing/materials	Mobility/freight	Food/land use
GHGs/warming agents			
CO_2	Cement, iron, steel, aluminum, ammonia, lime, urea, glass, soda ash, ferroalloy, titanium dioxide, ash, ferroalloy, titanium dioxide, magnesium, and use of carbonates		X
CH_4	Various (but small)		X
N_2O	Nitric acid, adipic acid, other		X
CO	X		X
Volatile organic compounds	X		X
F-gases	Aluminum, magnesium		
Contrails		X	
Hydrogen	X	X	
Cooling agents			
Particulates		Brakes + tires	X
NO_x			X
SO_2			X

In this chapter we focus on reducing the pollutants in table 6.1, which have either warming or cooling effects but that are not directly the result of fossil fuel consumption and use. For example, particulate emissions from brakes and tires come from cars that mainly use fossil fuels now, but the associated emissions still exist for electrified vehicles, so those are not fossil fuel related. Similarly, methane and N_2O emissions from food production and land use similarly would exist even if agricultural machinery were fully electrified, so those are not fossil fuel related either.

The main categories of interest for this chapter are non-fossil methane, fluorinated gases, non-fossil nitrous oxides, carbon dioxide from industrial processes, and carbon dioxide from land-use changes. In each case, understanding emissions sources at a fine level of detail makes creating an emissions reduction plan easier. Each country or state varies in the sources of these warming agents, but we show examples for the US below to illustrate the range of emissions drivers in a large, diverse, developed economy.

6.2.1 Base-year emissions by human activity

Figure 6.1 shows fossil versus non-fossil emissions by warming agent for the US in 2020 using 20 year GWPs, slicing the data from figure 4.4 in a slightly different way.

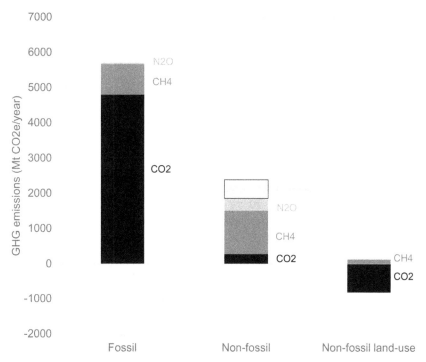

Figure 6.1. US fossil and non-fossil greenhouse gas emissions in 2020 by warming agent (20 year GWPs). Source: US EPA [6]. GWPs from table 7.SM.7 in IPCC [1].

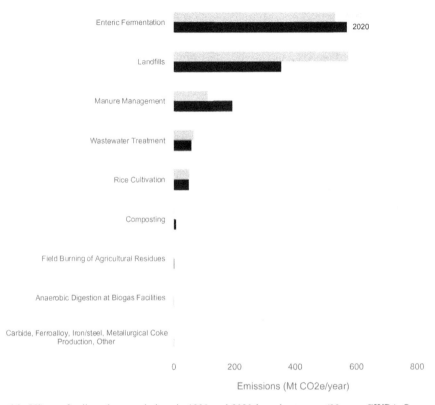

Figure 6.2. US non-fossil methane emissions in 1990 and 2020 by subcategory (20 year GWPs). Source: US EPA [6]. GWPs from table 7.SM.7 in IPCC [1].

Fossil emissions and non-fossil land-use emissions are dominated by carbon dioxide and methane. Non-fossil is split by warming agent, with methane being the most important contributor.

Figure 6.2 shows details on non-fossil methane sources in the US, calculated as carbon dioxide equivalent using 20 year GWPs, ranked from highest to lowest emitting in 2020. Enteric fermentation is methane emitted from the digestive processes of ruminant animals. Landfills, manure, wastewater treatment, rice cultivation, and biogas all generate emissions from anaerobic decay of organic matter.

Figure 6.3 shows emissions of fluorinated gases, which are almost all associated with electronics and electrical equipment manufacturing, as well as materials production (mainly aluminum and magnesium). The most important exception is HFCs needed to replace ozone depleting substances. Use of electrical equipment and electronics manufacturing will no doubt continue to grow in a zero-emissions world, but there are often ways to substitute for greenhouse gas-intensive processes if emissions of some warming agents grow in importance.

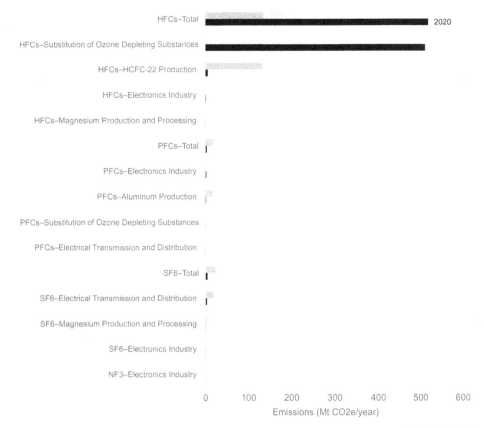

Figure 6.3. US F-gas emissions in 1990 and 2020 by subcategory (20 year GWPs). Source: US EPA [6]. GWPs from table 7.SM.7 in IPCC [1].

Figure 6.4 shows non-fossil nitrous oxide emissions, which are dominated by agricultural soil management. Most of the rest is associated with wastewater treatment, manure management, and manufacturing.

Figure 6.5 shows non-fossil carbon dioxide emissions for the US, excluding land use (which is treated separately below). The biggest category is 'non-energy use of fuels', which include reducing agents and solvents (any use of fuels resulting in sequestration of carbon in long-lived products, like some plastics derived from petrochemicals, is not included in this category). Most of the rest is associated with production of materials such as cement, iron and steel, ammonia, lime, urea, glass, aluminum, soda ash, ferroalloy, titanium dioxide, zinc, phosphoric acid, lead, carbide, magnesium, and process use of carbonates.

Figure 6.6 shows land-use change emissions, including CO_2, CH_4, and N_2O. Methane and nitrous oxide emissions are relatively small. The biggest effect by far is from the net negative emissions of CO_2, which result from regrowth of US forests after several centuries of large-scale land clearing beginning in the seventeenth

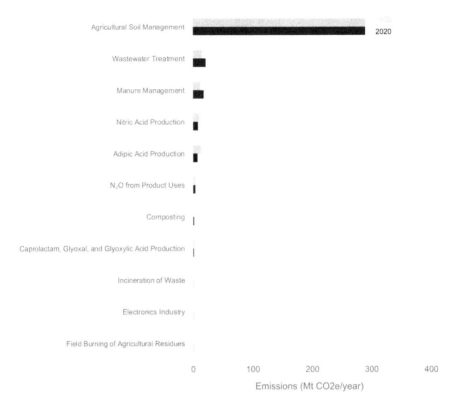

Figure 6.4. US non-fossil nitrous oxide emissions in 1990 and 2020 by subcategory (20 year GWPs). Source: US EPA [6]. GWPs from table 7.SM.7 in IPCC [1].

century. This regrowth is recapturing carbon emitted when many US forests were cleared, and the effects are large enough to significantly affect the total emissions budget. For many other countries, particularly in the developing world, net CO_2 emissions from land use are positive because of rapid deforestation.

6.3 Summary of non-fossil emissions by major category

Figure 6.7 shows non-fossil emissions by gas and major category. More than half of methane emissions come from agriculture and food production, with modest amounts associated with land use and the rest coming from waste management, mainly anaerobic decay in landfills and sewage treatment plants. Almost all nitrous oxide emissions are from agriculture and food production, and almost all non-fossil emissions from F-gases and carbon dioxide (excluding land-use change) are related to industrial production. We address the key drivers of warming in the following sections, including ecosystem disruptions, food and agriculture, industrial processes, aviation contrails, and (for future scenarios) the potential warming effects of green hydrogen.

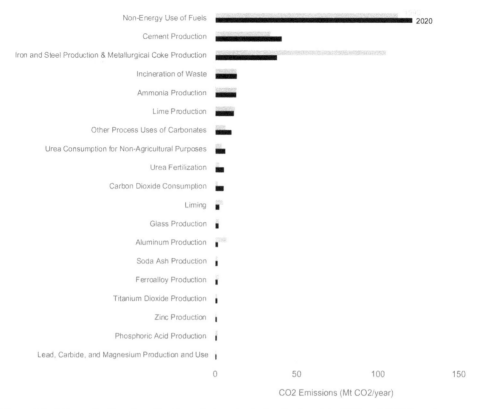

Figure 6.5. US non-fossil carbon dioxide emissions (excluding land use) in 1990 and 2020 by subcategory. Source: US EPA [6]. GWPs from table 7.SM.7 in IPCC [1]. GWPs for carbon dioxide equal 1.0 by definition for both 20 and 100 year periods.

6.3.1 Ecosystems disruptions

For all life on Earth, CO_2 is food. Carbon dioxide is first extracted from the atmosphere by plants and algae through photosynthesis, which in turn feed nearly every ecosystem along with all of humanity. Wherever there is life, ecosystems are storing and cycling carbon, and our planet's lands and seas still store much more carbon than our atmosphere.

Nearly every ecosystem is now being disrupted by human activity, generally through some combination of direct land conversion and less direct anthropogenic forces, such as temperature and precipitation shifts from climate change. Human-linked ecosystem disruptions often result in large near-term GHG emissions, as well as a reduction in the rates biological systems can pull CO_2 out of the atmosphere and store carbon in biomass and soils.

6.3.1.1 Deforestation and land-use change

Many of the world's most carbon-dense ecosystems are forests, and almost four billion tonnes of carbon dioxide each year still comes from deforestation, as trees

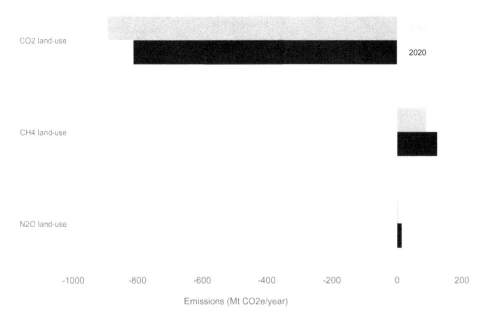

Figure 6.6. US greenhouse emissions from land-use change in 1990 and 2020 by category (20 year GWPs). Source: US EPA [6]. GWPs from table 7.SM.7 in IPCC [1]. GWPs for carbon dioxide equal 1.0 by definition for both 20 and 100 year periods.

are cut for timber or burned to make way for livestock grazing livestock or agricultural crops [2]. In the Amazon, Congo, and other forests, the path to deforestation often starts with road construction, followed by commercial logging, livestock grazing, and then row crop cultivation and urban development where economically feasible. Many government policies have historically encouraged deforestation as a means to claim land rights and achieve economic development, and some governments continue to encourage deforestation by overlooking violations of conservation boundaries or indigenous land rights, with some government officials even profiting from deforestation through corrupt relationships or official investments.

Peat swamp forests are especially large carbon sinks, as the waterlogged soil in these tropical ecosystems prevents dead leaves and wood from fully decomposing, building up layers of carbon-rich organic matter that can be 20 meters deep (about the height of a four story building) [15]. Unfortunately, many of the world's remaining peat swamp forests are still being logged, burned and drained to convert land to agriculture, resulting in some of world's largest sources of climate pollution beyond fossil fuels. Conversion to oil palm plantations has been the largest danger for peat forests, driven by rising demand for palm oil and palm kernel oil (often innocuously labeled 'vegetable oil') in a wide range of consumer products, from food to beauty products. Grasslands, wetlands, permafrost, mangrove swamps, and other

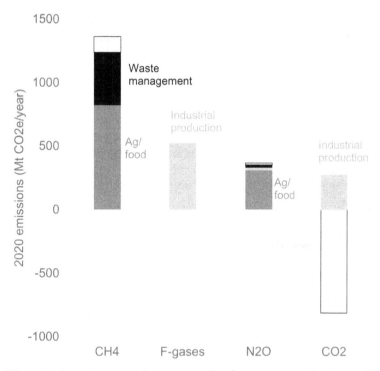

Figure 6.7. US non-fossil greenhouse emissions by gas and major category in 2020 (20 year GWPs). Source: US EPA [6]. GWPs from table 7.SM.7 in IPCC [1]. GWPs for carbon dioxide equal 1.0 by definition for both 20 and 100 year periods.

coastal ecosystems can also store large amounts of carbon, yet these systems are also still losing ground to agriculture, infrastructure and urban development in many parts of the world.

6.3.1.2 Marine ecosystems

Oceans cover over 70% of Earth's surface, and marine ecosystems store the vast majority of the world's carbon (over 90%), much more than the combined carbon in our atmosphere, terrestrial ecosystems, and fossil fuel reserves. Earth's oceans absorb carbon in three ways: direct CO_2 absorption from the atmosphere, storing carbon-containing sediments from rivers, and biological carbon storage in ocean ecosystems.

About a quarter of anthropogenic CO_2 emissions are absorbed by our oceans every year, as physics and chemistry push to balance the concentration of CO_2 in ocean water with the increasing concentration in the atmosphere, resulting in net CO_2 uptake as long as the gas concentration in air is higher than in water [2]. However, as CO_2 is absorbed by seawater it converts into carbonic acid (H_2CO_3), resulting in lower pH in the water and overall ocean acidification. This process is the reason why higher CO_2 pollution levels also make the oceans more acidic. The acceleration of ocean acidification from manmade CO_2 pollution is making it harder for many aquatic species to survive, especially corals, shellfish, other marine life with calcium carbonate shells that dissolve in acidic waters. Warming waters from climate change can compound the challenges from acidification, as well as add additional survival challenges for many other aquatic species. These marine ecosystem shifts can result in substantial biodiversity loss, alongside degradation of the ability to store biological carbon in many marine ecosystems.

Humans are also directly affecting the carbon cycle in ocean environments through fishing, with dredging and bottom trawling methods damaging seafloor ecosystems and stirring up carbon-rich sediments. Over 100 million metric tons of fish are taken out of oceans every year[1], removing organic carbon that would otherwise feed marine ecosystems and eventually be partially stored in ocean sediments. Nutrient runoff from agricultural overfertilization also often flows from rivers into seas, causing algae blooms and ocean 'dead zones' where decaying organic matter depletes oxygen levels and drives away or suffocates other marine life.

6.3.1.3 Climate feedbacks

Increasing impacts from climate change will likely accelerate GHG emissions from disruption of many ecosystems, which in turn will accelerate climate change. For example, climate-linked heat, pests, drought, and wildfires in many areas are causing near-term GHG emissions while also decreasing the long-term carbon storage capacity for many forests, which in turn makes our climate challenges worse. This type of vicious cycle is known as a 'positive feedback', not because it is positive for people or the planet, but because it accelerates the pace of a process such as climate change. There are 'negative feedbacks' in our climate system as well, such as the fact that higher atmospheric CO_2 concentrations can make some plants grow faster and absorb more carbon (an effect known as 'CO_2 fertilization' or 'carbon fertilization'), but research has typically found many more feedbacks that accelerate warming than feedbacks that slow it down [1, 16–19].

One particularly concerning feedback is the melting of arctic permafrost, which may result in billions of tons of CO_2 and methane released to the atmosphere from the decay of organic matter that has been frozen for thousands of years [17, 20]. We are still in the beginning stages of understanding other feedbacks, such as the long-term climate implications of ocean acidification. Given the pace of current climate change, most ecosystems are now experiencing shifts in decades that would have previously happened over millennia. Many species aren't equipped to handle these changes, and opportunistic species that thrive in changed conditions can become invasive and further disrupt ecosystem balance

6.3.2 Food and agriculture

What we eat and how we produce our food play an outsized role in our climate challenge [21]. The climate cost of food starts with the land we use to make it, which

---

[1] https://ourworldindata.org/fish-and-overfishing and https://www.ramlegacy.org

generally has been converted to agriculture or livestock grazing from a higher carbon ecosystem (such as forests, grasslands, or wetlands), whether this land-use change happened a few years or a few centuries in the past. Then there is GHG pollution linked to most parts of the food production process, from embodied GHG emissions in fertilizers, pesticides, and herbicides, to fossil fuel use in farm machinery, processing equipment, refrigeration, transportation, and cooking. While many of the GHGs linked to food production can be almost fully eliminated through an economy-wide shift to electrification and clean energy, there are also substantial sources of non-combustion GHGs in our global food system that require additional action.

6.3.2.1 Ruminant livestock

Livestock are a large source of anthropogenic methane emissions, second only to methane leakage from fossil fuel infrastructure (see figure 4.6). Cows are the biggest polluters, followed by sheep, goats, and other ruminants, which naturally burp out CH_4 as methanogenic bacteria in their rumens help these animals digest grasses and other fibrous feed. For these animals, methane isn't a problem, it's a natural by-product of their digestive system, but the sheer numbers of ruminant livestock make it a big problem for climate [22].

The other climate problem with such livestock is the land that's required to feed animals before they can feed us [21]. Livestock graze on over a quarter of Earth's ice-free land surface, and an additional third of agricultural land grows livestock feed[2]. Ruminant livestock are not very efficient at converting vegetable protein into meat and dairy products, so while grazing and animal feed collectively takes up nearly 80% of agricultural land, these animals supply under 20% of global calories and less than 40% of global protein[3]. Beyond the initial GHG emissions from converting natural ecosystems to livestock grazing and feed crops, the inefficient use of land for livestock is a large opportunity cost for climate. Reducing demand for beef and dairy products and improving the efficiency of livestock husbandry [21] could free up millions of hectares to shift back to carbon-storing natural ecosystems, or this land could be used to produce feedstocks for biofuels, BECCs, or bioplastics. Another strategy is to shift feedstuffs for livestock away from foods consumed by humans [23]. Still another is to shift more grazing and agriculture to organic practices [24], although that shift by itself is not a cure all.

Ruminant livestock may not always be detrimental to climate. Wild ruminants have grazed Earth's savannas and grasslands for millions of years, often grazing intensely in large herds and constantly moving to protect from predators. Researchers have started to consider the possibility that intensive rotational grazing that mimics natural systems could actually enhance root growth and soil carbon storage, though other research has cast doubt on whether soil carbon benefits can outweigh increased methane emissions from grazers [25]. Grazing livestock can also play a role in fire prevention, by eating understory plants to reduce the risk that fires

---

[2] https://www.fao.org/3/ar591e/ar591e.pdf and https://ourworldindata.org/global-land-for-agriculture
[3] https://ourworldindata.org/agricultural-land-by-global-diets

spread into tree canopies, thereby potentially reducing the size, intensity, and associated GHG emissions from wildfires. If we can find ways to substantially reduce methane emissions from ruminants then they may become a part of carefully managed climate-positive land-use solutions, though it's unlikely this can happen at scale with the billions of ruminants that currently feed humanity [26–28].

6.3.2.2 Crop production

The invention of chemical fertilizer helped spark the Green Revolution in agriculture, which has increased global food production enough to keep pace with population growth since the 1950s. But our increasing use of fertilizer to boost crop yields has also led to large increases in several sources of climate pollution. First, the production of the most common nitrogen fertilizer, ammonia (NH_3), requires hydrogen, which is typically stripped from natural gas or coal, resulting in both CO_2 emissions from hydrogen production and additional GHG emissions from the fossil fuels burned as heat for the process. More climate pollution happens after nitrogen fertilizer is applied to crops, as some portion always converts into the high-potency GHG nitrous oxide, which is 273 times more potent a warming agent than CO_2 (true for both GWP_{20} and GWP_{100} according to the latest IPCC Working Group I report [1]). N_2O emissions increase as more nitrogen fertilizer is applied, and this is true for both chemical fertilizer and organic fertilizers such as manure and compost, but other environmental factors also influence N_2O emissions, including soil moisture, temperature, pH, and microbial species [29].

6.3.2.3 Food waste

Poore and Nemecek [21] identify reducing food waste as one part of a comprehensive strategy to reduce greenhouse gas emissions. That effort can involve shifting the supply chain to more processed/frozen foods as well as identifying more efficient producers to allow consumers to choose them (analogously to 'Fair Trade' and organic labeling standards). Potential savings from improving the supply chain are likely lower than shifting consumer preferences to more plant-based products, but every emissions reduction helps.

6.3.3 Industrial processes

Greenhouse gas emissions from industrial processes are almost all carbon dioxide and F-gases. Methane and N_2O have only tiny emissions associated with industrial production

6.3.3.1 Carbon dioxide

Most non-fossil emissions of carbon dioxide (excluding land use) are from industrial processes that produce materials and manufactured goods. Two of the biggest segments are cement, and iron and steel, but their relative importance varies a lot by locale.

6.3.3.2 *F-gases*

All F-gas emissions are from industrial processes, and almost all of these are associated with the displacement of ozone-depleting substances. The rest result from manufacturing of electrical and electronic equipment.

6.3.4 Aviation contrails

Carbon dioxide emissions from long-haul aviation are significant and growing, but the indirect warming effects from the creation of contrails in the upper atmosphere are larger [30, 31]. In fact, Burkhardt [32] found that 'net radiative forcing due to contrail cirrus remains the largest single radiative-forcing component associated with aviation'. This important effect is often not included in emissions inventories but is worth tracking with aviation emissions projected to increase in coming decades [33].

The indirect effects of combustion in the upper atmosphere 'arise from emissions of greenhouse gases, aerosols and nitrogen oxides, and from changes in cloudiness in the upper troposphere' [32]. The direct emissions of greenhouse gases fall into our 'fossil' category, as do aerosols (e.g. soot), but most nitrogen oxides come from heated surfaces in contact with a nitrogen-rich atmosphere.

Nitrogen oxide emissions would result even from combustion of green hydrogen derived from zero-emissions sources. Water vapor, which also is emitted from residual water in jet fuel, is the main by-product of hydrogen combustion, and injecting water into the upper atmosphere may also have warming effects. Jacobson [34, section 2.2.3] analyzes these effects and finds a 70% reduction in contrail thickness when hydrogen is combusted instead of jet fuel.

6.3.5 Hydrogen

Some future scenarios suggest that hydrogen will play an increasing role in the world's energy system [35, 36], but there has also been justifiable skepticism about the feasibility of more widespread use of hydrogen [37, 38]. Hydrogen will play an important role in industrial production in a zero-emissions world [10], will likely be useful for energy storage for the power systems, and may be useful for long-haul aviation and shipping, but the dream of a hydrogen economy that reaches small consumers and light vehicles will almost certainly not be realized. The advantages of battery storage and electricity delivery are too great (and are growing), while hydrogen applications for small users have gained little traction in the past couple of decades, even as batteries have plummeted in cost.

The potential effects of an increase in the use of hydrogen depend on uncertainties about emissions rates, atmospheric chemistry, and warming effects (which depend on complex interactive factors) [39–41]. Research in the past two decades has found that hydrogen emissions do have a warming effect when all interactions are tallied. Early work found 100 year GWPs of around 4 [11, 12] but the most recent research shows 100 year GWPs around 11, while the 20 year GWP of hydrogen is about three times the 100 year value [7].

Better understanding of leakage rates and the complex interactions that lead to warming from hydrogen emissions will become increasingly important as hydrogen production increases. Hydrogen is a much smaller molecule than methane, so preventing leakage is harder, but just as for methane, eliminating leaks is in the interests of the producers (fuel that leaks can't be sold).

If hydrogen replaces jet fuel for long-haul aviation, emissions of nitrogen oxides and water vapor in the upper atmosphere may still create significant warming, even if the hydrogen is created using zero-emissions energy sources [34, section 2.2.3]. This area is not well studied, and the most recent work in the UK explicitly excluded the warming effects of use of hydrogen for aviation in their analysis [7].

6.4 Creating or adopting a business-as-usual scenario

Just as for fossil emissions, the first step is to create a business-as-usual (BAU) scenario, which requires understanding emissions associated with human activities. We can begin by splitting fossil and non-fossil emissions for each key greenhouse gas in a base year, and then drilling down further into each category to understand the key drivers.

Data availability will affect the level of detail possible for your BAU projection. In most cases, reaching the disaggregation represented in the EPA data for the US above is difficult, so you'll need to determine what your data allow.

6.4.1 Projecting base-year emissions to mid-century

As discussed in the previous two chapters, the BAU scenario involves projecting base-year emissions to the mid twenty-first century, using drivers by task like the ones shown in table 4.1 above. Those drivers can change over time as technology changes, and the best projections incorporate dynamic developments in a careful way.

6.5 Creating the climate-positive scenario

Just as for electrification, the climate-positive scenario uses the same drivers by task used in the BAU case. Examples for non-fossil emissions would include projections of food consumption, industrial production, and materials use. Of course, the climate-positive scenario can alter these trends to make emissions reductions easier, by (for example) changing eating preferences or altering the types of materials projected to be used by society in the future.

Most integrated assessment models account for feedbacks between scenario drivers and industrial production imperfectly or not at all, so that is an important area where those models could be improved [42]. Different zero-emissions technologies have different materials intensities and associated non-fossil emissions (such as for cement, steel, and aluminum) and tracking these as a function of scenario evolution would give policy makers insight into how 'embedded emissions' affect the scenario results. In most cases, 'use phase' emissions will dominate, but that's not always true, and understanding when it isn't true can yield policy insight.

Changing patterns of material use by consumers and businesses affect industrial process emissions associated with the production of those materials [43]. The potential for such emissions reductions is often ignored but is clearly worthy of more attention [44, 45]. The possibility of substituting information services for materials use ('dematerialization') is another option with great possibilities [4, 46].

On land use, one subtlety is that it's not enough to reforest after land is cleared, we need to stop all industrial-scale deforestation first, in particular that affecting old-growth forests and jungles. There must be no more destruction of these precious habitats from now on, but it has been singularly difficult to prevent countries from destroying them.

Just as countries continue to build new fossil infrastructure while claiming to support climate stabilization, countries continue to destroy old-growth forests while supporting reforestation projects. Stop industrial-scale deforestation immediately, then reforestation makes sense, otherwise it's a dangerous distraction from what climate stabilization requires, and it enables extinction of myriad species that once gone, cannot be replaced. Like many aspects of the climate problem, the moral implications of continued industrial-scale deforestation are deeply troubling to us, and that should be reason enough to stop it immediately.

6.6 Data sources

As shown above, climate change is not just an energy problem, and many new options are being deployed or are now under development to reduce non-fossil emissions. To the extent that these emissions can't be eliminated with technological changes, carbon capture will be needed to get to net zero [47], but for industrial processes to produce cement[4], steel[5], aluminum[6], lime, glass, and feedstocks there are options for substantially reducing or eliminating those emissions [9, 10, 35, 44, 48–54]. Same for food [21], general ecosystem management [55], forests [56], agricultural practices [57, 58], methane [59], nitrous oxide [60], black carbon [61], and fluorinated gases [62]. The IPCC [5] and [3] reviewed options in a comprehensive way.

For each of the categories for which data are available you'll need to create a BAU scenario and a pathway to zero emissions using data sources applicable to your chosen geography. In many cases you'll need to adapt previous work to suit your analysis, but as long as you follow physical and engineering principles and clearly document your assumptions in adopting that work, you'll be on solid analytical ground.

6.7 Chapter conclusions

Tracking non-fossil emissions separate from those associated with fossil fuel extraction and use is critically important. These emissions are often ignored or

---

[4] http://www.eng.cam.ac.uk/news/cambridge-engineers-invent-world-s-first-zero-emissions-cement
[5] https://www.hybritdevelopment.se/en/
[6] https://elysis.com/en

treated as an afterthought. Rigorous analysis of climate solutions requires that all major sources of greenhouse gas emissions be reduced or eliminated and just focusing on the energy sector (i.e. fossil-related emissions) is both insufficient and detrimental to identifying opportunities for whole-system redesign.

Further reading

IPCC 2022 *Climate Change 2022: Mitigation of Climate Change. Contribution of Working Group III to the Sixth Assessment Report of the Intergovernmental Panel on Climate Change* (Cambridge: Cambridge University Press) https://www.ipcc. ch/report/sixth-assessment-report-working-group-3/. Chapter 7 (land use/agriculture), chapter 11 (industry), and chapter 5 (demand, services and social aspects of mitigation), all have relevant insights and references for assessing non-fossil emissions reduction potentials.

Friedlingstein P *et al* 2022 Global carbon budget 2021 *Earth Syst. Sci. Data* **14** 1917–2005 https://doi.org/10.5194/essd-14-1917-2022. The most authoritative source on stocks and flows of carbon in the biosphere.

Poore J and Nemecek T 2018 Reducing food's environmental impacts through producers and consumers *Science* **360** 987–92 https://doi.org/10.1126/science. aaq0216. A comprehensive and authoritative view of the food system's environmental effects.

Project Drawdown (https://www.drawdown.org/solutions) gives significant detail on both fossil and non-fossil emissions reductions, focusing on land use, agriculture, and industrial processes.

Saunois M *et al* 2020 The global methane budget 2000–2017 *Earth Syst. Sci. Data* **12** 1561–623 https://doi.org/10.5194/essd-12-1561-2020. The most authoritative source on stocks and flows of methane in the biosphere.

Sovacool B K, Griffiths S, Kim J and Bazilian M 2021 Climate change and industrial F-gases: a critical and systematic review of developments, sociotechnical systems and policy options for reducing synthetic greenhouse gas emissions *Renew. Sust. Energy Rev.* **141** 110759 https://doi.org/10.1016/j. rser.2021.110759. The most authoritative source on flows of F-gases.

Tian H *et al* 2020 A comprehensive quantification of global nitrous oxide sources and sinks *Nature* **586** 248–56 https://doi.org/10.1038/s41586-020-2780-0. The most authoritative source on stocks and flows of nitrous oxides in the biosphere.

References

[1] IPCC 2021 *Climate Change 2021: The Physical Science Basis. Contribution of Working Group I to the Sixth Assessment Report of the Intergovernmental Panel on Climate Change* ed V Masson-Delmotte *et al* (Cambridge: Cambridge University Press) https://www.ipcc.ch/report/sixth-assessment-report-working-group-i/

[2] Friedlingstein P *et al* 2022 Global carbon budget 2021 *Earth Syst. Sci. Data* **14** 1917–2005

[3] IPCC 2022 *Climate Change 2022: Mitigation of Climate Change. Contribution of Working Group III to the Sixth Assessment Report of the Intergovernmental Panel on Climate Change*

(Cambridge: Cambridge University Press) https://www.ipcc.ch/report/sixth-assessment-report-working-group-3/

[4] Koomey J G, Matthews H S and Williams E 2013 Smart everything: will intelligent systems reduce resource use? *Annu. Rev. Environ. Resour.* **38** 311–43

[5] IPCC 2014 *Climate Change 2014: Mitigation of Climate Change. Contribution of Working Group III to the Fifth Assessment Report of the Intergovernmental Panel on Climate Change* ed O Edenhofer *et al* (Cambridge: Cambridge University Press) https://www.ipcc.ch/report/ar5/wg3/

[6] US EPA 2022 *Inventory of US Greenhouse Gas Emissions and Sinks: 1990–2020* (Washington, DC: US Environmental Protection Agency) https://www.epa.gov/ghgemissions/inventory-us-greenhouse-gas-emissions-and-sinks

[7] Warwick N, Griffiths P, Keeble J, Archibald A, Pyle J and Shine K 2022 *Atmospheric Implications of Increased Hydrogen Use* (Cambridge: University of Cambridge, NCAS, and University of Reading) https://assets.publishing.service.gov.uk/government/uploads/system/uploads/attachment_data/file/1067144/atmospheric-implications-of-increased-hydrogen-use.pdf

[8] Dreyfus G B, Xu Y, Shindell D T, Zaelke D and Ramanathan V 2022 Mitigating climate disruption in time: a self-consistent approach for avoiding both near-term and long-term global warming *Proc. Natl Acad. Sci. USA* **119** e2123536119

[9] Valentin V, Åhman M and Nilsson L J 2018 Assessment of hydrogen direct reduction for fossil-free steelmaking *J. Cleaner Prod.* **203** 736–45

[10] Griffiths S, Sovacool B K, Kim J, Bazilian M and Uratani J M 2021 Industrial decarbonization via hydrogen: a critical and systematic review of developments, socio-technical systems and policy options *Energy Res. Soc. Sci.* **80** 102208

[11] Derwent R, Simmonds P, O'Doherty S, Manning A, Collins W and Stevenson D 2006 Global environmental impacts of the hydrogen economy *Int. J. Nucl. Hydrogen Prod. Appl.* **1** 57–67

[12] Derwent R 2018 Hydrogen for heating: atmospheric impacts—a literature review *BEIS Research paper* Number 21 Department for Business, Energy and Industrial Strategy https://assets.publishing.service.gov.uk/government/uploads/system/uploads/attachment_data/file/760538/Hydrogen_atmospheric_impact_report.pdf

[13] Saunois M *et al* 2020 The global methane budget 2000–2017 *Earth Syst. Sci. Data.* **12** 1561–623

[14] Tian H *et al* 2020 A comprehensive quantification of global nitrous oxide sources and sinks *Nature* **586** 248–56

[15] Yule C M 2010 Loss of biodiversity and ecosystem functioning in Indo-Malayan peat swamp forests *Biodiversity Conservation* **19** 393–409

[16] Cox P M, Betts R A, Jones C D, Spall S A and Totterdell I J 2000 Acceleration of global warming due to carbon-cycle feedbacks in a coupled climate model *Nature* **408** 184–7

[17] Koven C D, Ringeval B, Friedlingstein P, Ciais P, Cadule P, Khvorostyanov D, Krinner G and Tarnocai C 2011 Permafrost carbon-climate feedbacks accelerate global warming *Proc. Natl Acad. Sci.* **108** 14769–74

[18] Zhang W, Miller P A, Jansson C, Samuelsson P, Mao J and Smith B 2018 Self-amplifying feedbacks accelerate greening and warming of the arctic *Geophys. Res. Lett.* **45** 7102–11

[19] Callaghan T V, Johansson M, Key J, Prowse T, Ananicheva M and Klepikov A 2011 Feedbacks and interactions: from the arctic cryosphere to the climate system *Ambio* **40** 75–86

[20] Walter Anthony K, Schneider von Deimling T, Nitze I, Frolking S, Emond A, Daanen R, Anthony P, Lindgren P, Jones B and Grosse G 2018 21st-century modeled permafrost carbon emissions accelerated by abrupt thaw beneath lakes *Nat. Commun.* **9** 3262

[21] Poore J and Nemecek T 2018 Reducing food's environmental impacts through producers and consumers *Science* **360** 987–92

[22] Hackmann T J and Spain J N 2010 Invited review: ruminant ecology and evolution: perspectives useful to ruminant livestock research and production *J. Dairy Sci.* **93** 1320–34

[23] Schader C *et al* 2015 Impacts of feeding less food-competing feedstuffs to livestock on global food system sustainability *J. R. Soc. Interface* **12** 20150891

[24] Muller A *et al* 2017 Strategies for feeding the world more sustainably with organic agriculture *Nat. Commun.* **8** 1290

[25] Roche L M, Cutts B B, Derner J D, Lubell M N and Tate K W 2015 On-ranch grazing strategies: context for the rotational grazing dilemma *Rangeland Ecol. Manag.* **68** 248–56

[26] Eagle A J and Olander L P 2012 Greenhouse gas mitigation with agricultural land management activities in the United States—a side-by-side comparison of biophysical potential *Adv. Agronomy* **115** 79–179

[27] Teague R, Provenza F, Kreuter U, Steffens T and Barnes M 2013 Multi-paddock grazing on rangelands: why the perceptual dichotomy between research results and rancher experience? *J. Environ. Manag.* **128** 699–717

[28] Norton B E, Barnes M and Teague R 2013 Grazing management can improve livestock distribution: increasing accessible forage and effective grazing capacity *Rangelands* **35** 45–51

[29] Wang C, Amon B, Schulz K and Mehdi B 2021 Factors that influence nitrous oxide emissions from agricultural soils as well as their representation in simulation models: a review *Agronomy* **11** 770

[30] Camero K 2019 Aviation's dirty secret: airplane contrails are a surprisingly potent cause of global warming *Science* 28 June

[31] Lee D S *et al* 2010 Transport impacts on atmosphere and climate: aviation *Atmos. Environ.* **44** 4678–734

[32] Burkhardt U and Kärcher B 2011 Global radiative forcing from contrail cirrus *Nat. Clim. Change* **1** 54–8

[33] Bock L and Burkhardt U 2019 Contrail cirrus radiative forcing for future air traffic *Atmos. Chem. Phys* **19** 8163–74

[34] Jacobson M Z 2020 *100% Clean, Renewable Energy and Storage for Everything* (Cambridge: Cambridge University Press)

[35] Davis S J *et al* 2018 Net-zero emissions energy systems *Science* **360** 6396

[36] Haegel N M *et al* 2017 Terawatt-scale photovoltaics: trajectories and challenges *Science* **356** 141

[37] Romm J J 2004 *The Hype about Hydrogen: Fact and Fiction in the Race to Save the Climate* (Washington, DC: Island)

[38] Plötz P 2022 Hydrogen technology is unlikely to play a major role in sustainable road transport *Nat. Electr.* **5** 8–10

[39] Tromp T K, Shia R-L, Allen M, Eiler J M and Yung Y L 2003 Potential environmental impact of a hydrogen economy on the stratosphere *Science* **300** 1740–2

[40] Kammen D M and Lipman T E 2003 Assessing the future hydrogen economy *Science* **302** 226–9

[41] Schultz M G, Diehl T, Brasseur G P and Zittel W 2003 Air pollution and climate-forcing impacts of a global hydrogen economy *Science* **302** 624–7

[42] Pauliuk S, Arvesen A, Stadler K and Hertwich E G 2017 Industrial ecology in integrated assessment models *Nat. Clim. Change* **7** 13–20

[43] Hertwich E G 2021 Increased carbon footprint of materials production driven by rise in investments *Nat. Geosci.* **14** 151–5

[44] IEA 2019 *Material Efficiency in Clean Energy Transitions* (Paris: International Energy Agency) https://www.iea.org/reports/material-efficiency-in-clean-energy-transitions

[45] D'Amico B, Pomponi F and Hart J 2021 Global potential for material substitution in building construction: the case of cross laminated timber *J. Cleaner Prod.* **279** 123487

[46] Hilty L, Lohmann W and Huang E M 2011 Sustainability and ICT—an overview of the field with a focus on socioeconomic aspects *Notiz. Politeia* **17** 13–28

[47] Paltsev S, Morris J, Kheshgi H and Herzog H 2021 Hard-to-abate sectors: the role of industrial carbon capture and storage (CCS) in emission mitigation *Appl. Energy* **300** 117322

[48] Rissman J *et al* 2020 Technologies and policies to decarbonize global industry: review and assessment of mitigation drivers through 2070 *Appl. Energy* **266** 114848

[49] IEA 2018 *Technology Roadmap: Low-Carbon Transition in the Cement Industry* (Paris: International Energy Agency) https://www.iea.org/reports/technology-roadmap-low-carbon-transition-in-the-cement-industry

[50] Favier A, Wolf C D, Scrivener K and Habert G 2018 *A Sustainable Future for the European Cement and Concrete Industry. Technology Assessment for Full Decarbonisation of the Industry by 2050* (Zurich: ETH Zurich)

[51] DECHEMA 2017 Low carbon energy and feedstock for the European chemical industry *Technology Study* German Society for Chemical Engineering and Biotechnology (DECHEMA) and released by the European Chemical Industry Council (Cefic) https://dechema.de/en/Low_carbon_chemical_industry.html

[52] Fischedick, M, Marzinkowski J, Winzer P and Weigel M 2014 Techno-economic evaluation of innovative steel production technologies *J. Cleaner Prod.* **84** 563–80

[53] van Ruijven B J, van Vuuren D P, Boskaljon W, Neelis M L, Saygin D and Patel M K 2016 Long-term model-based projections of energy use and CO_2 emissions from the global steel and cement industries *Resour., Conserv. Recycl.* **112** 15–36

[54] Allwood J M, Azevedo J, Cleaver C, Cullen J and Horton P 2022 *Materials and Manufacturing: Business Growth in a Transformative Journey to Zero Emissions* (Cambridge: UK Fires, Engineering and Physical Sciences Research Council, University of Cambridge)

[55] Girardin C A J, Jenkins S, Seddon N, Allen M, Lewis S L, Wheeler C E, Griscom B W and Malhi Y 2021 Nature-based solutions can help cool the planet—if we act now *Nature* **593** 191–4

[56] Agrawal A, Nepstad D and Chhatre A 2011 Reducing emissions from deforestation and forest degradation *Annu. Rev. Environ. Res.* **36** 373–96

[57] Gerber P J *et al* 2013 Technical options for the mitigation of direct methane and nitrous oxide emissions from livestock: a review *Animal* **7** 220–34

[58] Clark M A, Domingo N G G, Colgan K, Thakrar S K, Tilman D, Lynch J, Azevedo I L and Hill J D 2020 Global food system emissions could preclude achieving the 1.5 °C and 2 °C climate change targets *Science* **370** 705

[59] Nisbet E G *et al* 2020 Methane mitigation: methods to reduce emissions, on the path to the Paris Agreement *Rev. Geophys.* **58** e2019RG000675

[60] Winiwarter W, Höglund-Isaksson L, Klimont Z, Schöpp W and Amann M 2018 Technical opportunities to reduce global anthropogenic emissions of nitrous oxide *Environ. Res. Lett.* **13** 014011

[61] Schmidt C W 2011 Black carbon: the dark horse of climate change drivers *Environ. Health Perspect.* **119** A172–5

[62] Fenhann J 2000 Industrial non-energy, non-CO_2 greenhouse gas emissions *Technol. Forecast. Soc. Change* **63** 313–34

Solving Climate Change
A guide for learners and leaders
Jonathan Koomey and Ian Monroe

Chapter 7

Efficiency and optimization

There is an even cleaner form of energy than the Sun, more renewable than the wind: it's the energy we don't consume.

—Arthur H Rosenfeld

Chapter overview

- Efficiency and optimization mean optimizing energy, capital, and materials flows while reducing greenhouse gas emissions to zero.
- As always, we begin with understanding the tasks we want to accomplish and go from there.
- The three components of efficiency and optimization, defined as 'emissions efficiency' are
  - Moving to zero-emissions energy sources.
  - Improving end-use efficiency of accomplishing tasks for capital equipment and operations.
  - Altering the tasks we choose to accomplish.
- Information and communication technology is a key enabler of the shift to zero emissions. It's our 'ace in the hole'.
- End-use strategies can substantially reduce emissions but combined with supply-side shifts they enable even more sweeping change.
- Institutional change can be powerful because of increasing returns to scale, but it requires systematic application of whole system design and iterative risk management principles.

7.1 Introduction

The word 'efficiency' conveys the general notion of accomplishing human goals with a minimum of effort. 'Optimization' conveys a similar idea, with the additional nuance of continuous improvements.

Efficiency is defined as the ratio of some output (such as heat delivered to a room, distance driven, or economic activity) to some input (such as fuel input to a furnace, final energy, or materials use). In the climate solutions literature this term most often is used in the context of energy efficiency, focusing on accomplishing human goals with the least consumption of primary or final energy [6, 11, 13].

While energy efficiency and optimization are important components of stabilizing the climate, it is only part of what we mean when we talk about efficiency in this book. We use the concept of 'emissions efficiency' to capture the full range of options, following the economic literature on 'carbon efficiency' [25–27]. We define the problem more generally as optimizing energy, capital, and materials flows while reducing greenhouse gas emissions to zero.

At the highest level of abstraction, there are three ways of moving to zero emissions, all of which fall under the category of 'emissions efficiency':

1. Move to zero-emissions energy sources:
   - Demand-side: electrifying everything as described in chapter 4
   - Supply side: shifting to renewable electricity generation or nuclear power as described in chapter 5
2. Improve end-use efficiency of accomplishing tasks, including both technical efficiency of equipment and efficiency of operations.
3. Alter the tasks we choose to accomplish, such as changing the structure of cities to enable lower emissions living or choosing to eat fewer animal products.

After a shift to zero-emissions energy sources, efficiency and task definition still matter, because many of these zero-emissions sources are constrained in the near term. Lower energy use means lower emissions for a given level of zero-emission generation sources.

Reductions in end-use energy intensity can also make it possible to achieve emissions reductions with less widespread deployment of carbon capture and other supply side technologies. Hummel [28] called this a reduction in 'mitigation pressure' that allows for deeper emissions reductions than would be possible with accelerated supply-side options alone. Grübler et al [2] also allude to intensity reductions as an enabler of more rapid and more profound supply-side changes.

The concept of efficiency has a rich history in the energy literature [11, 13]. The most common usage is to define efficiency by applying the First Law of Thermodynamics:

$$\text{First Law efficiency} = \frac{\text{Energy out}}{\text{Energy in}} \qquad (7.1)$$

The First Law states that energy can be neither created nor destroyed, which implies that for a closed system it is impossible to extract more energy from the system than flows into it. For a gas-fired home furnace, that means we can't extract more heat from it than is contained in the fuel flowing in. The maximum First Law efficiency is 100%, but in practice we can never quite reach that goal. Typical furnaces in the US have First Law efficiencies of 80%–95%.

A less familiar metric is called Second Law efficiency, which is defined as follows:

$$\text{Second Law efficiency} = \frac{\text{Minimum energy use needed to accomplish a task}}{\text{Current energy use need to accomplish a task}} \quad (7.2)$$

This metric, also known as 'exergy efficiency', measures how efficient we are compared to the most efficient possible way to accomplish that task [11, 13]. For heat engines operating between two heat reservoirs, the maximum possible efficiency (i.e. the minimum energy use) is given by the Carnot limit, which is a function of the temperatures of the two reservoirs. Second Law efficiencies for the global economy are typically much lower than First Law efficiencies, more like 5%–15%, and much less than that for processes that are especially inefficient [11].

Second Law efficiencies provide a more accurate picture of the potential for efficiency improvements than do First Law efficiencies, which are tied to a specific process (such as a furnace) rather than the theoretical minimum energy needed to accomplish the task. Second Law efficiencies also point to the possibility of *redefining the task* and achieving even great energy savings. Our focus on linking emissions with associated tasks in earlier chapters emerges from this way of looking at the problem.

The full potential for energy efficiency and other demand-side actions has in general not been reflected fully in large-scale climate mitigation studies [29], although that situation has been improving in recent years. The most recent IPCC Working Group III report, for the first time, focused on the potential for shifts in end-use demands to reduce emissions [3: technical summary and chapter 5]. These findings are also summarized in [30]. The IPCC report found that end-use strategies affecting institutional and individual behavior (which fall squarely in what we define as efficiency and optimization) could reduce emissions by 40%–70% by 2050.

A key aspect of efficiency and optimization is tallying *all* societal costs and benefits. Just focusing on energy use or carbon emissions is a mistake because there are so many co-benefits of climate action. Optimizing efficiency in the societal sense means accounting for these co-benefits, which can offset some or all of the costs of reducing emissions [3, 7]. The societal perspective must always be primary when setting climate mitigation goals but understanding cost perspectives of market actors is important for designing effective policies.

One archetypal example is that of biomass cookstoves in the developing world. Many of these are inefficient and polluting, killing millions of people every year [19]. More efficient cookstoves using cleaner fuels could substantially reduce the toll of death and disease associated with combustion of biomass while also reducing operating costs, greenhouse gas emissions, and deforestation[1]. A narrow focus on the energy savings benefits from more efficient stoves would miss the health and quality of life benefits associated with solving this difficult problem.

7.2 Creating or adopting a business-as-usual scenario

Efficiency and optimization as discussed here may affect your business-as-usual (BAU) case. As discussed above, it is important to reflect recent trends in efficiency and energy intensity, and not all scenarios capture those trends accurately (historical trends tend to lag developments). You may need to tweak a BAU case adopted from another source.

For example, projections of energy use that relied on a fixed relationship between primary energy and economic activity, which were common in the 1960s and 1970s, were famously inaccurate when that relationship changed in the early 1970s [9, 31]. The relationship between electricity use and economic activity also changed around the same time, and changed again in the mid-1990s for the US [31]. Projections that missed those shifts in efficiency trends were way off the mark.

Changes in efficiency trends can affect the underlying drivers for your BAU scenario. For example, vehicle miles traveled (VMT) is a commonly used driver of light vehicle transportation emissions but shifting structural factors and clever use of information technology can change those trends. One salient example emerged during the COVID-19 pandemic: telecommuting, enabled by Internet connectivity, became much more widely accepted [32]. That means that a BAU case that assumes pre-COVID travel behaviors will likely be inaccurate (and overestimate VMT in the future).

7.3 Creating the climate-positive scenario

The climate-positive scenario that embodies the twin goals of electrification and shifting to non-fossil generation technologies needs to reflect the latest thinking on efficiency and optimization. That means adopting the broader view of 'emissions efficiency' rather than a narrow focus on energy efficiency (which is still important but is only part of a more comprehensive strategy).

Creutzig *et al* [30] provide a comprehensive view of end-use strategies to improve emissions efficiency, grouping actions into the major categories 'avoid', 'shift', and 'improve'. Avoiding means reducing 'unnecessary consumption' (such as avoiding food or plastic waste, extending product life span, improving materials efficiency, reducing flying), shifting means moving to already existing technologies and systems that reduce or eliminate emissions (such as electrification, reuse, and recycling), and improving means adopting higher efficiency technologies and actions than people would otherwise choose on their own.

There is a related literature on what has come to be called 'circular economy', which describes methods of reducing environmental impacts of human activities through reducing, reusing, recycling, and harnessing renewable energy and

---

[1] https://drawdown.org/solutions/improved-clean-cookstoves

materials flows. It grew out of efforts to 'close the loop' in manufacturing [33]. This way of addressing efficiency and optimization has its uses, but is not a panacea, and only a focus on absolute emissions reductions and what we call emissions efficiency will reliably achieve the desired results [34, 35]. Intermediate metrics such as percentage of materials recycled say little about cumulative emissions reductions, which are ultimately all the matters for climate, and inherent problems with recycling plastic in particular limit the usefulness of this approach in addressing certain materials [36].

Information and communication technologies (ICTs) are key cross-cutting drivers of greater efficiency and lower emissions [37]. They speed up our ability to collect data, distribute data to consumers, manage complexity, and more rapidly learn and adapt. That's why we call these technologies our 'ace in the hole'. They allow us to (a) move bits instead of atoms, (b) substitute smarts for parts, (c) dynamically control energy supply and demand, (d) collect high value data, and (e) help us design better gadgets and systems, all of these in ways that we never could before [1, 5].

The performance of electronic computers has shown remarkable and steady growth over the past eight decades, a finding that is not surprising to anyone with even a passing familiarity with computing technology [38]. What most folks don't know, however, is that the *electrical efficiency* of computing (the number of computations that can be completed per kilowatt-hour of electricity) has doubled at predictable rates since the dawn of the computer age [39, 40].

Figure 7.1 shows recent trends in computing efficiency relative to 1985. From the mid-1940s to about the year 2000, computations per kWh at peak output doubled every 1.6 years [39]. Around the year 2000, engineers started to hit physical limits on their ability to shrink transistors while controlling power use, which resulted in a slowing of the trend in computations per kWh to doubling every 2.6 years [40].

The existence of laptop computers, cellular phones, and personal digital assistants was enabled by these trends, which presage continuing rapid reductions in the power consumed by battery-powered computing devices, accompanied by new and varied applications for mobile computing, sensors, wireless communications, and controls.

The most important future effect of these trends is that the power needed to perform a task requiring a fixed number of computations will fall by half every 2.6 years, enabling mobile devices performing such tasks to become smaller and less power consuming, and making many more mobile computing applications feasible. Alternatively, the performance of some mobile devices will continue to double every 2.6 years while maintaining the same battery life (assuming battery capacity doesn't improve). Some applications (such as laptop computers) will likely tend towards the latter scenario, while others (such as mobile sensors and controls) will take advantage of increased efficiency to become less power hungry and more ubiquitous.

Innovations in low power operation (particularly when computing devices are in standby mode) [40], software design [41], and co-design of special purpose hardware

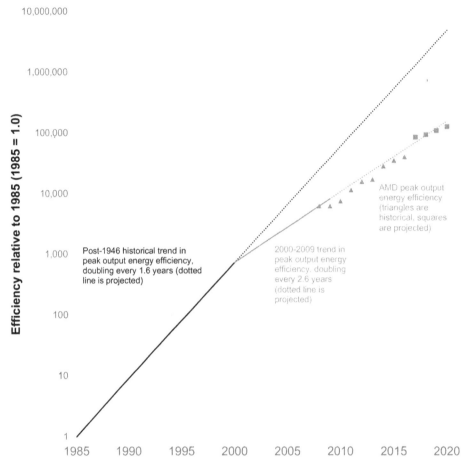

Figure 7.1. Trends in peak output efficiency of computing over time. Source: Efficiency trends to 2000 from [39]. Efficiency trends 2000 to 2009 from data in [39]. Trends 2008 to 2020 from data/analysis in [40]. 'Projected' points from AMD post 2016 match actuals from more recent data.

and software [42] will allow engineers (for a time) to exceed historical trends in improving peak output efficiency (which are derived from characteristics of general-purpose computing devices such as those highlighted in figure 7.1). We are still not at the physical limits but because these are approaching in the next few decades we will eventually need to rethink our computing technology from the ground up [5]. Fortunately, even if progress in computing efficiency stopped tomorrow there would be at least a couple of decades of benefits associated with applying current technology to applications for which it hasn't yet been used.

These technologies will allow us to better match energy services demanded with energy services supplied, and vastly increase our ability to collect and use data in real time. They will also help us minimize the energy use and emissions from accomplishing human goals, a technical capability that we sorely need. The environmental implications of these trends are profound and only recently becoming clear [5, 43–45].

As one of many examples of what is becoming possible using ultra low power computing, consider the wireless no-battery sensors created by Joshua R Smith's team at the University of Washington [46]. These sensors scavenge energy from stray television and radio signals, and they use so little power (tens of microwatts in this example) that they don't need any other power source. Stray light, motion, or heat can also be converted to meet slightly higher power needs, perhaps measured in milliwatts.

The contours of this design space are only beginning to be explored. Imagine wireless temperature, humidity, or pollution sensors that are powered by ambient energy flows, send information over wireless networks, and are so cheap and small that thousands can be installed where needed. Imagine sensors scattered throughout a factory so pollutant or materials leaks can be pinpointed rapidly and precisely. Imagine sensors spread over vast areas of glacial ice, measuring motion, temperature, and ambient solar insolation at very fine geographical resolution. Imagine tiny sensors inside products that tell consumers if temperatures while in transport and storage have been within a safe range. Imagine a solar powered outdoor trash can/compactor that notifies the dispatcher when it is full, thus saving truck trips (no need to imagine this one, it's real[2]). In short, these trends in computing will help us lower greenhouse gas emissions and allow vastly more efficient use of resources.

7.3.1 Moving to zero-emission technologies

Chapter 4 explored electrification as key to eliminating direct use of fossil fuels. Chapter 5 explored zero-emissions generation technologies for eliminating use of fossil fuels in the electricity sector. Both efforts can be accelerated by clever use of information technology.

For both demand and supply side technologies, ICT allows accurate data collection and real-time controls that enable us to more precisely match end-use demands with variable generation sources. It also allows more rapid and accurate real-time analysis.

Because of the rapid decline in the costs of monitoring technology (driven by improvements in computing and communications), our ability to understand the effects of our actions in real time is increasing rapidly. This means better control of processes, less waste, and better matching of energy services demanded with those supplied. In combination with battery, thermal, and (sometimes) hydrogen storage, these tools will allow the whole electricity system to work more reliably and cheaply.

Fortunately, the inrush of data from monitoring technologies has been accompanied by improvements in our ability to analyze and understand those data. Without new tools we'd have a hard time keeping up, which is why new data centers, industrial operations, and power system operators are increasingly demanding more powerful tracking software. These developments are important because energy-use data are available at increasingly fine levels of geographic and temporal disaggregation, which will enable vastly more responsive and reliable power systems.

---

[2] http://bigbellysolar.com

7.3.2 Improving end-use efficiency

While electrifying, we also need to make sure electric technologies are as efficient as they can be. In general, efficiency is cheaper than new supply, and saving energy at the end-use level captures significant additional savings upstream, by eliminating system losses [47]. The more efficient and more demand responsive we can make demand-side devices the easier and cheaper it will be to expand the electricity grid to accommodate an all-electric economy.

In the past two decades there have been substantial gains in the efficiency of electricity technologies for lighting, heating, cooling, water heating, electric motors, cooking, clothes washers, clothes dryers, and other equipment [48]. Some of these changes have been driven by technological improvements (especially for lighting, which has seen a large-scale shift to light emitting diode (LED) technology). Others have been driven by policy and changes in consumer preferences. The potential for efficiency improvements remains large and is constantly expanding as technology improves [8, 12, 47, 49]. Increasing returns to scale for granular technologies play an important role in the rapid progress of end-use efficiency technologies [10].

Adoption of higher efficiency technologies can be accelerated with a combination of widely used policies [50–59]. They include minimum efficiency standards for equipment, building efficiency standards, auto efficiency standards, energy labeling, and incentives for purchase of more efficient products. Less commonly used are mandatory end dates for sales of fossil fuel appliances and equipment, and incentives to scrap inefficient and fossil-fired equipment early, but these will grow in importance over time.

In the past, these programs were designed to be 'fuel neutral', but with increasing emphasis on electrification, the hesitance to promote electricity over natural gas has diminished. Ultimately, programs like ENERGY STAR, which promotes products in the top tier of efficiency, will need to deprecate natural gas appliances in their ratings (or eliminate them altogether). Efficiency standards will need to do the same.

The beauty of standards is that they eliminate the hassles, transactions, and computation costs associated with choosing the more efficient option. They are useful for cutting off the least efficient part of the market and they reach parts of the market (such as rental homes) that are most resistant to choosing more efficient products. Labels and incentives are useful for pushing adoption of the most efficient products. Early scrappage ('cash for clunkers') and end dates for sale of fossil fueled equipment can also be powerful tools for accelerating stock turnover.

With the proliferation of 'smart meters' that allow real-time metering of electricity use, our ability to understand energy use and emissions in all sectors will rapidly improve. In the early days of energy efficiency analysis (in the 1970s), researchers conducted market assessments using simple averages of costs and savings for a single average refrigerator model for the US (for example). Now we're able to monitor the response of millions of households to electricity prices in real time, and to disaggregate household electricity use into its component parts with unparalleled accuracy. That allows much more precise assessments of efficiency potentials and will give businesses the opportunity to target the biggest electricity users with energy saving innovations.

7.3.3 Redefining tasks

Chapter 3 contains one example of task redefinition, expanding our view from just the furnace to the actual task (keeping rooms at a given temperature). Another is creating durable materials based on calcium carbonate, which requires high temperatures and pressures using current technology (cement is the archetypal example). If an engineer focused on the First Law efficiency of making cement using existing processes, the possibilities would be limited. If she instead studied alternative means of making such materials (copying how chickens make eggshells at room temperature, for example [60]) the design space and potential emissions reductions expand considerably.

Changing patterns of material use by consumers and businesses affect industrial process emissions associated with the production of those materials [21, 61] as well as emissions from agricultural production and food consumption. The potential for such emissions reductions is often ignored but is clearly worthy of more attention [22, 23].

In the US, mail is delivered directly to everyone's home address. In many suburban areas of Canada, there are central mailboxes within a block or two of every house and the letter carrier only needs to drop mail at each central mailbox, making it possible to deliver mail to many more homes with much less labor and energy use (see figure 7.2 for an example). Redefining this task in that way would enable much greater efficiencies in multiple dimensions than just focusing on the efficiency of mail delivery vehicles.

The possibility of substituting information services for materials use ('dematerialization') is another promising option [5, 24]. It is usually possible to make products simpler in design using software and controls in the device itself, but we can also save energy and materials by avoiding the need to move physical objects and people from place to place.

The three archetypal examples of this effect are telecommuting [62], replacement of physical compact discs [63] or DVDs [64] with downloadable files, and video conferencing [65, 66]. It is not always true that moving bits instead of atoms reduces emissions, but it is often true.

To take full advantage of ICT's benefits, companies must reorganize themselves [67], which is another way of redefining tasks. ICT also makes such reorganization easier because it improves communication, coordination, and process controls, and creates the conditions under which complementary cost-reducing innovations can more rapidly be brought to market [68].

To reduce emissions in an institutional context, companies typically follow a set of steps such as the ones below, which are exactly analogous to steps needed for a complete and actionable emissions reduction plan for a state or country [1]:

Figure 7.2. A central mailbox in Canada circa 2022. Source: Copyright 2022 Jonathan Koomey.

1. Create a baseline inventory of corporate greenhouse gas emissions and track over time.
2. Project greenhouse gas emissions and commit to an aggressive improvement level in a future year to which the company will work towards.
3. Link greenhouse gas emissions to each business function.
4. Identify opportunities for transformational environmental improvements (using whole-systems integrated design) and create business plans for capturing them.
5. Assign responsibility for implementation.
6. Implement highest impact and most profitable changes in business processes.
7. Measure impacts over time.
8. Reward technical staff and managers for achieving improvement targets.
9. Re-evaluate opportunities each year and implement the highest impact and most profitable opportunities first.
10. Lather, rinse, repeat.

These 'steps to sustainability' will look familiar to companies that are already taking the climate issue seriously (they apply equally well to other environmental issues). Institutional innovations hold great promise for rapid and large-scale systems change because they can be easily replicated with the help of information technology, thus taking full advantage of increasing returns to scale [69]. This example also highlights the power of supply chains for achieving emission reductions goals at scale [70, 71].

Many companies use Six Sigma programs to institutionalize processes such as the steps above. In that case, the focus is broader than just on energy or emissions, but the idea is the same, assigning cross-departmental teams to identify opportunities, giving those teams responsibility for capturing those savings, and measuring the results. The teams are then rewarded for the real savings they produce, and in general, they find more and more.

Another historical example is from Dow Chemical in the 1980s and early 1990s. Ken Nelson, an engineer with Dow USA, created a contest among lower-level employees to root out waste and save energy. The first year of the contest they found dozens of projects with a measured return on investment (ROI) of 173% per year, and over the dozen-year life of the contest, the projects saved $110 million per year for an audited average ROI of about 200%. Those savings went straight to Dow's bottom line, and never petered out. The program stopped when Nelson retired, but it is the archetypal example of how opportunities for energy savings are a renewable resource [4].

7.3.4 Direct electricity used by ICT

ICT helps improve efficiency and optimize energy and materials flows, but it also uses electricity. There has been much confusion in the literature around electricity used by ICT, with the general tendency of many analysts to overestimate ICT electricity use [14, 15, 17, 20, 72].

The most accurate research shows that conventional data centers used less than 1% of the world's electricity in 2018, and data center electricity use grew only 6% from 2010 to 2018, while computations, network data flows, and data storage grew to 6.5 times, 11 times, and 26 times their 2010 levels during that period.

There is little recent work on total electricity used by all information technology equipment, but the most credible estimates for 2015 showed that all information technology globally, including data centers, networks, and computers, consumed about 3.6% of the world's electricity [73: table 6], and that total didn't grow from 2010 to 2015.

The most common error that leads to overestimates of electricity used by ICT is to assume that growth in service demands (such as computations or network data flows) will continue at historic rates but that efficiency will not improve at the same time. Why people continue to make consequential errors such as this is explored elsewhere [17], but the key thing is to view all astounding claims about ICT electricity use with great skepticism.

Sometimes ICT electricity demand does increase rapidly. Data center electricity use doubled between 2000 and 2005 [74], which is why the technology industry paid such attention to improving data center efficiency in the early to mid-2000s. They largely succeeded in putting the industry on a sustainable path [15].

An area of recent rapid growth in ICT is in cryptocurrency and associated technologies [16, 18]. Recent credible estimates (https://ccaf.io/cbeci/index) place electricity used by Bitcoin alone in the tenths of a percent of all global electricity consumption, and little to none of that is consumed in conventional data centers. This use of electricity has grown rapidly in the past five years and it is emblematic of how fast-moving information technologies can lead to rapid changes in end-use demands.

Such rapid growth can create issues for policy makers, but cryptocurrency is particularly problematic. Naïve extrapolation of near-term grown rates often leads to absurd conclusions [17, 72] and there is still significant uncertainty about the long-term economic value of these technologies. The electricity use is driven by the economics of 'mining' and if there are big changes in the price of cryptocurrencies (as there often are) the associated electricity use is volatile and particularly hard to track.

7.4 Chapter conclusions

Efficiency and optimization will play a key role in creating a climate-positive world, but a whole-systems view implies a focus on 'emissions efficiency', not just energy efficiency. Lower end-use demands enable faster emissions reductions and make supply-side changes less onerous. We can achieve lower emissions intensities by switching to non-fossil technologies (by electrifying end-uses and moving to zero-emissions generation), improving the efficiency and operational optimization of end-use equipment, and altering/redefining tasks to minimize total emissions. The IPCC has shown emissions reductions of roughly a factor of two associated with end-use strategies, and information and communication technologies can help us capture that potential.

Further reading

Creutzig F *et al* 2022 Demand-side solutions to climate change mitigation consistent with high levels of well-being *Nat. Clim. Change* **12** 36–46 https://doi.org/10.1038/s41558-021-01219-y. This article summarizes findings on demand-side emissions reduction options from the latest IPCC mitigation report (2022) as well as other research.

Energy Star is a voluntary labeling program (run by the US Environmental Protection Agency and US Department of Energy) that promotes the most efficient products on the market: http://www.energystar.gov. It is mainly focused on the US market but it influences standards and labels around the world.

Harvey L D 2015 *A Handbook on Low-Energy Buildings and District-Energy Systems: Fundamentals, Techniques and Examples* (New York: Routledge). This book shows how to create superefficient buildings and district energy systems.

IEA 2019 *Material Efficiency in Clean Energy Transitions* (Paris: International Energy Agency) https://www.iea.org/reports/material-efficiency-in-clean-energy-transitions. This report gives a high-level view of how materials efficiency can contribute to achieving a climate-positive world.

Koomey J G, Berard S, Sanchez M and Wong H 2011 Implications of historical trends in the electrical efficiency of computing *IEEE Ann. Hist. Comput.* **33** 46–54 http://doi.ieeecomputersociety.org/10.1109/MAHC.2010.28. The original article documenting long-term trends in the efficiency of general-purpose computers from the mid-1940s onwards.

Koomey J and Naffziger S 2016 Energy efficiency of computing: what's next? *Electr. Design* 28 November http://electronicdesign.com/microprocessors/energy-effi-ciency-computing-what-s-next. This article summarizes the latest data on trends in the efficiency of general-purpose computing devices, showing that the rate of improvement in peak output efficiency slowed after the year 2000.

Koomey J G, Matthews H S and Williams E 2013 Smart everything: will intelligent systems reduce resource use? *Annu. Rev. Env. Resour.* **38** 311–43 https://doi.org/10.1146/annurev-environ-021512-110549. This article gives a comprehensive overview of how to think about information technology and resource use.

Meys R, Kätelhön A, Bachmann M, Winter B, Zibunas C, Suh S and Bardow A 2021 Achieving net-zero greenhouse gas emission plastics by a circular carbon economy *Science* **374** 71–6 https://doi.org/10.1126/science.abg9853. This article plots a path to a zero fossil plastics industry, relying on carbon captured from combustion or the air and hydrogen created using zero-emissions electricity.

Poore J and Nemecek T 2018 Reducing food's environmental impacts through producers and consumers *Science* **360** 987 https://doi.org/10.1126/science.aaq0216. This meta-analysis explores tradeoffs in reducing emissions from the food system, showing that shifting to more plant-based diets is a key option for reducing food-related emissions.

The American Council for an Energy Efficiency Economy (ACEEE) has been fighting to improve appliance and equipment efficiency for decades in the US: https://www.aceee.org.

The Carbon Trust focuses on credible measurement and verification of carbon savings: https://www.carbontrust.com/our-projects.

The Collaborative Labeling and Appliance Standards Program (CLASP) studies and supports energy efficiency and quality of appliances and equipment around the world: https://www.clasp.ngo/.

References

[1] Koomey J G 2012 *Cold Cash, Cool Climate: Science-Based Advice for Ecological Entrepreneurs* (El Dorado Hills, CA: Analytics)

[2] Grübler A *et al* 2018 A low energy demand scenario for meeting the 1.5 °C target and sustainable development goals without negative emission technologies *Nat. Energy* **3** 515–27

[3] IPCC 2022 *Climate Change 2022: Mitigation of Climate Change. Contribution of Working Group III to the Sixth Assessment Report of the Intergovernmental Panel on Climate Change*

(Cambridge: Cambridge University Press) https://www.ipcc.ch/report/sixth-assessment-report-working-group-3/

[4] Lovins A B *et al* 2011 *Reinventing Fire: Bold Business Solutions for the New Energy Era* (White River Junction, VT: Chelsea Green) https://www.rmi.org/insights/reinventing-fire/

[5] Koomey J G, Matthews H S and Williams E 2013 Smart everything: will intelligent systems reduce resource use? *Annu. Rev. Environ. Res.* **38** 311–43

[6] Williams J H, Jones R A, Haley B, Kwok G, Hargreaves J, Farbes J and Torn M S 2021 Carbon-neutral pathways for the United States *AGU Adv.* **2** e2020AV000284

[7] Roberts D 2020 Air pollution is much worse than we thought: ditching fossil fuels would pay for itself through clean air alone *Vox* 12 August https://www.vox.com/energy-and-environment/2020/8/12/21361498/climate-change-air-pollution-us-india-china-deaths

[8] Lovins A B, Ürge-Vorsatz D, Mundaca L, Kammen D M and Glassman J W 2019 Recalibrating climate prospects *Environ. Res. Lett.* **14** 120201

[9] Craig P, Gadgil A and Koomey J 2002 What can history teach us? A retrospective analysis of long-term energy forecasts for the US *Annual Review of Energy and the Environment 2002* ed R H Socolow, D Anderson and J Harte (Palo Alto, CA: Annual Reviews) pp 83–118

[10] Wilson C, Grübler A, Bento N, Healey S, De Stercke S and Zimm C 2020 Granular technologies to accelerate decarbonization *Science* **368** 36

[11] Carnahan W, Ford K W, Prosperetti A, Rochlin G I, Rosenfeld A H, Ross M, Rothberg J, Seidel G and Socolow R H 1975 Efficient use of energy: art 1—a physics perspective, a report of the Research Opportunities Group of a summer study held in Princeton, New Jersey in July 1974 *AIP Conf. Proc.* **25** PB-24277 https://www.osti.gov/biblio/7134080

[12] Lovins A, Bendewald M, Kinsley M, Bony L, Hutchinson H, Pradhan A, Sheikh I and Acher Z 2010 *X10E Factor Ten Engineering Design Principles* (Old Snowmass, CO: Rocky Mountain Institute) https://www.rmi.org/our-work/areas-of-innovation/office-chief-scientist/10xe-factor-ten-engineering/

[13] Grübler A N, Johansson T B, Mundaca L, Nakicenovic N, Pachauri S, Riahi K, Rogner H-H and Strupeit L 2015 Energy primer *Global Energy Assessment—Toward a Sustainable Future* (Cambridge: Cambridge University Press) ch 1 https://iiasa.ac.at/web/home/research/Flagship-Projects/Global-Energy-Assessment/Chapter1.en.html

[14] Koomey J 2017 *Turning Numbers into Knowledge: Mastering the Art of Problem Solving* 3rd edn (El Dorado Hills, CA: Analytics)

[15] Masanet E, Shehabi A, Lei N, Smith S and Koomey J 2020 Recalibrating global data center energy-use estimates *Science* **367** 984

[16] Koomey J 2019 *Estimating Bitcoin Electricity Use: A Beginner's Guide* (Washington, DC: Coin Center) https://coincenter.org/entry/bitcoin-electricity

[17] Koomey J and Masanet E 2021 Does not compute: avoiding pitfalls assessing the Internet's energy and carbon impacts *Joule* **5** 1625–8

[18] Lei N, Masanet E and Koomey J 2021 Best practices for analyzing the direct energy use of blockchain technology systems: review and policy recommendations *Energy Policy* **156** 112422

[19] Smith K R and Pillarisetti A 2017 *Household Air Pollution from Solid Cookfuels and Its Effects on Health* (Washington, DC: The International Bank for Reconstruction and Development/The World Bank) http://europepmc.org/abstract/MED/30212117 http://europepmc.org/books/NBK525225 https://www.ncbi.nlm.nih.gov/books/NBK525225

[20] Koomey J, Calwell C, Laitner S, Thornton J, Brown R E, Eto J, Webber C and Cullicott C 2002 Sorry, wrong number: the use and misuse of numerical facts in analysis and media

reporting of energy issues *Annual Review of Energy and the Environment 2002* ed R H Socolow, D Anderson and J Harte (Palo Alto, CA: Annual Reviews) pp 119–58

[21] Hertwich E G 2021 Increased carbon footprint of materials production driven by rise in investments *Nat. Geosci.* **14** 151–5

[22] IEA 2019 *Material Efficiency In Clean Energy Transitions* (Paris: International Energy Agency) https://www.iea.org/reports/material-efficiency-in-clean-energy-transitions

[23] D'Amico B, Pomponi F and Hart J 2021 Global potential for material substitution in building construction: the case of cross laminated timber *J. Cleaner Prod.* **279** 123487

[24] Hilty L, Lohmann W and Huang E M 2011 Sustainability and ICT—an overview of the field with a focus on socioeconomic aspects *Notiz. Politeia* **17** 13–28

[25] Xu Q, Zhong M and Cao M 2022 Does digital investment affect carbon efficiency? Spatial effect and mechanism discussion *Sci. Total Environ.* **827** 154321

[26] Li S, Wang W, Diao H and Wang L 2022 Measuring the efficiency of energy and carbon emissions: a review of definitions, models, and input-output variables *Energies* **15** 962

[27] Trinks A, Mulder M and Scholtens B 2020 An efficiency perspective on carbon emissions and financial performance *Ecol. Econ.* **175** 106632

[28] Hummel H 2006 Interpreting global energy and emission scenarios: methods for understanding and communicating policy insights *Thesis* Interdisciplinary Program on Environment and Resources, Stanford University, CA https://profiles.stanford.edu/holmes-hummel

[29] Wilson C, Grübler A, Gallagher K S and Nemet G F 2012 Marginalization of end-use technologies in energy innovation for climate protection *Nat. Clim. Change.* **2** 780–8

[30] Creutzig F *et al* 2022 Demand-side solutions to climate change mitigation consistent with high levels of well-being *Nat. Clim. Change* **12** 36–46

[31] Hirsh R F and Koomey J G 2015 Electricity consumption and economic growth: a new relationship with significant consequences? *Electr. J.* **28** 72–84

[32] Qian X 2022 *Telecommuting During COVID-19: How Does It Shape the Future Workplace and Workforce?* (St Paul, MN: Minnesota Department of Transportation) https://www.cts.umn.edu/publications/report/telecommuting-during-covid-19-how-does-it-shape-the-future-workplace-and-workforce

[33] Frosch R A and Gallopoulos N E 1989 Strategies for manufacturing: wastes from one industrial process can serve as the raw materials for another, thereby reducing the impact of industry on the environment *Sci. Am.* **261** 144–52

[34] Mayers K, Davis T and Van Wassenhove L N 2021 The limits of the 'sustainable' economy *Harvard Business Rev.* 16 June https://hbr.org/2021/06/the-limits-of-the-sustainable-economy

[35] Mayers K, Davis T and Van Wassenhove L N 2021 Circular economy: a slippery step on the path to sustainability? Seven lessons to address resource consumption *INSEAD* Working paper No. 2021/65/TOM https://ssrn.com/abstract=3960908

[36] Enck J and Dell J 2022 Plastic recycling doesn't work and will never work *Atlantic* 30 May https://www.theatlantic.com/ideas/archive/2022/05/single-use-plastic-chemical-recycling-disposal/661141/

[37] Wilson C, Kerr L, Sprei F, Vrain E and Wilson M 2020 Potential climate benefits of digital consumer innovations *Annu. Rev. Environ. Res.* **45** 113–44

[38] Nordhaus W D 2007 Two centuries of productivity growth in computing *J. Econ. History* **67** 128–59

[39] Koomey J G, Berard S, Sanchez M and Wong H 2011 Implications of historical trends in the electrical efficiency of computing *IEEE Ann. History Comput.* **33** 46–54

[40] Koomey J and Naffziger S 2016 Energy efficiency of computing: what's next? *Electronic Design* 20 November http://electronicdesign.com/microprocessors/energy-efficiency-comput-ing-what-s-next

[41] Leiserson C E, Thompson N C, Emer J S, Kuszmaul B C, Lampson B W, Sanchez D and Schardl T B 2020 There's plenty of room at the top: what will drive computer performance after Moore's law? *Science* **368** eaam9744

[42] Shalf J, Quinlan D and Janssen C 2011 Rethinking hardware-software codesign for exascale systems *Computer* **44** 22–30

[43] Koomey J 2012 The computing trend that will change everything *Technol. Rev.* 9 April 76–7 https://www.technologyreview.com/2012/04/09/186841/the-computing-trend-that-will-change-everything/

[44] Greene K 2011 A new and improved Moore's law *Technol. Rev.* 12 September https://www.technologyreview.com/2011/09/12/191382/a-new-and-improved-moores-law/

[45] The Economist 2011 A deeper law than Moore's? *Economist* 10 October http://www.economist.com/blogs/dailychart/2011/10/computing-power

[46] Parks A N, Sample A P, Zhao Y and Smith J R 2013 A wireless sensing platform utilizing ambient RF energy *Proc. IEEE Topical Conf. on Biomedical Wireless Technologies, Networks, and Sensing Systems (20–23 January)*

[47] Lovins A B 2005 *Energy End-Use Efficiency* (Amsterdam: Rocky Mountain Institute for InterAcademy Council) https://rmi.org/insight/energy-end-use-efficiency/

[48] IEA 2020 *Energy Efficiency 2020* (Paris: International Energy Agency, Organization for Economic Cooperation and Development (OECD)) https://www.iea.org/reports/energy-efficiency-2020

[49] Lovins A 2010 *Integrative Design: A Disruptive Source of Expanding Returns to Investments in Energy Efficiency* X10-09 (Old Snowmass, CO: Rocky Mountain Institute) http://www.rmi.org/rmi/Library/2010-09_IntegrativeDesign

[50] Meyers S, Cubero E, Williams A, Rosenquist G, Blum H, Iyama S, Iyer M and Kantner C 2021 *Energy and Economic Impacts of US Federal Energy and Water Conservation Standards Adopted from 1987 through 2020* (Berkeley, CA: Lawrence Berkeley National Laboratory) https://ees.lbl.gov/publications/energy-and-economic-impacts-us-5

[51] IEA 2016 *Achievements of Appliance Energy Efficiency Standards and Labelling Programs: A Global Assessment In 2016* (Paris: International Energy Agency) https://www.iea.org/reports/achievements-of-appliance-energy-efficiency-standards-and-labelling-programs

[52] Greene D L, Greenwald J M and Ciez R E 2020 US fuel economy and greenhouse gas standards: what have they achieved and what have we learned? *Energy Policy* **146** 111783

[53] National Research Council 2003 Effectiveness and impact of corporate average fuel economy (CAFE) standardsCommittee on the Effectiveness and Impact of Corporate Average Fuel Economy (CAFE) Standards Board on Energy and Environmental Systems Division on Engineering and Physical Sciences Transportation Research Board, National Research Council, National Academy Press

[54] Livingston O V, Cole P C, Elliott D B and Bartlett R 2014 *Building Energy Codes Program: National Benefits Assessment 1992–2040* (Richland, WA: Pacific Northwest National Laboratory) PNNL-22610 https://www.osti.gov/biblio/1756522

[55] Tyler M, Winiarski D, Rosenberg M and Liu B 2021 *Impacts of Model Building Energy Codes—Interim Update* (Richland, WA: Pacific Northwest National Laboratory) https://www.energycodes.gov/sites/default/files/2021-07/Impacts_of_Model_Energy_Codes_2010-2040_Interim_Update_07182021.pdf

[56] Cunningham L J and Eck R J 2021 *Renewable Energy and Energy Efficiency Incentives: A Summary of Federal Programs* (Washington, DC: Congressional Research Service) R40913 https://crsreports.congress.gov/product/pdf/R/R40913

[57] Glatt S and Shields G 2010 *State Energy Efficiency Tax Incentives for Industry* (Washington, DC: US Department of Energy) https://www1.eere.energy.gov/manufacturing/states/pdfs/state_ee_tax_incentives_for_industry.pdf

[58] Greene D L, Patterson P D, Singh M and Li J 2005 Feebates, rebates, and gas guzzler taxes: a study of incentives for increased fuel economy *Energy Policy* **33** 757–75

[59] Langer T 2005 *Vehicle Efficiency Incentives: An Update on Feebates for States* (Washington, DC: American Council for an Energy Efficient Economy) T051 https://www.aceee.org/sites/default/files/publications/researchreports/t051.pdf

[60] Oyen M 2021 Multiscale mechanics of eggshell and shell membrane *JOM* **73** 1676–83

[61] Hertwich E G *et al* 2019 Material efficiency strategies to reducing greenhouse gas emissions associated with buildings, vehicles, and electronics—a review *Environ. Res. Lett.* **14** 043004

[62] Atkyns R, Blazek M and Roitz J 2002 Measurement of environmental impacts of telework adoption amidst change in complex organizations: AT&T survey methodology and results *Resour. Conserv. Recycl.* **36** 267–85

[63] Weber C, Koomey J G and Matthews S 2010 The energy and climate change impacts of different music delivery methods *J. Industr. Ecol.* **14** 754–69

[64] Mayers K, Koomey J, Hall R, Bauer M, France C and Webb A 2014 The carbon footprint of games distribution *J. Industr. Ecol.* **19** 402–415

[65] The Climate Group 2008 *Smart 2020: Enabling the Low Carbon Economy in the Information Age* The Climate Group, on behalf of the Global eSustainability Initiative (GeSI) https://gesi.org/research/smart-2020-enabling-the-low-carbon-economy-in-the-information-age

[66] Tao Y, Steckel D, Klemeš J J and You F 2021 Trend towards virtual and hybrid conferences may be an effective climate change mitigation strategy *Nat. Commun.* **12** 7324

[67] Brynjolfsson E and Hitt L 1996 Paradox lost? Firm-level evidence on the returns to information systems spending *Manag. Sci.* **42** 541–58

[68] Brynjolfsson E and Hitt L M 2000 Beyond computation: information technology, organizational transformation and business performance *J. Econ. Perspect.* **14** 23–48

[69] McAfee A and Brynjolfsson E 2008 Investing in the IT that makes a competitive difference *Harvard Business Rev.* July–August p 98–108

[70] Liu M L, Li Z H, Anwar S and Zhang Y 2021 Supply chain carbon emission reductions and coordination when consumers have a strong preference for low-carbon products *Environ. Sci. Pollut. Res. Int.* **28** 19969–83

[71] The Carbon Trust 2018 Big benefits await those who tackle supply chain emissions *The Carbon Trust* 24 August https://www.carbontrust.com/news-and-events/insights/big-benefits-await-tackle-supply-chain-emissions

[72] Masanet E, Shehabi A, Lei N, Vranken H, Koomey J and Malmodin J 2019 Implausible projections overestimate near-term bitcoin CO_2 emissions *Nat. Clim. Change* **9** 653–4

[73] Malmodin J and Lundén D 2018 The energy and carbon footprint of the global ICT and E&M sectors 2010–2015 *Sustainability* **10** 3027

[74] Koomey J 2008 Worldwide electricity used in data centers. *Environ. Res. Lett.* **3** 034008

Chapter 8

Remove carbon

Even if we get to net zero, we still need to get carbon dioxide out of the atmosphere.
—John Kerry[1]

Chapter overview

- We need to get emissions to zero as quickly as we can to stabilize the climate, as discussed in previous chapters, but that's not the end of the story.
- We also need to scale up carbon removal to drawdown past climate pollution and return atmospheric GHG concentrations to pre-industrial levels.
- Carbon removal should not be considered an alternative to immediate, rapid direct emissions reductions, it is a supplement needed for a climate-positive planet.
- We'll need billions of tonnes per year of carbon removal to make a measurable dent in atmospheric carbon dioxide concentrations in the next few decades.
- Enhancing the ability of land and marine ecosystems to remove and store carbon often represents the most cost-effective approach in the near term, but additionality and permanence concerns must be addressed.
- Technological options, such as direct-air capture of carbon dioxide, must also play significant roles as soon as possible, but for that to happen we need to start funding and deploying technology now.

8.1 Introduction

In his splendidly clear book, *The 100% Solution*, Solomon Goldstein-Rose points out a critical fact that distinguishes climate change from other societal problems:

---

[1] https://www.huffpost.com/entry/john-kerry-climate_n_6081c355e4b05c4290738500

The interesting thing about climate change, which makes it different from most other global issues is that we can't expect any lessening of impacts on the way to a full solution. Most social problems in the world, such as poverty or diseases, cause a certain amount of harm each year, and if we make a little progress on a given issue, it causes a little less harm the following year.

Climate change is different. The impacts are caused by higher average global temperatures, which are driven by higher-than-average levels of greenhouse gases in the atmosphere. Reducing emissions is not enough to avoid, or even reduce, climate change effects—reducing the amount of CO_2 in the atmosphere is the only way to do that, which requires totally eliminating emissions and then removing some portion of the CO_2 that's already in the atmosphere [1].

We discussed the dependence of global warming risks on cumulative emissions in chapter 1, but it's worth reviewing it in the context of Goldstein-Rose's quote above.

First, temperatures will continue to increase until we get nearly to zero emissions [2]. Cutting emissions in half will just slow the increases in temperatures. We need to get to zero emissions as quickly as we can.

Second, reducing emissions to zero is only the start of mitigating harm from climate change, we also need to extract carbon dioxide from the atmosphere. CO_2 is the most important greenhouse gas, and one that stays in the atmosphere for centuries unless we actively work to clean up our climate trash. What are some practical ways to do that?

The most feasible near-term path to such 'drawdown' is increasing carbon uptake from the biosphere [3], which is the path used in the low energy demand scenario [4]. However, enhancing natural sinks alone is unlikely to quickly remove enough carbon to return us to a pre-industrial climate, particularly if managing land to maximize carbon storage conflicts with other needs such as food production. We may soon also need to recapture large amounts of carbon dioxide from the atmosphere using **direct-air capture** (DAC) technology. We need to research and scale such technologies, but it will likely be over a decade before the world will be able to accomplish significant **negative emissions** using such technological means.

In the near term, most carbon removal will come from **nature-based solutions** such as stopping deforestation, planting trees (afforestation in new areas and reforestation in previously forested areas), rebuilding coastal wetlands and kelp forests, enhancing ocean CO_2 uptake, and accelerating weathering of certain minerals. Some innovators are also starting to deploy **hybrid nature-based and engineered solutions**, where plants or algae do the work of removing carbon from the atmosphere, then engineered chemical or mechanical systems convert the carbon into a stable form for long-term storage or conversion into useful products. Whether fully nature-based or hybrid systems, the effects of enhancing biospheric carbon sinks may take many decades to show results and won't always be cheap [5–9], but we need to get started now.

The innovations needed to make such efforts successful are as much social and institutional as they are technical, but we largely know how to increase biospheric carbon uptake now. We also know there are biophysical and economic constraints on such activities [10].

Minx [11] suggests a natural progression of effort, with an early focus on enhancing land-based carbon sinks, moving later to technological options that might scale better and could capture cost reductions from manufacturing economies of scale and technological innovation. We generally agree. Since we will likely need both nature-based and engineered carbon removal systems to fully solve our climate challenges we should pilot a wide portfolio of potential carbon removal solutions now to find approaches that maximize pollution drawdown while minimizing the unforeseen social and environmental tradeoffs that can come from scale.

Finally, we need to move away from the idea that increasing uptake of carbon in the atmosphere is somehow an alternative to reducing emissions to zero [12]. We need to do both, as quickly as we can. That is one reason why 'carbon offsets' are a dead end. There is no alternative to rapid and immediate emissions reductions [3]. None.

8.2 Understanding carbon removal

Carbon dioxide removal (or CDR) through use of **negative emissions technologies** is removing carbon from the atmosphere after it has already been emitted, to reduce concentrations of carbon dioxide and reverse the warming associated with previously emitted carbon [13].

To put carbon removal options in context, it's first important to understand the global carbon cycle. The authoritative source on that topic is Friedlingstein *et al* [14]. Figure 8.1 summarizes the results of that study, averaged over the period 2011 to 2020.

This figure is a rich source of insight. The stocks and flows are expressed in terms of carbon because carbon exists in the atmosphere as carbon dioxide but is often in other forms in other parts of the cycle. To convert carbon to carbon dioxide multiply by the ratio of total molecular weight of carbon dioxide to the molecular weight of carbon (44/12 or 3.667).

The annual net influx of carbon to the atmosphere from burning fossil fuels is about 9.5 billion tonnes (gigatonnes), which is equivalent to about 35 gigatonnes of carbon dioxide per year. Permafrost releases about 1 Gt C per year, vegetation takes up about 3 Gt C per year and the ocean takes up about 3 Gt C per year. A little more than half of the fossil fuel plus permafrost inflows to the atmosphere are taken up by the oceans and land-based vegetation, with that uptake split roughly 50:50 between oceans and land.

The stocks of carbon also tell some important stories. First, the stock of carbon in the atmosphere (in the form of CO_2) is almost 900 Gt C. Current reserves of fossil fuels, coincidentally, also contain about 900 GtC, so if all this fuel were burned, that would increase carbon dioxide concentrations by about 50% (because half of the emissions would be taken up by the oceans and land-based vegetation). That would

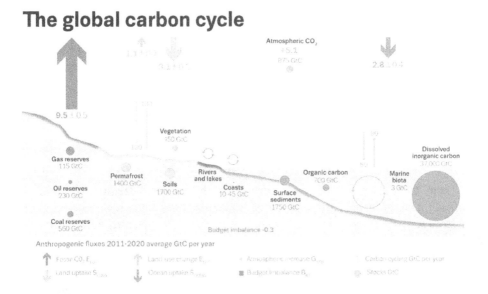

The global carbon cycle

Figure 8.1. The global carbon cycle, stocks and average annual flows, 2011 to 2020. Source: Friedlingstein *et al* [14]. Results are averaged 2011 to 2020. Each gigatonne of carbon emitted equals 3.667 Gt CO_2 and 2.124 parts per million by volume of CO_2 in the atmosphere. Reproduced under Creative Commons Attribution 4.0 International License: https://creativecommons.org/licenses/by/4.0/.

push concentrations to over 600 parts per million (from just over 400 million ppm in the past decade).

Recall from the discussion in chapter 1 that fossil reserves are the stocks of fuels that can be extracted at current prices using current technology. As technology changes, more fossil resources can be converted to reserves, so there will a lot more fossil fuel available over the next century than just the reserves shown in figure 8.1.

The biggest stock of carbon is dissolved inorganic carbon in the oceans, at 37 000 Gt C, with surface sediments at 1750 Gt C, soils at 1700 Gt C, and permafrost at 1400 Gt C. Even small changes in these stocks would have noticeable effects on carbon dioxide concentrations in the atmosphere, which is why climate scientists are worried about positive feedbacks as the climate warms (e.g. warming increases carbon releases from permafrost) but also why carbon removal could make a significant dent in atmospheric concentrations in carbon dioxide over time.

8.3 Carbon removal is not a silver bullet

Removing gigatonnes of carbon from the atmosphere can have beneficial effects, it is not an alternative to emissions reductions, it will be slow and costly [5], and it won't reverse all the effects of greenhouse gases in the atmosphere [15, 16]. The stored heat in the oceans somewhat offsets the cooling effect of lower greenhouse gas concentrations from carbon removal. In addition, outgassing of CO_2 from biospheric sinks reduces the effectiveness of carbon dioxide removal after reaching zero emissions.

As Tokarska and Zickfeld [15] put it carbon dioxide removal may be a helpful tool in conjunction with other efforts aimed at reducing the rise in atmospheric CO_2, it is not a silver bullet to restore the climate system to a desirable state on timescales relevant to human civilization once the impacts of climate change turn out to be 'dangerous'.

Carbon removal is only useful when it is additive to rapid emissions reductions, and by itself will not solve the climate problem, despite the fond hopes of the fossil fuel industry. We need rapid emissions reductions AND carbon removal.

8.4 Carbon capture and storage is not the same as carbon removal

We treat technology to remove carbon from exhaust streams of fossil fuel combustion or industrial process emissions (commonly known as carbon capture and storage (CCS)) as a way of reducing emissions. In general, eliminating these emissions is a better way to tackle the problem.

CCS is expensive (in terms of both capital and operating costs) and it doesn't reduce emissions to zero, even in the best case. It also requires a large carbon tax or subsidy to make the economics work for producers, but such economic engineering generally also helps promote non-fossil energy sources that are much cheaper than CCS already, so it's unclear how CCS can compete for fossil energy sources. And the use of CCS keeps fossil plants operating longer than they would otherwise, so it's counterproductive to the goal of ending fossil fuels. CCS may play a useful role in reducing emissions from industrial processes, such as cement, but even in those cases process innovations may be more cost effective.

8.5 Carbon removal options

It's important to treat the uptake of carbon from the atmosphere as distinct from the storage of that carbon. Some options both take up carbon and store it in one step (such as afforestation and reforestation) others just take it up (such as direct-air capture), so storage costs are additional.

There are also different types of storage, with different levels of permanence. Forest carbon can be released by forest fires, invasive pests, or diseases, which are all increasingly likely in a warming world. There are similar 'permanence' concerns for ocean uptake—as the ocean warms, more methane and carbon dioxide may be released. Above ground storage in weathered minerals or underground geological storage can in principle be more permanent than carbon stored in forests, but the process of capturing the carbon can be much more expensive and difficult for those options.

Co-benefits from different carbon dioxide removal options vary a lot, but in general, nature-based solutions have lots of co-benefits beyond carbon removal, while more technological options (such as direct-air capture or biomass with carbon capture) have few or none [17]. Stopping deforestation and protecting oceans from exploitation have obvious co-benefits, as is often also true for reforestation, afforestation, improving soil carbon management, enhanced weathering on land, and expanding mangrove swamps and other marine adjacent ecosystems (also

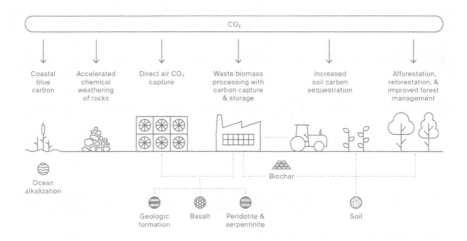

Figure 8.2. Carbon removal options. Source: Wilcox *et al* [13]. Reproduced under Creative Commons Attribution 4.0 International License (CC BY 4.0): https://creativecommons.org/licenses/by/4.0/.

known as 'blue carbon'). Co-benefits make the economics of those options a lot better, even if the value of some of them can't be quantified (such as avoiding extinction of species).

Precise boundary definitions and accurate accounting are essential for understanding costs and potentials for carbon removal [6–8]. Figure 8.2, from Wilcox *et al* [13], summarizes the options at a high level, and IPCC [14] performs a more detailed review. We describe each of them in turn.

8.5.1 Protecting ecosystems

Improving forest management and stopping the destruction of existing forests is the first priority for many emissions reduction plans, both because the co-benefits of existing ecosystems are large and because existing forests (and their associated species) are irreplaceable. Project Drawdown calls this **forest protection**[2], which is an apt name, and the United Nations and others often use the acronym **REDD**, an abbreviation for **reducing emissions from deforestation and degradation**, as well as **REDD+**, with the 'plus' referring to the role of conservation, sustainable management of forests and enhancement of forest carbon stocks in developing countries. We define these task more broadly as stopping destruction of existing forest and marine ecosystems, since there are many carbon-rich ecosystems beyond forests. Protecting these natural systems both avoids emissions and preserves the abilities for already carbon-dense ecosystems to remove even more carbon from our atmosphere.

However, protecting forests and other carbon-rich ecosystems is not a panacea. The **permanence** of carbon removed by forests and other nature-based ecosystem

---

[2] https://drawdown.org/solutions/forest-protection

solutions is a concern, since a wildfire, hurricane, bark beetle infestation, disease outbreak, or change in forest management can quickly reverse decades of carbon conservation, and a warming world makes many of these challenges more likely. The **additionality** of forest preservation carbon credits has also often been called into question, particularly when forest conservation organizations sell carbon offsets from supposed avoided deforestation on already protected land. Finally, there are **leakage** concerns for forest conservation efforts that decrease timber production from one location without reducing overall timber demand, since reduced wood supply from one forest may result in higher deforestation pressures in other forests. These concerns certainly don't mean that forests shouldn't be conserved, but it's important to understand that the climate impacts of ecosystem conservation aren't as straight-forward as calculating all the direct carbon stored in each location, since nearly every place is connected to our global economy and our changing climate.

8.5.2 Reforestation and afforestation

Reforestation means re-growing forests on previously forested land and **afforestation** refers to planting forests on land not previously known to support them [9, 13], although some researchers lump afforestation and regrowth of existing forests [18] under reforestation as a shorthand [19]. There are complex factors affecting the direct emissions benefits of forestry options, but they are increasingly well understood [20]. Full life cycle accounting of project climate benefits should also consider any potential permanence, additionality, and leakage concerns, as discussed above.

8.5.3 Increased soil carbon sequestration

Strictly speaking, forestry options are part of the larger category of carbon sequestration in soils, which also includes changes in land-use, but usually carbon sequestered from land and agricultural practices is treated in their own category, under **soil carbon sequestration**. This includes the use of **biochar** to enhance soil carbon storage and moisture retention, alongside other changes in agricultural practices, such as **low-till agriculture**, **crop switching**, **compost fertilization**, and **rotational grazing techniques** [13]. While many studies have shown short-term carbon removal benefits from soil management choices, there are still debates about the permanence of carbon stored in soils, since interventions may be reversible, and additionality and leakage concerns are also real.

8.5.4 Blue carbon

Coastal blue carbon refers to carbon sequestered at the interface between ocean and land, often in mangrove swamps, tidal marshes, kelp forests, and similar ecosystems [13]. Macreadie *et al* [21] and Hilmi [22] discuss potential emissions reductions and carbon removal opportunities from blue carbon, while Macreadie *et al* [23] discusses the future of blue carbon science. While coastal blue carbon solutions are often at an earlier stage than many land-based carbon removal options, some projects are already showing promising potential for both substantial carbon removal alongside fisheries and biodiversity co-benefits. There has also been some research on blue carbon removal techniques for the open ocean, including ocean fertilization techniques designed to create large algae blooms that store carbon when they die and sink to the ocean floor, although these initial experiments have come with concerns about ecosystem disruptions and haven't demonstrated lasting climate benefits. Other researchers have suggested that enhanced whale conservation could lead to meaningful carbon storage from millions of multi-ton whale bodies, as well as enhanced phytoplankton growth from whale-waste fertilization [24].

8.5.5 Enhanced weathering

The uptake of CO_2 by mineralizing carbon from weathering rocks is Earth's largest natural climate cooling system, but this process typically takes thousands of years to make a dent in atmospheric CO_2. Some researchers and entrepreneurs are now exploring **enhanced weathering** options to speed up this **CO_2 mineralization** process, which generally involves alkaline materials interacting with the carbon dioxide in the atmosphere to create stable storage [13]. It can occur underground, at the land surface, or in the ocean. To accelerate this process, alkaline material can be ground up to increase the reactive surface area. To do this at gigatonne scale would likely involve significant costs and energy use, but in some cases by-products of existing mining processes may be applied with minimal additional processing. Given that carbon is very stable once mineralized, permanence is much less of a concern relative to other nature-based solutions.

8.5.6 Biomass processing with carbon capture and storage

For the past decade or two, many mitigation scenarios have relied heavily on **biomass energy carbon capture and storage (BECCS)** systems to create negative emissions [25]. Plants recapture the carbon dioxide as they grow and then when those plants are burned to produce energy, the resulting carbon dioxide is assumed to be captured and stored [13]. While this technology is potentially feasible, there are complexities and constraints that can limit its potential [6]. The inefficiencies inherent to photosynthesis, serious constraints on land and water use, challenges with biomass collection and processing, and thermodynamic and practical constraints on how much carbon dioxide can be re-captured from combusted biomass will make scaling this technology difficult beyond a certain point, although it may have uses in niche applications.

Some ventures are now doing away with the energy part of the equation and instead converting biomass primarily to biochar and **bio-oil**, the latter of which can be injected underground for very long-term carbon storage, essentially creating a reverse oil industry. Such systems are climate positive when **waste biomass** is used, such as crop or forestry residues that would otherwise be burned. BECCS and other engineered biomass carbon capture systems have the potential advantage of more permanent carbon storage compared to nature-based solutions, but leakage issues can still arise if biomass is grown on land that would have been used for other commercial purposes.

8.5.7 Direct-air capture with storage

Direct-air capture technology (DAC) extracts carbon dioxide from the atmosphere and stores it in safe way, either as mineralized rocks or in underground caverns, or in useful carbon-based products [6–8, 26–31]. DAC technology is in its infancy, but initial commercial facilities are already operational, and it is clear that massively scaling up production of machines that extract carbon dioxide would reduce the capital cost of that equipment substantially (just as mass production always does). As with enhanced weathering and engineered biomass carbon storage, DAC has the potential advantage for very long-term carbon storage without additionality concerns.

The energy use associated with running DAC machines will still be an issue, however, because carbon dioxide exists in modest concentrations in the atmosphere and thermodynamics is challenging in such situations. Storing billions of tons of material per year in solid, liquid, or gaseous form will likely also be a difficult and costly challenge.

This doesn't mean we shouldn't develop DAC technology, just that it will likely take more than a decade to be important enough to make a measurable difference. The sooner we get started the sooner we can identify the most promising avenues for continued development.

8.6 Potentials and costs for carbon removal

How much carbon removal do we need? The short answer is 'as much as we can manage'. Making a measurable dent in carbon dioxide concentrations means removing billions of tonnes of carbon dioxide per year from the atmosphere. Exactly how much we can capture in practice is a complicated question of implementation logistics, social complexity, land-use constraints, and technology.

The first step is to get to net zero emissions from the biosphere by stopping industrial-scale deforestation and any other large-scale disturbance of marine and terrestrial ecosystems. Scaling other kinds of carbon removal can happen in parallel but stopping biospheric destruction of all types needs to be the highest priority. These efforts will have modest direct costs [32, 33] but will yield so many co-benefits that they should easily pay for themselves from society's perspective.

That effort by itself can result in at least a gigatonne of carbon dioxide reductions annually, as per Friedlingstein *et al* [14], but there are still big uncertainties in net emissions from land-use changes globally [34]. In addition, that global net number includes carbon uptake in some places (such as the US), so that stopping industrial-scale deforestation while maintaining carbon uptake in places that already show net negative emissions would reduce net global emissions by much more than 1 gigatonne per year.

The *Carbon Removal Primer* [13] and these other key sources [6–8] summarize the costs and potentials for a range of carbon removal options, but it's a frustrating body of research. The uncertainties are huge and the dependence of most options on scaling up to achieve cost reductions makes it hard to create firm estimates for the future.

Costs of carbon removal are often expressed in $ per tonne removed. An option that removes one tonne of CO_2 for one dollar (a bargain!) would cost society one billion dollars for every gigatonne removed. If the cost is one hundred dollars per tonne, every gigatonne would cost one hundred billion dollars, and ten gigatonnes per year would cost $1 trillion per year. In a global economy that generates almost $100 trillion per year that's probably affordable, but it is still real money, and if there are ways to remove carbon more cheaply, we'll need to find them.

Roe *et al* [35] find 'transforming the land sector and deploying measures in agriculture, forestry, wetlands and bioenergy could feasibly and sustainably contribute about … 15 billion tonnes of carbon dioxide equivalent ($GtCO_2e$) per year' in global emissions reductions. Such large emissions reductions would require substantial changes in management of the biosphere but would also make a measurable difference in carbon dioxide concentrations in the medium to longer term.

Fuss *et al* [6] find 'our best estimates for sustainable global NET potentials in 2050 are 0.5–3.6 $GtCO_2$/year for afforestation and reforestation, 0.5–5 $GtCO_2$/year for BECCS, 0.5–2 $GtCO_2$/year for biochar, 2–4 $GtCO_2$/year for enhanced weathering, 0.5–5 $GtCO_2$/year for DAC, and up to 5 $GtCO_2$/year for soil carbon sequestration'.

Fuss also estimates low and high costs for each of these options. We combine Fuss's low costs with high potentials, reasoning that only if costs were low could higher potentials be achieved. Figure 8.3 shows those results. We also combine high costs with low potentials in figure 8.4, reasoning potentials for each option would be reduced if higher costs prevailed. While a more comprehensive sensitivity analysis would no doubt yield insights, we think this crude summary is useful at a high level.

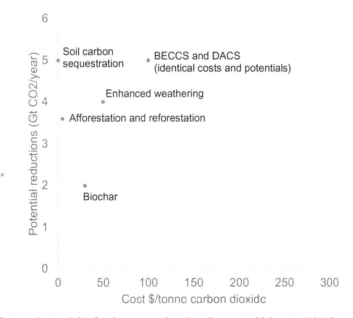

Figure 8.3. Costs and potentials of carbon removal options (low costs, high potentials). Source: Fuss *et al* [6], combining low cost estimates with high potentials estimates.

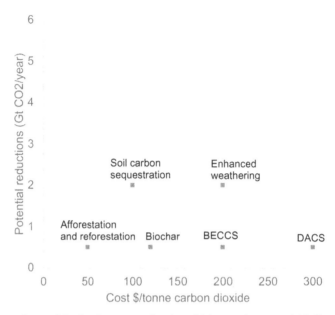

Figure 8.4. Costs and potentials of carbon removal options (high costs, low potentials). Source: Fuss *et al* [6], combining high cost estimates with low potentials estimates.

In the low cost, high potentials case, the options that enhance biospheric carbon uptake (such as soil carbon sequestration, afforestation, reforestation, and biochar) are a lot cheaper than $100/t CO_2, about 10 Gt CO_2/year reduction potentials in total. In the high cost, low potentials case, the savings are about 3 Gt CO_2/year at costs between $50 and $120/tonne.

Enhanced weathering adds another 2 Gt to 4 Gt CO_2/year to the potential depending on the sensitivity case, at costs between $200 and $50/tonne, respectively. More technological solutions (BECCS and DACs) offer another 10 Gt CO_2/year in the low cost case (both together) but in the high cost case that potential drops to 1 Gt CO_2/year in total. Estimated costs for these options range between $100 and $300/tonne. Note that these costs don't factor in the projected permanence of carbon storage or the impact of climate change on carbon storage duration, which could increase costs and decrease potential for some carbon removal technologies.

8.7 Creating or adopting a BAU scenario

Net emissions from land-use vary greatly in different places, so you'll need to collect estimates, data, and projections for your chosen geography. In some places, such as the Amazon, deforestation has accelerated in recent years, so it's particularly important to use current data when it's available.

8.8 Creating the climate-positive scenario

The name of the game is, as always, reducing cumulative emissions and atmospheric concentrations, so the sooner deforestation can be stopped and carbon removal ramped up, the better off the world will be. For your chosen geography, you'll need to begin by creating a concrete plan for stopping industrial-scale deforestation, which is by far the cheapest option from society's perspective.

Next, you'll need an inventory and potentials analysis for carbon removal that incorporates local resources and constraints. If one doesn't yet exist, you'll need to create one, ideally modeling it on well-constructed analyses for other geographies. Project Drawdown is a great place to start, and they have developed detailed plans for many different carbon removal options (see the further reading below for the link). For large countries, total potentials should be at or close to the gigatonne per year scale, but countries with smaller land areas will have smaller potentials.

8.9 Chapter conclusions

Carbon removal is the only way to reduce the damage from historical carbon dioxide emissions, which otherwise would warm the Earth for centuries to millennia. Reducing carbon dioxide concentrations in the atmosphere will take unprecedented effort, will sometimes be expensive, and will take many years to scale up. That means we need to get started immediately.

The most promising options in the near term are those that enhance the ability of land-based and marine ecosystems to remove carbon and store it. Many of these options have significant co-benefits, which makes their economics more attractive than more technological options for carbon removal, but many nature-based options also have substantial challenges with additionality, permanence, and leakage that may become worse in a warmer world.

Technological carbon removal options, such as direct-air capture, may be able to contribute significant carbon removal by mid-century, but are not likely to be relevant for climate-positive plans targeting the next decade. Some technological options may reach meaningful scale sooner, particularly hybrid carbon removal systems that combine nature-based carbon capture with engineered long-term carbon storage. We need to start developing and deploying all promising nature-based, technological, and hybrid options now, as we will likely need to scale a wide portfolio of carbon removal solutions to reach a climate-positive future.

Further reading

Canadell J G and Raupach M R 2008 Managing forests for climate change mitigation *Science* **320** 1456–7 https://doi.org/10.1126/science.1155458. Excellent high level summary of potential forest contributions to carbon removal.

Cook-Patton S C *et al* 2020 Mapping carbon accumulation potential from global natural forest regrowth *Nature* **585** 545–50 https://doi.org/10.1038/s41586-020-2686-x. Forest regrowth is one of the most cost-effective options for carbon removal, with lots of co-benefits.

Hausfather Z and Flegal J 2022 We need to draw down carbon—not just stop emitting it *MIT Technol. Rev.* 5 July https://www.technologyreview.com/2022/07/05/1055322/we-need-to-draw-down-carbon-not-just-stop-emitting-it/. This summary article addresses the mistaken belief, commonly held in the sustainability community, that carbon removal is an excuse for delaying rapid direct emissions reductions.

Hilmi N, Chami R, Sutherland M D, Hall-Spencer J M, Lebleu L, Benitez M B and Levin L A 2021 The role of blue carbon in climate change mitigation and carbon stock conservation *Frontiers Clim.* **3** https://doi.org/10.3389/fclim.2021.710546. Blue carbon is stored in ecosystems at the boundary of ocean and land, it is of increasing interest to researchers.

IPCC 2021 *Climate Change 2021: The Physical Science Basis. Contribution of Working Group I to the Sixth Assessment Report of the Intergovernmental Panel on Climate Change* V Masson-Delmotte *et al* (ed) (Cambridge: Cambridge University Press) https://www.ipcc.ch/report/sixth-assessment-report-working-group-i/, in particular chapter 5: Global carbon and other biogeochemical cycles and feedbacks. This chapter of the latest IPCC report on the science of climate change summarizes the key issues and available data in a comprehensive way.

Macreadie P I, Costa M D P, Atwood T B, Friess D A, Kelleway J J, Kennedy H, Lovelock C E, Serrano O and Duarte C M 2021 Blue carbon as a natural climate solution *Nat. Rev. Earth Environ.* **2** 826–39 https://doi.org/10.1038/s43017-021-00224-1. More on the potential for carbon reduction at the ocean/land interface.

Matthews H D, Zickfeld K, Dickau M, MacIsaac A J, Mathesius S, Nzotungicimpaye C-M and Luers A 2022 Temporary nature-based carbon removal can lower peak warming in a well-below 2 °C scenario *Commun. Earth Environ.* **3** 65 https://doi.org/10.1038/s43247-022-00391-z. This study explores the effect of early nature-based carbon removal that is subsequently reversed in the second half of the twenty-first century.

National Research Council 2015 *Climate Intervention: Carbon Dioxide Removal and Reliable Sequestration* (Washington, DC: The National Academies Press) https://doi.org/10.17226/18805. Important summary of CDR and sequestration, but a bit dated now.

National Academies of Sciences, Engineering, and Medicine 2019 *Negative Emissions Technologies and Reliable Sequestration: A Research Agenda* (Washington, DC: The National Academies Press) https://doi.org/10.17226/25259. A useful summary of the most important research topics on negative emissions.

National Academies of Sciences, Engineering, and Medicine 2021 *A Research Strategy for Ocean-based Carbon Dioxide Removal and Sequestration* (Washington, DC: The National Academies Press) https://doi.org/10.17226/26278. A useful source on current research topics for ocean-based CDR.

Project Drawdown analysis contains many options for carbon removal: https://www.drawdown.org/solutions/table-of-solutions.

Tokarska K B and Zickfeld K 2015 The effectiveness of net negative carbon dioxide emissions in reversing anthropogenic climate change *Environ. Res. Lett.* **10** 094013 https://doi.org/10.1088/1748-9326/10/9/094013. This study models the temperature and other effects of carbon removal over time, showing the complex interactions between carbon sources and sinks as well as heat flows from the ocean.

Wilcox J, Kolosz B and Freeman J (ed) 2021 *Carbon Dioxide Removal (CDR) Primer* https://cdrprimer.org/. A treasure drove of clear thinking and data on carbon removal, built on the expertise of dozens of researchers.

References

[1] Goldstein-Rose S 2020 *The 100% Solution: A Plan For Solving Climate Change* (New York: Melville House)

[2] Matthews H D and Caldeira K 2008 Stabilizing climate requires near-zero emissions *Geophys. Res. Lett.* **35** L04705

[3] Wilcox J, Kolosz B and Freeman J (ed) 2021 *Carbon Dioxide Removal (CDR) Primer* (Burnaby, BC: Hemlock Printers) https://cdrprimer.org/

[4] Grübler A *et al* 2018 A low energy demand scenario for meeting the 1.5 °C target and sustainable development goals without negative emission technologies *Nat. Energy* **3** 515–27

[5] Horton H 2022 Greenhouse gas removal 'not a silver bullet to achieve net zero': UK scientists say carbon capture is 'hard and expensive' and focus must be on reducing emissions *Guardian* 30 May https://theguardian.com/environment/2022/may/30/greenhouse-gas-removal-not-a-silver-bullet-to-achieve-net-zero?CMP=share_btn_tw

[6] Fuss S *et al* 2018 Negative emissions—part 2: costs, potentials and side effects *Environ. Res. Lett.* **13** 063002

[7] Minx J C *et al* 2018 Negative emissions—part 1: research landscape and synthesis *Environ. Res. Lett.* **13** 063001

[8] Nemet *et al* 2018 Negative emissions—part 3: innovation and upscaling *Environ. Res. Lett.* **13** 063003

[9] Arora V K and Montenegro A 2011 Small temperature benefits provided by realistic afforestation efforts *Nat. Geosci.* **4** 514–8

[10] Smith P *et al* 2016 Biophysical and economic limits to negative CO_2 emissions *Nat. Clim. Change* **6** 42–50

[11] Hausfather Z and Flegal J 2022 We need to draw down carbon—not just stop emitting it *MIT Technol. Rev.* 5 July https://technologyreview.com/2022/07/05/1055322/we-need-to-draw-down-carbon-not-just-stop-emitting-it/

[12] Harvey F 2022 We cannot adapt our way out of climate crisis, warns leading scientist *Guardian* 1 June https://theguardian.com/environment/2022/jun/01/we-cannot-adapt-our-way-out-of-climate-crisis-warns-leading-scientist?CMP=Share_iOSApp_Other

[13] Friedlingstein P *et al* 2022 Global carbon budget 2021 *Earth Syst. Sci. Data.* **14** 1917–2005

[14] Tokarska K B and Zickfeld K 2015 The effectiveness of net negative carbon dioxide emissions in reversing anthropogenic climate change *Environ. Res. Lett.* **10** 094013

[15] Solomon S, Plattner G-K, Knutti R and Friedlingstein P 2009 Irreversible climate change due to carbon dioxide emissions *Proc. Natl Acad. Sci.* **106** 1704–9

[16] Fuss S *et al* 2018 Negative emissions—part 2: costs, potentials and side effects *Environ. Res. Lett.* **13** 063002

[17] Arora V K and Montenegro A 2011 Small temperature benefits provided by realistic afforestation efforts *Nat. Geosci.* **4** 514–8

[18] Cook-Patton S C *et al* 2020 Mapping carbon accumulation potential from global natural forest regrowth *Nature* **585** 545–50

[19] Canadell J G and Raupach M R 2008 Managing forests for climate change mitigation *Science* **320** 1456–7

[20] Jackson R B *et al* 2008 Protecting climate with forests *Environ. Res. Lett.* **3** 044006

[21] Macreadie P I *et al* 2021 Blue carbon as a natural climate solution *Nat. Rev. Earth Environ.* **2** 826–39

[22] Hilmi N, Chami R, Sutherland M D, Hall-Spencer J M, Lebleu L, Benitez M B and Levin L A 2021 The role of blue carbon in climate change mitigation and carbon stock conservation *Frontiers Clim.* **3** 710546

[23] Macreadie P I *et al* 2019 The future of blue carbon science *Nat. Commun.* **10** 3998

[24] Yeo S 2021 How whales help cool the Earth *BBC Future* 20 January https://bbc.com/future/article/20210119-why-saving-whales-can-help-fight-climate-change

[25] Köberle A C 2019 The value of BECCS in IAMs: a review *Curr. Sust./Renew. Energy Rep.* **6** 107–15

[26] Keith D W, Holmes G, St Angelo D and Heidel K 2018 A process for capturing CO_2 from the atmosphere *Joule* **2** 1573–94

[27] APS 2011 *Direct Air Capture of CO_2 with Chemicals: A Technology Assessment for the APS Panel on Public Affairs* (Washington, DC: American Physical Society) https://aps.org/policy/reports/assessments/upload/dac2011.pdf

[28] Sanz-Pérez E S, Murdock C R, Didas S A and Jones C W 2016 Direct capture of CO_2 from ambient air *Chem. Rev* **116** 11840–76

[29] Buck H J 2020 Should carbon removal be treated as waste management? Lessons from the cultural history of waste *Interf. Focus* **10** 20200010

[30] Grant N, Hawkes A, Mittal S and Gambhir A 2021 The policy implications of an uncertain carbon dioxide removal potential *Joule* **5** 2593–605

[31] Madhu K, Pauliuk S, Dhathri S and Creutzig F 2021 Understanding environmental trade-offs and resource demand of direct air capture technologies through comparative life-cycle assessment *Nat. Energy* **6** 1035–44

[32] Grieg-Gran M 2006 *The Cost of Avoiding Deforestation: Report Prepared for the Stern Review of the Economics of Climate Change* (London: International Institute for Environment and Development) https://cbd.int/financial/finplanning/g-costdeforestion.pdf

[33] Busch J and Engelmann J 2017 Cost-effectiveness of reducing emissions from tropical deforestation, 2016–2050 *Environ. Res. Lett.* **13** 015001

[34] Popp A *et al* 2017 Land-use futures in the shared socio-economic pathways *Global Environ. Change* **42** 331–45

[35] Roe S *et al* 2019 Contribution of the land sector to a 1.5 °C world *Nat. Clim. Change* **9** 817–28

Chapter 9

Align incentives

Give me a lever long enough and a fulcrum on which to place it, and I shall move the world.

—Archimedes

Chapter overview

- Aligning incentives enables rapid societal change.
- By incentives we mean any action, policy, program, or institutional change that helps achieve climate positivity quickly (it's more than just 'getting prices right').
- We need to make it easy for people and institutions to make climate-positive choices.
- We need to change the game so that the economic, legal, and social systems all support rapid movement toward climate stability.
- We need to fix the rules so that old and outmoded practices don't impede the transition to zero emissions.
- Many conventional analyses erroneously assume that incentives won't change, or won't change enough, to reach zero emissions quickly. Instead, we can *choose* to change incentives and make a climate-positive world a reality, but it's up to us. Those incentives won't change on their own.
- The next chapter, 'Mobilizing money', zooms in on money-related incentives like pricing pollution, subsidizing the transition, removing fossil money from politics, and moving money from planned fossil investments into climate-positive investments.

9.1 Introduction

The global economy is now powered mostly by fossil fuels, and that's been true for more than a century. The fossil fuel industry is the most powerful in human history,

doi:10.1088/978-0-7503-4032-8ch9

with more than $5 trillion US in revenues in 2022 (see appendix F for details) and it has shaped countless tax, regulatory, and other policies in its favor at all levels of government. We must level the playing field by aligning broader societal incentives with the goal of creating a climate-positive world. We also need to combat fossil-funded corruption everywhere it exists.

Griffith [1] correctly points to outmoded regulations as a major impediment to change, but the problem is even bigger than that. It also means more than just putting a price on carbon dioxide emissions. Making carbon polluters pay would be a salutary step, but it's not the end of the story, as we discuss in more detail in our next chapter [2].

We need to make it easy for people and institutions to move to zero emissions quickly, and we use the broad term 'incentives' to describe any action, policy, program, or institutional change that helps achieve that goal. There are many effective ways to change incentives and promote the transition to net zero emissions, and we should use them all.

Aligning incentives can mean everything from maximizing convenience of climate solutions to taxing pollution, imposing mandates, reorganizing subsidies, imposing legal liability on polluters, changing property rights, exposing and prosecuting corruption, enhancing social pressure to shift from polluting choices, and fixing outmoded rules, practices, and regulations. We'll likely to do all of this and more to properly align incentives with the goal of creating a climate-positive planet.

This chapter reviews these and other important ways to re-design non-financial, focusing on making it easy, changing the game, and fixing the rules. The next chapter, 'Mobilizing money', focuses on those incentives that are money related.

9.2 Making it easy

Most folks don't have time in their busy lives to worry about climate on a regular basis, so we need to make it easy for them to make the right choices. Time and attention have value and good solutions embrace that reality, otherwise adoption will be slower than we want.

It's hard to value time in our personal lives but we all know it is limited, in the near term by life's complexity and ultimately by our finite lifespan. That's true for businesses too, whose biggest cost is usually payroll (people's time as an expense). Readily available climate solutions that save time AND reduce pollution can practically sell themselves.

9.2.1 Make climate-friendly options the most convenient

Governments and businesses should promote climate-friendly options, make them readily available, and make them more convenient than the fossil fueled options they displace. The charging network for all electric vehicles (EVs) needs to be as widely available and as reliable as the current Tesla supercharger network, for example. In the early stages of adoption, EVs should be given preferential registration fees and access to high occupancy vehicle lanes, with these preferences phasing out once target adoption levels are achieved.

9.2.2 Make climate action normal good behavior

Public service campaigns on littering, seatbelts, drunk driving, and distracted driving have all driven changes in public perceptions of what 'good behavior' is, and we need similar efforts to make responsible climate choices 'the right thing to do' [3]. Governments can also publicize successful emission reductions efforts. Individuals can state their social expectations about climate action and share their successes. Companies and investors can reinforce those norms by rewarding employees who help reach climate goals and publicizing climate leadership by their employees. Non-profits can help individuals, companies, and communities accomplish these goals.

9.2.3 Make the invisible visible

New technology and analysis tools can help highlight how to 'do the right thing', as well as which climate choices matter most. For example, recent analysis of total life-cycle emissions of oil and gas has demonstrated significant variation in emissions depending on where these fuels are found, how they are extracted, how they are processed, and how they are used [4–6]. Remote sensing of methane emissions from fossil fuel production has enabled more accurate estimates of these emissions [7, 8], and recent developments in information technology portend much wider use of mobile sensors and communication equipment for tracking real time emissions [9].

Governments can mandate standardized disclosure of full scope 1–3 greenhouse gas emissions for companies, as well as for individual products, services, and investments, much like most governments now require food labels to include calorie counts and nutrient contents. Standardization can prevent 'greenwashing' and ensure that climate action results in real emissions reductions. Companies can participate in creating those disclosure requirements and disseminating those disclosures for their own operations, although pollution disclosures should always be verified by governments or independent third-party auditors.

Combining life-cycle climate footprint data with innovative apps and other information technology tools can help individuals, businesses and other organizations make more climate-friendly choices, as well as track the cumulative climate impacts of their own climate choices. The Oroeco app (of which Ian was a co-creator) and the UC Berkeley CoolClimate Network tools pioneered many techniques for 'gamifying' personal and business climate footprints more than a decade ago, and now many similar tools help improve climate choices by combining life-cycle climate data with real and virtual rewards.

Banks and credit card companies have also started to offer automated assessments of individual climate footprints based on financial transaction data, and some also offer the option to automatically offset the climate impacts of purchases. In addition, many individual retailers and consumer portals are adding climate data to point-of-sale decisions, from restaurants adding climate footprints to food items to Google adding CO_2e calculations to airline flights. As regulations such as France's Climate and Resilience Law increasingly make climate footprint data both standardized and required, innovative technologies will likely do even more to nudge choices towards climate positivity.

9.2.4 Tighten regulations, mandate solutions, ban polluting technologies

Most nations have regulations on the energy efficiency and emissions from energy-using equipment while states and localities have regulations affecting local building design and construction. These regulations are helpful in removing the least efficient and most polluting products and structures from the market. One of their benefits is that they eliminate transaction and hassle costs from that choice because consumers can no longer purchase the less efficient or more polluting devices. Another is that such regulations are always benefit–cost tested, and their savings are virtually always many times greater than their costs to society[1].

Each country creates its own processes to update these standards, and all need to re-examine them to speed up these updates. They are an important part of making it easy.

More recently, countries, states, and localities have been passing laws to phase out sales of inefficient incandescent light bulbs and fossil fuel-fired equipment. Other laws mandate the purchase of electrified equipment, most importantly for vehicles and natural gas in homes and businesses.

Many cities are now phasing in bans on natural gas connections for buildings while many countries are phasing in bans on purchasing internal combustion vehicles. Companies and communities can also move faster than governments by enacting their own policies that mandate climate solutions and ban polluting technologies, as many are now doing with internal mandates for 100% renewable energy and 100% 'fossil free' investment portfolio targets. Regulations and private policies move the market towards lower emissions more rapidly than would happen otherwise.

9.3 Changing the game

System change is the name of the game. The choices of individuals (particularly wealthy ones with large carbon footprints) can be helpful for cutting emissions, but it is even more important to change the system in which people make those choices, because that enables more rapid changes than would the actions of individuals on their own.

9.3.1 Make a stable climate a core right and principle

The continued orderly development of human civilization requires a stable climate, and that means getting to zero emissions as quickly as possible. Stating this assumption clearly should be the first step towards 'changing the game' and can help alter how policy makers, investors, companies, and consumers discuss this issue.

The United Nations has already voted to define a 'clean, healthy, and sustainable' environment as a human right [10, 11], which implies a stable climate. Constitutions

---

[1] Here is a recent example: https://www.reuters.com/business/autos-transportation/us-epa-proposing-rules-cut-emissions-heavy-trucks-2022-03-07/.

of national governments should do the same. Mission statements for companies and non-governmental organizations should state their goal of getting to zero emissions as soon as possible. Trade agreements should prioritize rapid emissions reductions and remove impediments to emissions reductions embedded in such agreements [12]. Countries should sign on to the Fossil Fuel Non-Proliferation Treaty and enshrine the goal of keeping fossil fuels in the ground [13–16].

In addition, anyone can make climate action a top priority without waiting for national and international policy makers. Companies and other organizations can make achieving climate positivity a core part of their operating missions, and individual households and communities can integrate climate action into every decision they make.

9.3.2 Restructure property rights

Government defines property rights [17] and can choose how these will be structured and enforced [18], which can strongly affect the incentives individuals and institutions have to protect the commons and reduce emissions [19]. Many of these choices are in some ways arbitrary, which means we can make choices that favor long-term sustainability, and there's nothing wrong with that.

For example, the property rights associated with a battery for many years transferred fully from the manufacturer to the purchaser of that battery. In recent times, as society has become more aware of the toxic effects of chemicals in groundwater, some countries (notably those in Europe) have moved towards a modification of property rights, under the rubric of 'extended producer responsibility'. The manufacturer is still responsible for the costs of safely recycling that battery, even after it is sold to the customer.

This change is not much different from property rights for land in the US, where rights for mining, water, wind generation, and air (in the sense of the right to fly above land) can be split off and sold separately, depending on the laws for the state in which the property resides. Redefining property rights is a powerful but often neglected tool for changing incentives to achieve a climate-positive society (and it is yet another case where economic models, which almost always assume fixed consumer preferences and property rights, underestimate the possible speed and scope of change towards zero emissions).

9.3.3 Harness learning effects, economies of scale, and spillovers

Most climate-positive products start out as niche products that have low production volumes and high markups. Minimum standards, tax credits, government purchasing, and corporate purchasing can help move niche products to become more widely available, reducing their prices to everyone (at the same time as economies of scale, learning effects, and network externalities reduce costs still further). We've seen such 'spillover effects' when first Germany and then China aggressively promoted purchasing and development of solar photovoltaics [20], and such policies can help move society more rapidly towards zero emissions.

9.3.4 Harness social incentives

Shaming is counterproductive [21], but arguing in a principled and informed way that wealthy people should change their behavior is a terrific idea. The wealthy have a disproportionate effect on emissions and also have the money and influence to effect big changes. The wealthy are often thought leaders and can set an example for others to follow, validating the idea that climate is a problem worth solving.

Social influences can have dramatic effects on people's behavior, but whether they result in lasting change is still a subject of debate and varies a great deal from one person to the next [22]. One powerful tool is comparing individual energy use and emissions with those of peers and neighbors, either by demographics[2] or location [23]. Technology tools that visualize and gamify climate impacts can make climate action socially rewarding, and thoughtfully designed climate tracking and gamification tools can foster virtuous cycles of improvement, particularly when social elements are combined with leaderboards and reward systems that encourage collaboration and competition to reach climate targets.

9.4 Fixing the rules

The rules of most modern economies were set during an era of fossil fuel abundance, but they need to be rewritten to reflect the need to get to zero emissions as soon as possible.

9.4.1 Rationalize regulations

Fixing existing regulations should be a high priority to facilitate the transition to zero emissions. Griffith [24] describes the shocking statistic that rooftop solar costs *one-third* as much in Australia as it does in the US (others [24] say it is more than a factor of two). As far as we know, the laws of physics are the same in both Sydney and San Francisco, and solar panels are sold in a global market, so the difference is in the regulatory regime that makes installation of US solar systems much more expensive [24]. For example, permitting is a laborious process that can take weeks or months in the US, but if it were standardized and streamlined, need not take more than a day. Building and electrical codes also need to be modernized to reflect changing technologies, and tariffs on solar panels and other clean energy technologies should be eliminated.

Simply mandating solar panels for new homes, as California has done, is another way to get rid of these regulatory costs and make residential solar really cheap [25]. The inspector must come through the building site at regular intervals anyway, the walls are open, the electrician is on site, and the roof can be constructed with solar panels in mind from the very beginning. The mandate also expands the market for solar installations and pushes costs down through learning effects. It is yet another example of how mandates remove transaction and hassle costs from the equation and drive societal costs down.

---

[2] https://coolclimate.org/calculator

The design of the US Corporate Average Fuel Economy Standards (CAFE) originally had one fleet average efficiency standard for cars and a lower one for light trucks [26]. This policy created the perverse incentive for the auto manufacturers to redefine more and more vehicles as bigger and heavier light trucks, thus partly circumventing the intent of the standards of increasing efficiency over time. While this policy has improved over time, such perverse incentives are pervasive.

In Europe, the Energy Charter Treaty (ECT) allows energy companies to sue governments for threatening their profits. The treaty was originally established to compensate companies for property appropriated by treaty signatories, but now it is being weaponized by the fossil fuel industry to fight climate action [27–29]. Unfortunately, trade and investment agreements like this one are binding on countries, but emissions reduction agreements (such as the Paris Agreement) technically are not. This structural asymmetry is one that needs to be fixed but the more urgent need is to prevent the ECT from being used to hamper European climate action.

Lead in aviation gasoline (for small planes) is still allowed in the US, even though lead's terrible health effects have long been known [30] and lead has been phased out of all other fuels. Fortunately, the US Environmental Protection Agency plans on releasing an Endangerment Finding on lead in aviation gasoline in 2022, which is the first step towards banning it, and that will create a strong incentive to electrify general aviation[3].

As Griffith points out, fixing inefficient regulations such as these should be a high priority when kickstarting the transition to zero emissions. It is not just about writing new laws, it is also about removing obsolete ones.

In addition, we need to stop the fossil fuel industry and the 'not in my back yard' (NIMBY) movement from gumming up the regulatory process, which is one of their preferred strategies for what the futurist Alex Steffen correctly calls 'predatory delay'. One prominent recent example in the US relates to the fossil fuel industry influencing building code revisions to inhibit electrification [31], but there no doubt are many others.

The US in particular needs to build things faster and prevent people and industries from slowing things down for no good reason other than their own perceived self-interest or whim, and poorly designed bureaucratic systems in the US make many types of climate solution infrastructure substantially more expensive relative to Europe and Asia [32]. Decarbonizing the electricity grid and getting to zero emissions will require us to build a lot of new equipment, including a vastly expanded electricity transmission network. We need to change the rules so we can get that done quickly and cheaply.

9.4.2 Change laws, rules, and regulations

The legal system has been rigged by the fossil fuel industry for decades in both subtle and not-so-subtle ways. For example, bidding in a federal auction for oil and gas

---

[3] https://www.epa.gov/regulations-emissions-vehicles-and-engines/regulations-lead-emissions-aircraft

drilling rights is illegal if you want to set that land aside and not develop it. The activist Tim DeChristopher spent two years in prison after bidding on 22,500 acres in a Bureau of Land Management auction in 2008 with no intent to produce fossil fuels [33]. But why shouldn't we allow people and institutions to bid on fossil fuel leases if they want to keep those fuels in the ground? Isn't that just private actors expressing their preferences using their wealth as they see fit in an ostensibly free market?

The bonding requirements for oil and gas wells are far too low to pay for clean-up after the wells stop producing [34–37]. In recent years, many fossil fuel companies declared bankruptcy, leaving orphaned wells for others to clean up, costing US taxpayers billions [36, 38]. The industry captured profits from producing oil and gas from those wells, then skipped town and forced others to pay to clean them up. It's long past time for that loophole to be fixed.

Other structural issues relate to the incentives of incumbent actors (such as fossil fuel companies and the utility industry) to resist encroachment on their core businesses. For example, fossil fuel industries block efforts to ban natural gas in buildings and electric utilities frequently fight against rooftop solar and energy storage. One fix to this overarching issue relates to incentives and market structure (making the zero emissions path more profitable), another to removing money from politics and ending the revolving door of influence (which we discuss more in our next chapter). Stronger laws prohibiting policy makers from working for industry (and vice versa) would also be helpful here.

While not every chemical release is related to climate change, many chemicals are derived from fossil fuels. The current lax regulation of chemicals (and plastics in particular) reflects broader regulatory dysfunction that also affects climate [39] and the profitability of fossil fuel refineries. Hormone-disrupting chemicals have become widespread and are beginning to affect the fertility and development of organisms in a measurable way [40, 41]. Tiny microplastic particles can circle the globe in our oceans and atmosphere. Plastics are now found throughout the food chain in every ecosystem, even in Antarctica, wreaking havoc with wildlife and in some cases affecting human health [42].

Most countries allow companies to use whatever chemical they please, and only consider regulations later if obvious problems emerge [42], often relying on imperfect analogues when these chemicals are analyzed at all [43]. We need to make companies demonstrate safety *before* releasing a chemical or a plastic into the environment, just as drug manufacturers must do for pharmaceuticals [44, 45]. Such requirements are particularly important for rapidly growing activities with currently unknown impacts such as rocket launches [46], but many more mundane products have broad effects on the environment. We will need legally binding international agreements to tackle this problem, such as the recent agreement on plastic pollution [47].

9.4.3 Make polluters liable for climate damages

The number and scope of legal challenges attempting to hold fossil fuel companies accountable for creating climate change in the first place have increased substantially in the past decade [48–51]. The fossil fuel industry has known since the 1950s that climate change driven by fossil fuel combustion would create major risks for society and that their products were causing it [52, 53], but they downplayed the risks, funded fake experts, and worked in myriad ways to delay action to reduce emissions [54–58]. This area of law is evolving rapidly, but it's fair to say that fossil fuel companies will face increasing legal pressure and may face liability for lying about climate change for decades, as well as some degree of financial or even criminal liability for climate-related disasters, such as heat waves, droughts, wildfires, floods, and other extreme events [59]. COP 27's focus on creating a "Loss and Damage Fund" is yet another expression of increasing interest in this issue. Governments should hasten this process by clearly designating legal liabilities for large climate polluters, while empowering regulators, non-profits, and citizen groups to win lawsuits for climate damages.

9.4.4 Make companies liable for failure to disclose and prepare for climate risks

Recently, legal actions against companies for failing to prepare for climate change have created a real business risk for energy companies [60]. Most investors now understand that climate change is real and dangerous for many types of businesses, and financial regulators are increasingly requiring companies to report material climate risks in standardized reporting to investors. Companies that fail to report and prepare for climate risks will increasingly find themselves liable to regulators and investors for climate-related damages. The more that market regulators require standardized climate risk and preparation disclosures, the greater the incentives for forward-thinking companies to better respond to climate risks to steer clear of legal liabilities.

9.5 Building your scenarios

The most important part of the scenario-building process is generating plausible and evocative stories and supporting them with data, analysis, and exhaustive documentation [61, 62]. Re-designing societal incentives and detailing the required changes will teach you important lessons about the constraints and opportunities associated with creating a climate-positive world.

9.5.1 Creating or adopting a business-as-usual scenario

Most business-as-usual (BAU) scenarios assume little change in the next few decades, but there still may be some obvious drivers to tweak. When a fuel looks likely to be phased out anyway because of technological and regulatory trends already in motion (such as lead in aviation gas, for example) it makes sense to incorporate that change into your current trends continued case. A policy that has been implemented but is in its early stages should also be included in your BAU scenario as an adjustment to a scenario that doesn't yet include it.

9.5.2 Creating the climate-positive scenario

Changing incentives is a powerful lever for speeding up climate action. We explored many possible changes to existing incentives above, but only some will apply to your chosen geography. Your base-year emissions inventory and the associated projection into the future is the first place to start in prioritizing policy action, because it allows you to focus first on what's important.

The political environment of the state or country you are analyzing also matters. If a policy has already been tried but been reversed in that geography, other options might be more appealing. The key is to create and argue for a plausible mix of changes in existing incentives that have individually in other places or at other times been powerful enough to shift emissions significantly, combining them in ways that are internally consistent and likely to yield the desired results.

In our class we don't require students to assess all the costs of the policy and programs they propose, because time is short, but ideally a national plan would attempt such cost accounting. Koomey [62] contains a plan with an exemplary level of detail for the US building sector, so that's a good place to start (the data are now old but it's a great example of how to create detailed and well-documented intervention scenarios).

The best you can do is argue from historical program experience and document your analysis exhaustively. Then 'iterative risk management' kicks in and the plan would be adjusted over time to reflect how well each program or policy is working, doing more of what works and less of what doesn't.

9.6 Chapter conclusions

Changing incentives can enable rapid shifts towards climate positivity. Conversely, not aligning incentives with our climate goals will ensure that we won't reach zero emissions in time to avoid the worse effects of climate change.

Many models and policy studies convey the mistaken impression that we can't solve the climate problem quickly enough, but that erroneous result often follows from the assumption that we won't change existing incentives. If we modify incentives in the right way and to the right degree, society can and will change quickly, much like the world responded to (and recovered from) World War II.

Further reading

Axelrod R S and VanDeveer S D (ed) 2019 *The Global Environment: Institutions, Law, and Policy* (Washington, DC: CQ Press). This compilation gives a comprehensive view on global environmental issues from a wide range of experts.

Cullenward D and Victor D G 2020 *Making Climate Policy Work* (New York: Polity Press). A critical look at market-based instruments for solving the climate problem.

Edwards L and Cox S 2020 Cap and adapt: failsafe policy for the climate emergency *Solutions* 1 September https://thesolutionsjournal.com/2020/09/01/cap-and-adapt-failsafe-policy-for-the-climate-emergency/. A policy framework for enabling the transition to zero emissions.

Harvey H, Orvis R and Rissman J 2018 *Designing Climate Solutions: A Policy Guide for Low-Carbon Energy* (Washington, DC: Island). Useful comprehensive summary of policy solutions for the energy sector.

Moore F C, Lacasse K, Mach K J, Shin Y A, Gross L J and Beckage B 2022 Determinants of emissions pathways in the coupled climate–social system *Nature* **603** 103–11 https://doi.org/10.1038/s41586-022-04423-8. A sophisticated recent must-read analysis of key drivers of change related to emissions.

Vig N J, Kraft M E and Rabe B G (ed) 2021 *Environmental Policy: New Directions for the Twenty-First Century* (Washington, DC: CQ Press). Now in its eleventh edition, this classic book summarizes the history and current status of US environmental policy.

References

[1] Cullenward D and Victor D G 2020 *Making Climate Policy Work* (New York: Polity)

[2] Goldberg M H, Gustafson A and van der Linden S 2020 Leveraging social science to generate lasting engagement with climate change solutions *One Earth* **3** 314–24

[3] Koomey J, Gordon D, Brandt A and Bergerson J 2016 *Getting Smart about Oil in a Warming World* (Washington, DC: Carnegie Endowment for International Peace) http://carnegieendowment.org/2016/10/04/getting-smart-about-oil-in-warming-world-pub-64784

[4] Gordon D 2022 *No Standard Oil: Managing Abundant Petroleum in a Warming World* (New York: Oxford University Press) https://nostandardoil.com

[5] Gordon D, Koomey J, Brandt A and Bergerson J 2022 *Know Your Oil and Gas: Generating Climate Intelligence to Cut Petroleum Industry Emissions* (Boulder, CO: Rocky Mountain Institute) https://rmi.org/insight/kyog/

[6] Schneising O, Burrows J P, Dickerson R R, Buchwitz M, Reuter M and Bovensmann H 2014 Remote sensing of fugitive methane emissions from oil and gas production in North American tight geologic formations *Earth's Future* **2** 548–58

[7] Kuze A, Kikuchi N, Kataoka F, Suto H, Shiomi K and Kondo Y 2020 Detection of methane emission from a local source using GOSAT target observations *Remote Sensing* **12** 267

[8] Koomey J G, Matthews H S and Williams E 2013 Smart everything: will intelligent systems reduce resource use? *Annu. Rev. Environ. Resour.* **38** 311–43

[9] UN 2022 *The Human Right to a Clean, Healthy and Sustainable Environment* (New York: United Nations) https://digitallibrary.un.org/record/3982508?ln=en

[10] UNEP 2022 In historic move, UN declares healthy environment a human right *United Nations Environment Programme* 28 July https://unep.org/news-and-stories/story/historic-move-un-declares-healthy-environment-human-right

[11] Klein N 2015 *This Changes Everything: Capitalism Versus The Climate* (New York: Simon and Schuster)

[12] Green F and Denniss R 2018 Cutting with both arms of the scissors: the economic and political case for restrictive supply-side climate policies *Clim. Change* **150** 73–87

[13] Asheim G B, Fæhn T, Nyborg K, Greaker M, Hagem C, Harstad B, Hoel M O, Lund D and Rosendahl K E 2019 The case for a supply-side climate treaty *Science* **365** 325

[14] Newell P and Simms A 2019 Towards a fossil fuel non-proliferation treaty *Clim. Policy* **20** 1043–54

[15] Erickson P, Lazarus M and Piggot G 2018 Limiting fossil fuel production as the next big step in climate policy *Nat. Clim. Change* **8** 1037–43

[16] de Soto H 2003 *The Mystery of Capital: Why Capitalism Triumphs in the West and Fails Everywhere Else* (New York: Basic)

[17] McMillan J 2003 *Reinventing the Bazaar: A Natural History of Markets* (New York: Norton)

[18] Koomey J G 2012 *Cold Cash, Cool Climate: Science-Based Advice for Ecological Entrepreneurs* (El Dorado Hills, CA: Analytics)

[19] Nemet G 2019 *How Solar Became Cheap: A Model for Low-Carbon Innovation* (New York: Routledge) https://howsolargotcheap.com

[20] Sovacool B K *et al* 2017 New frontiers and conceptual frameworks for energy justice *Energy Policy* **105** 677–91

[21] Goldberg M H, Gustafson A and van der Linden S 2020 Leveraging social science to generate lasting engagement with climate change solutions *One Earth* **3** 314–24

[22] Jones C and Kammen D M 2014 Spatial distribution of US household carbon footprints reveals suburbanization undermines greenhouse gas benefits of urban population density *Environ. Sci. Technol.* **48** 895–902

[23] Griffith S 2021 *Electrify: An Optimist's Playbook for Our Clean Energy Future* (Cambridge, MA: MIT Press)

[24] Birch A 2018 *How to Halve the Cost of Residential Solar in the US: Former Sungevity CEO Andrew Birch Breaks Down Why American Consumers Are Being Charged Two Times More For Solar Than Their Peers Overseas* (Boston, MA: Greentech Media) https://greentechmedia.com/articles/read/how-to-halve-the-cost-of-residential-solar-in-the-us

[25] Weaver J F 2018 California residential solar power headed toward $1/W and 2.5¢/kWh *PV Mag.* 14 May https://pv-magazine-usa.com/2018/05/14/california-residential-solar-power-headed-to-1-12-w-2-5¢-kwh/

[26] Greene D L, Greenwald J M and Ciez R E 2020 US fuel economy and greenhouse gas standards: what have they achieved and what have we learned? *Energy Policy* **146** 111783

[27] Rankin J 2021 Secretive court system poses threat to Paris climate deal, says whistleblower: treaty allows energy corporations to sue governments for billions over policies that could hurt their profits *The Guardian* 3 November https://theguardian.com/environment/2021/nov/03/secretive-court-system-poses-threat-to-climate-deal-says-whistleblower

[28] Ambrose J 2019 Energy treaty 'risks undermining EU's green new deal': Calls for ECT to be scrapped to stop fossil fuel firms using it to take governments to court *The Guardian* 9 December https://www.theguardian.com/business/2019/dec/09/energy-treaty-risks-undermining-eus-green-new-deal

[29] Provost C and Kennard M 2015 The obscure legal system that lets corporations sue countries *The Guardian* 10 June https://theguardian.com/business/2015/jun/10/obscure-legal-system-lets-corportations-sue-states-ttip-icsid

[30] Wolfe P J, Giang A, Ashok A, Selin N E and Barrett S R H 2016 Costs of IQ loss from leaded aviation gasoline emissions *Environ. Sci. Technol.* **50** 9026–33

[31] Kaufman A C 2022 A battle over building codes may be the most important climate fight you've never heard of *Huffington Post* 3 March https://huffpost.com/entry/building-codes-climate_n_621e4b69e4b0afc668c68e59

[32] Klein E 2022 What America needs is a liberalism that builds *The New York Times* 29 May https://nytimes.com/2022/05/29/opinion/biden-liberalism-infrastructure-building.html

[33] Goodell J 2011 A Rosa Parks moment: climate activist Tim DeChristopher sentenced to prison *Rolling Stone* 27 July https://rollingstone.com/politics/politics-news/a-rosa-parks-moment-climate-activist-tim-dechristopher-sentenced-to-prison-237801/

[34] Ho J S, Shih J-S, Muehlenbachs L A, Munnings C and Krupnick A J 2018 Managing environmental liability: an evaluation of bonding requirements for oil and gas wells in the United States *Environ. Sci. Technol.* **52** 3908–16

[35] Raimi D, Nerurkar N and Bordoff J 2020 *Green Stimulus for Oil and Gas Workers: Considering a Major Federal Effort to Plug Orphaned and Abandoned Wells* (New York: Columbia University Center on Global Energy Policy and Resources for the Future) https://energypolicy.columbia.edu/research/report/green-stimulus-oil-and-gas-workers-considering-major-federal-effort-plug-orphaned-and-abandoned

[36] Jenkins D 2022 Orphaned oil wells costing taxpayers a bundle *The Hill* 2 July https://thehill.com/opinion/energy-environment/593183-orphaned-oil-wells-costing-taxpayers-a-bundle/amp/

[37] Carbon Tracker 2022 Event horizon: a case study of holdback and the point of no return for decommissioning upstream oil and gas 'assets' *Carbon Tracker* 29 July https://carbontracker.org/reports/event-horizon-a-case-study-of-holdback-analysis/

[38] Groom N and Mason J 2022 US unveils $1.15 bln for abandoned oil and gas well clean-up *Reuters* 31 January https://www.reuters.com/world/us/us-unveils-115-bln-abandoned-oil-gas-well-clean-up-2022-01-31/

[39] Kysar D A 2010 *Regulating from Nowhere: Environmental Law and the Search for Objectivity* (New Haven, CT: Yale University Press)

[40] Colborn T, Dumanoski D and Myers J P 1997 *Our Stolen Future: Are We Threatening Our Fertility, Intelligence, and Survival? A Scientific Detective Story* (New York: Plume)

[41] Persson L *et al* 2022 Outside the safe operating space of the planetary boundary for novel entities *Environ. Sci. Technol.* **56** 1510–21

[42] Freinkel S 2011 *Plastic: A Toxic Love Story* (New York: Houghton Mifflin Harcourt)

[43] Kelly J M *et al* 2021 Global cancer risk from unregulated polycyclic aromatic hydrocarbons *GeoHealth* **5** e2021GH000401

[44] Carusi A, Wittwehr C and Whelan M 2022 *Addressing Evidence Needs in Chemicals Policy and Regulation* (Luxembourg: Publications Office of the European Union) https://publications.jrc.ec.europa.eu/repository/handle/JRC126724

[45] Zimmerman J B, Anastas P T, Erythropel H C and Leitner W 2020 Designing for a green chemistry future *Science* **367** 397–400

[46] Piesing M 2022 The pollution caused by rocket launches *BBC Future* 15 July https://bbc.com/future/article/20220713-how-to-make-rocket-launches-less-polluting

[47] UNEP 2022 *What You Need to Know about the Plastic Pollution Resolution* (Nairobi: United Nations Environment Program) https://unep.org/news-and-stories/story/what-you-need-know-about-plastic-pollution-resolution

[48] Burger M and Wentz J 2018 Holding fossil fuel companies accountable for their contribution to climate change: where does the law stand? *Bull. Atomic Sci.* **74** 397–403

[49] Setzer J and Higham C 2021 *Global Trends in Climate Change Litigation: 2021 Snapshot* (London: Grantham Research Institute on Climate Change and the Environment and Centre for Climate Change Economics and Policy, London School of Economics and Political Science) https://lse.ac.uk/granthaminstitute/publication/global-trends-in-climate-litigation-2021-snapshot/

[50] Bhargava A, Franta B, Toral K M, Tandon A, Benjamin L and Setzer J 2022 *Policy Briefing: Climate-Washing Litigation–Legal Liability for Misleading Climate Communications* (London: The Climate Social Science Network) https://lse.ac.uk/granthaminstitute/publication/climate-washing-litigation-legal-liability-for-misleading-climate-communications/

[51] Clark L 2022 Big Oil tees up next Supreme Court climate showdown *E&E News ClimateWire* 9 June https://eenews.net/articles/big-oil-tees-up-next-supreme-court-climate-showdown/

[52] Exxon 1982 CO$_2$ greenhouse effect *Technical Review* Exxon Corporation 1 April https://insideclimatenews.org/wp-content/uploads/2015/09/1982-Exxon-Primer-on-CO2-Greenhouse-Effect.pdf

[53] Franta B 2018 Early oil industry knowledge of CO$_2$ and global warming *Nat. Clim. Change* **8** 1024–5

[54] Franta B 2021 Weaponizing economics: big oil, economic consultants, and climate policy delay *Environ. Politics* **31** 555–75

[55] Franta B 2021 Early oil industry disinformation on global warming *Environ. Politics* **30** 663–8

[56] Oreskes N and Conway E M 2010 *Merchants of Doubt: How a Handful of Scientists Obscured the Truth on Issues from Tobacco Smoke to Global Warming* (New York: Bloomsbury)

[57] Oreskes N 2015 The fact of uncertainty, the uncertainty of facts and the cultural resonance of doubt *Phil. Trans. R. Soc.* A **373** 20140455

[58] Bonneuil C, Choquet P-L and Franta B 2021 Early warnings and emerging accountability: total's responses to global warming, 1971–2021 *Global Environ. Change* **71** 102386

[59] Eubanks S Y 2022 I led the US lawsuit against big tobacco for its harmful lies. Big oil is next *The Guardian* 5 July https://theguardian.com/commentisfree/2022/jul/05/us-lawsuit-big-tobacco-big-oil-fossil-fuel-companies

[60] Gundlach J 2020 Climate risks are becoming legal liabilities for the energy sector *Nat. Energy* **5** 94–7

[61] Ghanadan R and Koomey J 2005 Using energy scenarios to explore alternative energy pathways in California *Energy Policy* **33** 1117–42

[62] Schwartz P 1996 *The Art of the Long View: Planning for the Future in an Uncertain World* (New York: Doubleday)

Chapter 10

Mobilize money

Just as we argued in the 1980s that those who conducted business with apartheid South Africa were aiding and abetting an immoral system, we can say that nobody should profit from the rising temperatures, seas, and human suffering caused by the burning of fossil fuels.

—Desmond Tutu[1]

Chapter overview

- The world now spends more than $20 trillion US every year on fossil fuels, related infrastructure, and associated environmental costs.
- Re-allocating these costs and expenditures towards climate mitigation infrastructure will be more than enough to solve the problem.
- Re-allocating these capital flows will save money over time in direct cost terms, as well as reduce and eventually eliminate at least $5 trillion US per year in human health, environmental damage, and climate risks.
- Policies (such as pricing pollution, subsidizing solutions, mandates, and bans) can ensure that climate-positive technologies are convenient and always cheaper to buy and maintain than the polluting technologies they replace.
- Scaling up carbon removal technologies to return our climate to a stable state will require new funding mechanisms (perhaps even a new global currency) but financial and policy innovations point to some possible paths forward.
- The world will also need to allocate much more funding to climate adaptation and resilience, as well as reparations for climate damages. Such funding should be allocated based on financial need and social equity, and the largest climate polluters should be the most responsible for paying to alleviate climate damages.

[1] https://www.theguardian.com/commentisfree/2014/sep/21/desmond-tutu-climate-change-is-the-global-enemy

- Ending the corrupting role of climate-polluter money in politics is essential, as is rooting out myriad sources of perverse financial incentives and market distortions.
- The political challenges of re-allocating capital are not trivial, but the end goal is a climate-positive world that is simply better than the one it replaces, with net negative emissions, lower total costs, less economic volatility, and many other co-benefits for society that will improve the quality of life for nearly everyone.

10.1 Introduction

Solving climate change ultimately requires shifting trillions of dollars every year into climate-positive alternatives. Climate spending is mostly investing, since the upfront capital costs of renewable energy, electrification, and efficiency technologies more than pay for themselves over time with savings from substantially lower operating costs.

Most climate solution spending also leads to substantial co-benefits, from dramatically improved air and water quality to avoided damages from direct and indirect climate risks. When we properly tally the costs and benefits from society's perspective, as we should, it becomes clear that solving climate change will lead to net financial benefits for most people around the world. This chapter gives context related to financial incentives, presents the benefits and pitfalls of pricing pollution, discusses incentives to promote technology adoption and innovation, and explains the different ways capital can and should be re-allocated to facilitate the transition to a climate-positive world.

10.2 Context

This section gives overall context for the discussion in this chapter, quantifying rough magnitudes of monetary flows and presenting key issues to consider in analyzing how to use financial incentives of all sorts to create a climate-positive world.

10.2.1 How much money needs to shift into climate solutions?

According to the IPCC's latest mitigation analysis, the world should invest an additional $3.2 trillion per year on average in zero-emissions technologies between now and 2030 to start the world on the path to zero emissions [1]. Figure 10.1 shows two views of those investments, first split by sector (energy efficiency, transportation, electricity, and agriculture, forestry, and land-use) and the other by major world region (developed versus developing countries). In contrast, annual investment in zero-emissions technologies has only averaged around $600 billion since 2017 [2], as shown in figure 10.2.

This seven-fold gap between what's needed and what's now being deployed demonstrates the scale of the climate financing challenge we face. These large funding gaps exist in every part of the world, and for every type of zero-emissions technology researched by IPCC [1].

In addition to mitigation, the world must also start investing billions (and likely trillions) of dollars per year in carbon direct removal (CDR) technologies. While

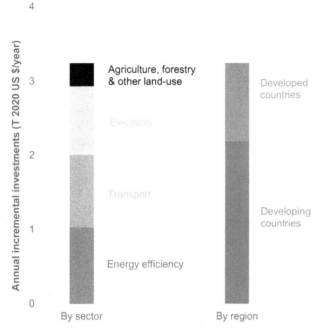

Figure 10.1. Average annual incremental mitigation investment requirements for climate stabilization through 2030. Source: Figure 15.4 in IPCC [1] (calculated as the average of low and high values by category, then subtracting out current expenditures by category).

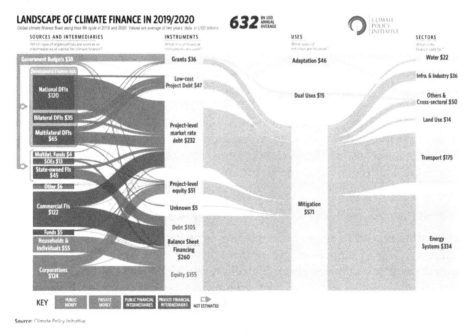

Figure 10.2. Landscape of climate finance circa 2019/2020 (billion US $/year). Source: Reproduced with permission from Buchner *et al* [4]. Copyright 2021 Climate Policy Initiative.

many climate mitigation technologies pay for themselves (mostly or fully) through operational savings, CDR solutions generally have a net cost. Of course, in the broader sense, CDR funding can also be seen as an investment, with every ton of CO_2 removed resulting in less climate-related damages for the hundreds of years that pollution would have otherwise remained in our atmosphere, but these climate benefits are diffused around the planet over time and space at scales beyond the bounds of traditional financial profitability analysis. Due to these spatial and temporal gaps, government mandates that guarantee CDR profitability are likely the only mechanisms that will drive enough capital to scale carbon removals at the necessary scale.

As the world scales up funding for climate pollution mitigation and CDR pollution drawdown, much more money will also need to be mobilized for climate adaptation and resilience, as well as reparations for losses in many communities already suffering from climate disruptions. The UN estimates that climate adaptation and resilience funding requirements will reach about $150 billion/year in 2030 and $300 billion/year in 2050 (2020 US $), even in an aggressive mitigation scenario [3], and delayed climate action will result in even higher costs.

Equity and social justice concerns are particularly acute for adaptation, resilience, and reparation funding decisions, since the communities most in danger from climate disruptions are often those who can least afford to prepare for the climate-amplified disasters that the world is already experiencing. There have long been tensions over how society should equitably value human lives and human suffering compared with infrastructure losses and damages to our supporting ecosystems. These tensions will likely amplify as finite climate funding is prioritized. Should limited climate impact funding go to seawalls that protect expensive homes on the coast, or to relocating poorer communities in flood zones, or to helping rural wildfire victims, or to better managing forests to minimize wildfire risks despite droughts, winds, and rising temperatures? The answers are not easy.

Government policies will play the largest roles in moving climate markets in the right directions, creating new capital, and equitably distributing funding to minimize future climate losses while compensating those already suffering from climate impacts. Effective governance is again an essential element. If we are serious about equity and climate justice all stakeholders must help direct decision-making, not just stakeholders with the most financial resources.

10.2.2 Shifting off fossil fuels can pay for getting to zero emissions

The spending needed to get to zero emissions is on par with what the world already spends every year to support the global fossil fuel industry. Annual upstream oil and gas production revenues are about $5 trillion/year (2020 US dollars), while global coal mining revenues are around $600 billion per year (see appendix F for details). Even more revenue is generated from downstream fossil fuel products. Society also directly subsidizes fossil fuels to the tune of about $500 billion per year.

In addition, there are pollution costs associated with continued use of fossil fuels. The International Monetary Fund (IMF) estimates societal greenhouse gas pollution costs at \$60/tonne of carbon dioxide equivalent [5], which implies total costs of about \$3.6 trillion US per year circa 2020 if applied to all greenhouse gas emissions, and this is a lower-bound estimate from the literature on the social cost of carbon [6]. There are also other pollution costs associated with the health and environmental effects of fossil air pollution, which the IMF estimates at \$2.4 trillion per year [5]. Finally, there are investments in equipment and structures outside of the fossil fuel industry (whose investments are captured in the fossil production revenues) of more than \$10 trillion per year, taken from the US Bureau of Economic Analysis (see caption for figure 10.3 for details)[2].

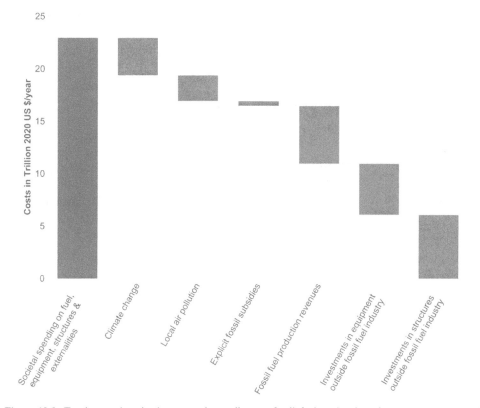

Figure 10.3. Total annual societal costs and spending on fossil fuels, related equipment, structures, and externalities, circa 2020. Sources: Parry *et al* [5] for local air pollution and explicit subsidies. Climate change damages estimated using \$60/tonne of carbon dioxide equivalent from Parry *et al* [5] and 60 Gt CO_2 equivalent total emissions for 2020 from Koomey *et al* [7]. Fossil fuel production revenues from appendix F. Investments in equipment and structures taken from US Bureau of Economic Analysis data for 2020 for the US (https://apps.bea.gov/itable/index.cfm) scaled to the world using the ratio of world to US GDP, adjusted up to reflect a slight difference between the percentage of non-financial investments for the world (1.39%) and the US (1.32%). The GDP and the percentage of non-financial investments are taken from World Bank data for 2020 and 2016, respectively (https://databank.worldbank.org/home.aspx), which are the latest years available. Total global GDP in 2020 was about \$89 T/year (2020\$).

---

[2] https://apps.bea.gov/itable/index.cfm

Figure 10.3 shows total costs associated with fossil fuels circa 2020 of about $22 trillion/year along with the components that add to that total. About a quarter of this $22 trillion consists of direct expenditures on fossil fuels. As society moves to phase out fossil fuels, more and more of these direct expenditures, as well as related subsidies, will be allocated instead to zero-emissions alternatives, which will also reduce societal costs of pollution. Purchasers of fossil fuels at market prices are now paying less than half of the total societal costs of these fuels.

In addition, figure 10.3 includes estimates for capital expenditures outside the fossil fuel industry for fossil powered equipment and structures, which total about $5 T/year and $6 T/year, respectively. Since most of these capital expenditures are for energy-using equipment, at least part of these capital flows will need to be re-allocated to non-emitting alternatives.

Even if climate solutions alternatives cost a bit more than fossil technologies up front (which they often will until the world embraces climate solutions in earnest and fully captures increasing returns to scale), the avoided climate change and other pollution costs will more than compensate for higher upfront costs. A world without fossil fuels will be better and cheaper for society than preserving the status quo. One can quibble with details of such calculations, but the broad picture has been confirmed by many independent analyses, and the estimates of fossil fuel related pollution costs are almost certainly lower-bound numbers.

Of course, shifting trillions of dollars out of fossil fuels and into climate solutions is much easier said than done. While this transition should have overwhelmingly net positive impacts for most people and parts of the world, there will likely be substantial economic losses for both private and public fossil fuel companies, as well as the governments and local economies that depend upon revenues from fossil fuel extraction and processing. Although fossil fuel revenues disproportionately benefit fossil executives and a handful of (often corrupt) government officials, millions more people work in fossil fueled economies and benefit from fossil-funded government programs, and billions more people have come to expect readily available fossil fuels at prices that don't reflect the harm caused by fossil pollution.

Shifting what we collectively pay to maintain our fossil addiction could fund most of the climate solution technology the world needs. But there are powerful vested interests that will continue to delay this transition if delay increases their profits and power, and there are also legitimate social equity and distributional challenges to navigate. The devils are always in the details, but if we're serious about solving climate, this massive financial transition must take place, and there are many potential levers that we can pull to mobilize capital in the most effective way.

10.2.3 Speeding up innovation in finance and business models

Solving our climate problems, at the highest level, involves displacing fossil fuels and non-fossil emissions with capital investment [8, 9]. A climate-positive world is one in which spending on energy services shifts heavily to capital investments and away from variable costs such as fuel and operating expenses.

The shift to a more capital-intensive energy system highlights the importance of financial innovation for smoothing the path to zero emissions, particularly for homes and smaller businesses (many big companies are already expert at it). Business model innovation is something all companies should master, both to reduce emissions and increase profitability [10]. Both are examples of institutional/ societal innovation that are often ignored by assumption in economic models of climate mitigation options.

Griffith [8, 9] points out that the modern car and home loans were invented in the US in the first half of the 20th century. These loans solved a cash flow problem for households who were wealthy enough in principle to buy new cars and homes but needed to spread those costs over time.

Griffith argues that to accelerate the transition of households to zero emissions, we need a financial innovation such as that (he calls it a 'climate loan'). Electrifying appliances, vehicles, and homes and installing solar panels requires capital expenditures that would be difficult for many households to afford without a loan, and we need to make such loans both widely available and inexpensive if we hope to rapidly reduce emissions in the residential sector.

There are other kinds of financial and business model innovations that could play an important role. Some rooftop solar installers now offer options to finance systems over time with no money down, and some community-oriented utilities allow customers to pay off zero-emissions investments over time on their monthly utility bills. Companies are also exploring programs to help employees fund programs to reduce and then eliminate emissions at home, including Carbon Savings Account programs that would function similarly to tax-advantaged Health Savings Account (HSA) plans.

10.2.4 Greater wealth = greater climate responsibility

Greater wealth almost always leads to higher climate pollution levels, both directly from more resource-intensive wealthy lifestyles, and through financed emissions in investment portfolios. Analysis from Oxfam International, summarized in figure 10.4, has found that the wealthiest 1% of humanity has been responsible for more climate pollution than the poorest 50% of our global population, and the wealthiest 10% of the planet is responsible for more than half of global GHG emissions [12].

While climate solutions need to reach everyone, those with greater wealth (and consequently more resources) should take more responsibility for solving the climate crisis. Those with the most wealth also wield considerable power and influence, which they should use to help fix this problem.

International climate policy debates are often framed in terms of richer countries who have long profited from polluting industries, and poorer countries that strive to catch up. Historic inequities in national climate pollution should of course be considered when assigning climate responsibility, but the debate about national inequities often ignores the relatively small groups of wealthy individuals around the world who are collectively responsible for an enormous share of climate pollution. In 1990, the wealthy in North America and Europe dominated global emissions, but

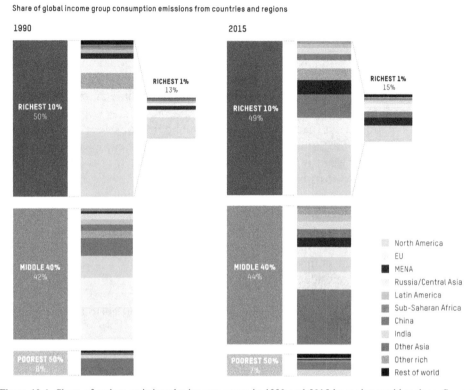

Figure 10.4. Share of carbon emissions by income group in 1990 and 2015 by major world regions. Source: Reproduced with permission from Gore [16].

now emissions per capita are nearly as high from the wealthiest citizens in Asia, the Middle East, and other parts of the world, as illustrated in figure 10.4.

When we frame our climate challenges in terms of the economic inequality of emissions, it makes sense that assigning economic responsibility for climate solutions should roughly align with household income. We all must redesign our homes, our transportation, our diets, our jobs, and our banking and investments to make them compatible with a climate-positive planet. The wealthier we are the more we should reduce emissions, particularly if that wealth spans multiple homes, private jets, substantial investments in polluting companies, or cultural influence over others' consumption choices.

Even those of us who don't think of ourselves as particularly wealthy or influential are still more responsible for solving climate than we think. The average household in wealthy countries has much higher emissions than the global average, and household GHGs generally increase with every increase in household income [13].

While the wealthiest among us should be doing the most to mobilize the capital needed to solve climate, we all have roles to play. Whether it's through our own spending and investing decisions, how we influence the choices that others make, or how we engage with our political and corporate leaders, we all must work to move money out of pollution and into solutions.

10.2.5 Helping power sector market structures evolve

The shift to a more capital-intensive energy system has other implications, particularly in the electricity sector. The economics of electricity systems has for many decades been strongly influenced by the short run operating costs (marginal costs) of fossil-fired electricity generators [14]. Rate design and utility business models have been structured around the idea of marginal costs being one of the primary drivers of utility system dispatch and investments.

The shift to a system dominated by capital investment means that market designs for the electricity system need to evolve [15–27]. The end state of having a large proportion of electricity generation being low or nearly zero marginal costs with significant storage in the system [28] will force system operators to develop alternative heuristics, improved contracts, and new market structures for dispatching power plants, paying for the construction of new plants, and determining how demand-size aggregators and storage resources will participate in these markets. This will also mean developing new analytical methods for assessing the *value* of different operating rules and power system investments, not just assessing costs [29–31].

10.3 Price pollution

While getting prices right won't solve the climate problem on its own, that step would help a lot. This section describes the benefits and complexities of pricing carbon, adding additional subtleties about pricing other warming agents and other kinds of pollution. As described in the previous chapter, other policy and business model changes are also needed, but to the extent we can make the price of fuels better reflect their true societal costs, we should do it.

10.3.1 Price carbon pollution

Pricing climate pollution has been widely used to mobilize money for climate solutions. The basic logic is that putting a price on greenhouse gas pollution will incorporate social and environmental externalities into market decisions, providing incentives for polluters to reduce GHGs, while zero-emission products and services become more cost-competitive. Revenues raised from pricing pollution can also be used to fund zero-emissions technologies, further tilting markets towards climate solutions.

Greenhouse gas emissions pricing can either come in the form of direct taxes on climate pollution, or through 'cap and trade' carbon compliance markets where maximum pollution allocations are determined and ratcheted down each year, forcing polluters to purchase pollution credits from carbon trading markets or carbon offset projects [32, 33]. Economists generally favor direct carbon taxes, which are simpler to design and implement, but many governments have instead created cap and trade carbon emissions trading scheme (ETS) markets, which have often been more politically feasible and favored by regulated industries. Cap and trade ETS systems play the largest role in pollution pricing, with the largest markets in Europe, China, and several US states [34].

Unfortunately, current climate pollution pricing systems still don't cover enough of the world's GHG emissions or set high enough pollution prices to drive substantial emissions reductions. About 22% of global GHG emissions are now covered by some form of climate pollution pricing, with about $53 billion raised in revenues from pollution pricing in 2020 [34]. Meanwhile, over three quarters of global GHGs are not priced into markets at all, and less than 4% of global climate pollution is priced above the $40 per metric ton CO_2e level recommended as the lowest price to be compliant with the Paris Agreement [34].

What is the right price for climate pollution? The economics literature provides little clear guidance, giving huge ranges for the prices needed to achieve certain emissions reductions or the social cost of carbon [1, 4, 35–38]. The practical answer is 'as high as practically possible'. There is virtually no danger of setting a price that's too high because the politics of pricing make that almost impossible and the need for rapid emissions reductions gets more and more urgent as time passes.

While current pollution pricing systems are a step in the right direction (particularly compared with the de facto price of zero), it's clear that climate pollution needs to be priced much higher than it is in most current systems, while pollution pricing must also expand to include the majority of the world's GHGs. Appropriately pricing greenhouse gas pollution would make fossil fuels, deforestation, and other climate pollution sources much less economically viable, while potentially raising trillions of dollars that could fund climate solutions in the process.

If the world's approximately 60 Gt CO_2 equivalent in annual greenhouse gas emissions were taxed at a rate of $60/t CO_2 equivalent (a lower-bound value used in IMF analysis [31], for example), around $3.6 trillion per year could be raised and redirected into climate solutions. Pollution revenues could be maintained around this level if CO_2 equivalent prices continue to rise as annual emissions fall, as most economists recommend.

10.3.2 Understand the limits of carbon pricing

A price on carbon would no doubt be helpful on reducing emissions but understanding how and where it would be most helpful depends on knowing the structure of energy systems. To deliver energy to small users, a large portion of energy costs comes from the capital costs of infrastructure, such as the distribution lines for electricity or the supply chains for retail gasoline. That means that carbon prices on fuels get diluted for small users because the carbon prices are a relatively small component of energy prices.

The story is quite different for most large users of fossil fuels, where fuel is a dominant cost of production and high carbon fuels are an important fuel source. Electric utilities, for example, have existing fossil-fired generators where fuel is the biggest part of their costs, and new generators where fuel costs for carbon-based fuels are still significant as a percent of the total.

Consider the effect of a $10/tonne of CO_2 price on the cost of an existing bituminous coal plant. At typical combustion efficiency that price implies 1 cent/kWh on the running cost of an existing coal plant (see appendix G). Current (June

2022) prices for CO_2 in Europe are about \$90/tonne and in California they are about \$30/tonne.[3] Those prices imply adders of 9 cents/kWh in Europe and 3 cents/kWh in California, respectively, compared to coal plant operating costs that are typically 3–4 cents/kWh. Modest carbon prices can have a huge effect on utility dispatch and retirement decisions for coal plants at these prices.

For the small electricity consumer, the effect of carbon taxes is diluted. First, coal isn't the only source of generation in most power systems, there are also often natural gas and zero-emissions nuclear, hydro, wind and solar plants. If we consider a simplified utility system with half coal and half zero-emissions sources, that means the 1 cent/kWh is cut in half for the consumer, to 0.5 cents/kWh.

Second, electricity prices are composed of fuel costs, operating and maintenance (O&M) costs, and most importantly, capital costs. Capital plus O&M costs are typically two thirds (or more) of the price of electricity in developed countries like the US[4]. That fact reduces the effect of the \$10 per tonne CO_2 tax by another two thirds, to 0.17 cents/kWh, or to less than one fifth of the cost the utility sees for operating its coal plant.

This example explains why carbon taxes can be a strong economic driver of emissions reductions for utilities and industrial consumers (who use a lot of high carbon fuels). These users have much lower fuel prices than small consumers, so the percentage impact of any given carbon tax (which typically only depends on the carbon content of the fuels) is much higher for bigger users, and the percentage change is what mostly drives economic decision-making.

Carbon taxes are weak drivers of consumer behavior but strong drivers of emissions reductions for utilities and industrial consumers. The same \$10 per tonne CO_2 charge only adds about 10 cents per gallon on the price of gasoline (see appendix G), which is tiny compared to the price of gasoline almost everywhere.

Passing carbon taxes can be politically difficult, but the alternative of emissions trading has its own political issues [39]. Emissions trading systems in the European Union and in California have in the past suffered from overallocation of permits and 'leakage', so that market prices for carbon emissions are much lower than would exist in better designed market regimes. The problem is that powerful market actors can (and often do) warp the design of such regimes to favor their interests, rendering these policy instruments far less effective at reducing emissions in a verifiable way.

The characteristics of technological systems responsible for emissions also create practical difficulties for properly pricing carbon emissions. For example, most proposals for taxing emissions from petroleum fuels begin with the assumption that the tax should be based on the *carbon content* of the fuel when it is burned, but 'upstream' (exploration and production) and 'midstream' (refining) emissions are big enough to matter [40–44].

For oil, these latter two components can increase the total life-cycle emissions per barrel 40%–50% compared to combustion emissions alone [45, 46]. In addition, the 'owners' of upstream and midstream emissions are in many cases far removed from

---

[3] https://carboncredits.com/carbon-prices-today/
[4] https://www.eia.gov/energyexplained/electricity/prices-and-factors-affecting-prices.php

the consumers actually burning the fuels. These complexities make designing what an economist would call an 'optimal' emissions price for these fuels almost impossible, but few economists have wrestled with these complexities. Of course, any price higher than zero would be an improvement over the status quo, but understanding the complexities is essential for designing the best policies.

One potential benefit of pricing emissions (either using taxes or auctioning permits in an emissions trading system) is that it generates revenues that can then be distributed to share the benefits and promote a more rapid transition to zero emissions. In principle, this can help the politics of pricing emissions, and it sometimes does, but people's general aversion to taxes of all sorts can make it challenging to argue for emissions pricing in many places.

Countries that price carbon can also implement border adjustments to reflect the warming externalities associated with manufactured goods imported from countries that don't have a carbon tax. The mechanics of such systems are complicated, but it's an area that has been well studied [38].

10.3.3 Price other GHGs

Carbon emissions are only part of the problem, but only rarely have pricing mechanisms acknowledged the other warming agents. The technical complexities of estimating greenhouse gas equivalences using Global Warming Potentials are well-known [56, 57], and in fact there is no 'right answer' when comparing the warming effect of emissions of carbon dioxide to that from methane, nitrous oxide, or other greenhouse gases. Warming effects compared to carbon dioxide vary by gas and by the time period of analysis, in most cases substantially [49], as we show in chapter 2.

Despite this complexity, we know that setting a price of zero for these warming agents is wrong. Better to price them as best we can than to ignore their warming effects. The alternative is to regulate such emissions, which in some cases will be the preferred option, but either way, we can't ignore these warming agents.

10.3.4 Price other fossil pollutants

Fossil fuel combustion causes air pollution beyond just greenhouse gases, and those pollutants cause significant health and environmental effects that are only partially controlled by existing regulations, resulting in millions of deaths every year linked to poor air quality [50–65]. Air pollution is an 'externality' that needs to be internalized. The costs of such pollution are large enough to matter and internalizing those costs through taxes (and regulations) will shift incentives substantially and make it easier to create a climate-positive society.

Another significant category of fossil-related pollution that's largely unpriced is uncontrolled plastic waste, with millions of tons of plastics (and small particles called 'microplastics') flowing into our air, soils, oceans, and biological systems every year. Emissions from washing synthetic fibers (such as polyester) and the wear of synthetic rubber vehicle tires contribute the largest shares of microplastic pollution [66]. It's still unclear exactly how much harm such pollution is causing to humans, wildlife, and ecosystems, but there are many potential concerns, and some microplastics are small enough to enter the human bloodstream. A recent Dutch study found detectable microplastics in the blood of 77% of individuals tested in the Netherlands [67].

Beyond plastics, there are many other potentially hazardous chemicals made from oil, coal, and natural gas, and these fossil fuel co-products impose societal costs and add profitability to the fossil industry. Pricing all forms of non-climate fossil pollution can make the prices of fossil fuels much closer to full societal costs and make the transition away from fossil fuels easier and quicker.

10.4 Subsidize investments and innovation

One way to smooth the transition to a climate-positive world is to subsidize early retirement and replacement of fossil equipment with zero-emissions alternatives. This time-tested technique has been used widely for decades and we'll need to expand its use. We'll also need to ramp up funding of research and development to fill the pipeline with zero-emissions options for decades to come. Finally, we should explore new ways for companies to reward process innovations that result in emissions reductions.

10.4.1 Defray capital costs with rebates

Because the transition to zero emissions is mainly one that substitutes capital for fuels, individuals and institutions will see real upfront costs for scrapping existing equipment early and buying new electrified equipment. We can ease the pain of that transition by subsidizing the purchase costs for new equipment and paying bounties for those who choose to retire fossil capital early and replace it with electrical equipment. There is a long history of such rebates in the utility industry as well as for tax credits from state and national governments.

Smoothing the transition is reason enough to subsidize climate-positive technologies, but such incentives also create innovation because they allow products to scale up and move down the learning curve. The end game is making the climate-positive options cheaper in direct cost terms than the high emissions options they displace, which argues for high initial subsidies (to spur adoption) that ramp down as adoption targets are reached. Combining such subsidies with institutional purchase requirements can spur even more rapid changes.

10.4.2 Invest in climate tech R&D

We already have most of the technologies needed to fully move the global economy to zero emissions, but investments in R&D can uncover new technologies, materials, and processes that further push down costs for existing products. R&D investment is also key to discovering entirely new solutions that achieve goals in better ways.

Companies, investors, and governments all play important roles in the climate tech R&D cycle. Government funding is especially crucial for funding the primary research at universities and national labs that leads to breakthrough discoveries, while companies and investors bring discoveries into markets, then fund further

R&D cycles to make new technologies scale and continuously improve. As we discuss above, most funding should support deployment of renewable energy, electrification, and other solution technologies that already exist, but funding climate tech R&D at much higher than current levels is also essential.

10.4.3 Pay for performance

Innovation means more than just developing and deploying new technologies. Process innovation is a way for companies to generate continuous improvements, but it requires aligning the incentives of employees and managers with driving emissions to zero.

Once standardized measurements of full-life-cycle emissions become commonplace, it will then be possible for companies to reward employee efforts to reduce greenhouse gas emissions with direct financial incentives. If rapid direct emissions reductions are a corporate goal, then giving bonuses to employees who help achieve that goal is natural and sensible.

One precedent for such efforts is the Six Sigma program in many large companies as discussed in chapter 7, which has the task of finding opportunities for cost reductions and new products. Six Sigma practitioners are often rewarded based on whether they achieve the cost savings and revenue increases projected in their original business plans, and the same kind of process would work here.

10.5 Redirect capital

Most of the capital needed to create a climate-positive future needs to be redirected from other activities. We already spend at least $5 trillion every year in direct expenditures on fossil fuels, and at least $6 trillion per year in unpriced greenhouse gas and other pollution externalities from those fuels. In addition, capital expenditures for equipment and structures total more than ten trillion dollars every year. Redirecting part of those funds will be more than enough to solve the problem.

This section explores ways to redirect those funds. It's more than just a technical or financial change, it also involves getting polluter money out of politics, ending pro-pollution subsidies, modifying central bank policies to facilitate climate action, and experimenting with new money focused particularly on carbon removal. The structural changes needed won't be easy to implement but are essential to accelerating the transition to a climate-positive world.

10.5.1 Equitably distribute pollution revenues

If pollution is priced above the social cost of carbon, then many polluting technologies and businesses will no longer be profitable, and at least some portion of pollution prices will be passed on to consumers in the form of higher prices for some sources of fuel, electricity, food, and materials. If not properly managed, these economic shifts could worsen economic inequities. For this reason, many advocates for better pollution pricing also advocating returning most or all of the revenues raised directly to citizens, with poorer households receiving more money back in direct payments than they are expected to pay in higher pollution prices.

The 'Carbon Fee and Dividend' system proposed by Citizens Climate Lobby (CCL) is one potential model for returning pollution revenues to citizens[5]. A few pollution markets return a portion of revenues to households but none now return all such revenues.

Of course, if more pollution revenues are returned as dividends to citizens, less will be available to subsidize climate-positive technologies or pay for climate adaptation, resilience, or reparations, although the degree each household is suffering from climate damages could also factor into the payment amounts that each household receives.

Widespread political support is key for using pollution pricing to drive substantial emissions reductions over time. Successful pollution pricing systems will allow the majority of citizens to receive net economic benefits, while the most vulnerable are protected and companies and individuals who can afford it pay more. Fossil giants and others who profit from pollution will likely continue to fight any pollution pricing that hurts their bottom line but putting a higher price on climate pollution will also force polluters to diversify, and several other policy and financial tools can hasten this transition.

10.5.2 End pro-pollution subsidies

There are incentives beyond emissions pricing that matter greatly. For example, the existing economy includes myriad subsidies for the global fossil fuel industry [68–78], many of which are hidden. The International Institute for Sustainable Development [72] estimated that G20 nations alone allocated $277 B US/year to oil and gas production from 2017 to 2019, which is of the same order of magnitude as profits from the global oil and gas industry[6]. An estimate for the OECD shows lower numbers of about $50 B US/year in 2019 for fossil fuel production subsidies [79].

Exact comparisons are complicated, but it's fair to say that eliminating such subsidies would substantially reduce oil and gas industry profits, thus increasing the cost of capital for that industry [80]. A transition to zero emissions must include efforts to root out and remove all such biases and subsidies everywhere in the world, otherwise the speed and extent of the transition will be hindered.

Five examples of hidden subsidies are the historically below-market leasing fees for fossil fuel leases on federal lands in the US [81], low bonding requirements for oil and gas drilling [82–84], misplaced incentives for shutting down abandoned wells [85], the lax structure of US bankruptcy and other laws that allow companies to privatize profits and socialize costs [4, 86], and subsidies that lower the price of fossil fuels to consumers [87, 88]. There are many other similar hidden subsidies elsewhere.

---

[5] https://citizensclimatelobby.org/basics-carbon-fee-dividend/
[6] The Fortune Global 500 contained 85 oil and gas companies that generated $223 B and $404 B of profits in 2018 and 2019, respectively. https://www.globaldata.com/oil-and-gas-sector-continues-to-rule-2019-fortune-global-500-list-in-revenue-generation-finds-globaldata

10.5.3 Donate to climate action

Given the scope of our climate crisis, it's remarkable that less than 2% ($6 to $10 billion/year) of global philanthropy funds climate action [89]. Figure 10.5 shows a subset of global philanthropic giving to climate action, that from major foundations, broken out by regions and sectors. Climate philanthropy must become a higher priority for the world to reach Net Zero by 2040. While better climate policies and investments can drive many climate solutions, climate philanthropy is necessary for some types of climate solutions, particularly solutions that don't yield immediate or direct financial returns.

Donations are particularly needed for climate communications and public policy engagement, which are essential for mobilizing the political will to push policy-makers to implement effective climate policies. By bridging our climate communi-cation gaps, climate philanthropy can help break political inertia by generating the constituent pressure politicians need to act. Donor-led technology efforts can also enhance climate transparency, pressure companies and investors to improve, better inform the public about climate impacts and solutions, and empower individuals everywhere to take action.

10.5.4 Get polluter money out of politics

Effective climate policymaking is difficult when money tied to polluting interests is allowed to disproportionately influence political systems. Historically, political

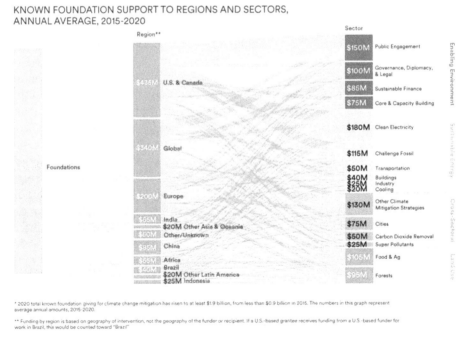

Figure 10.5. Major foundation support for climate action by regions and sectors (annual average 2015 to 2020). Source: Reproduced with permission from ClimateWorks Foundation, *Funding Trends* 2021 [97].

donations supporting climate action have been dwarfed by donations and lobbying from groups tied to climate polluting industries (particularly fossil fuels, utilities, automobiles, chemicals, forestry, and agriculture). In the United States, over $2 billion was spent on lobbying to block climate regulations between 2000 and 2016, more than ten times what was spent by proponents of legislative climate action [90]. Fossil fuel companies in particular have spent decades investing in politicians that deny climate science and hold up climate regulations, with the largest share of fossil campaign donations going to lawmakers with the longest track records of blocking climate action [91].

The problem of climate-polluter money in politics is a worldwide issue, with the wealthy few who profit from climate pollution using cash to tip political scales from local governments to international climate negotiations. Over 500 lobbyists connected with the fossil fuel industry attended the United Nations climate negotiations in Glasgow, Scotland [92], a fossil-funded delegation that was larger than any single country, and larger than the combined delegations from the eight countries worst affected by climate change in the last two decades: Puerto Rico, Myanmar, Haiti, Philippines, Mozambique, Bahamas, Bangladesh, and Pakistan.

According to the watchdog InfluenceMap [93], the fossil giants ExxonMobil and Chevron have the largest negative impacts on global climate policy, followed by Toyota Motor, Southern Company, Sempra Energy, BASF, ConocoPhillips, Glencore International, BP, and OMV to round out the top ten political polluters. Often these fossil giants publicly claim to support climate action while simultaneously funneling money to climate-obstructing trade groups, such as the American Petroleum Institute and US Chamber of Commerce, allowing companies to pool lobbying resources while distancing their brands from publicly unpopular positions [94]. InfluenceMap also stresses that a large share (potentially the majority) of fossil political influence still happens without public transparency, and that efforts to track the impacts of polluting interests likely miss large amounts of dark money that flows behind the scenes, as well as the full policy influence of state-owned fossil giants in Russia, Asia, the Middle East, and Latin America.

There is also a 'revolving door' problem in many political systems, where individuals cycle between roles as government regulators and roles as highly-paid consultants for the same polluting industries they were supposed to have regulated. The nonprofit Corporate Europe Observatory identified over 70 cases of European government regulators cycling between government jobs and positions with six fossil fuel giants and five fossil fuel lobby groups between 2015 and 2020 [95]. Former government officials are often sought after as corporate lobbyists since they already have personal connections in governments and know how to effectively pull political strings. There are even some government officials who maintain large personal investments in fossil fuel assets while continuing to act as regulators, creating obvious conflicts of interests that are surprisingly still allowed in many political systems.

Perhaps more surprising than fossil fuel lobbying is the fact that nearly all types of companies give more money to politicians who block climate action than they donate to politicians who champion climate solutions. Analysis by Bloomberg found that the average 'climate obstructionist' lawmaker in the US received about $1.84 in corporate donations for every $1 donated to lawmakers who support climate legislation [96]. Bloomberg's analysis found that every company in the S&P 100 has been giving to climate obstructionist politicians, including donations from many of the world's largest technology, finance, and healthcare companies. With most of the world's largest companies rewarding climate inaction with their political donations, as well as well-funded fossil lobbying campaigns threatening to fund the opponents of politicians who dare to advocate for aggressive climate action [97], it's no wonder that climate policies have fallen far short of what climate science says is necessary.

How do we get polluter money out of politics? The simple answer is that we need much better rules that keep polluting interests from corrupting the political process at all levels. Ideally three rules would apply to all levels of government: (1) ban polluting industries from making political donations; (2) ban politicians from having direct investments in fossil fuels and any other industries they regulate; and (3) ban government regulators from taking lucrative positions with companies or industry groups they once regulated. Enacting these rules would go a long way towards cleaning up climate corruption and aligning lawmaking with climate science.

10.5.5 Factor climate into spending decisions

While much of what we've discussed in this chapter covers tactics to make the supply side of climate solutions more economically attractive, the money we spend also sends powerful demand-side signals. Spending decisions ultimately dictate the flow of capital into pollution or solutions, and every level of spending is effectively a continuous vote on our climate future, whether cash is flowing from governments and companies or households and individuals. Implementing policies to ensure that climate-positive options are the cheapest, most convenient, and default options can go a long way towards shifting capital in the right directions, but consumer choice will always be the final deciding factor in free market economies.

Information technology can play a large role in calculating the full climate impacts of products and services, as well as making these previously invisible climate impacts visible and actionable for consumers. Some companies are already adopting climate labeling standards that calculate the full Scope 1-3 GHG footprint of products and services, empowering potential buyers to add climate criteria into their choices. As climate labeling becomes standardized, verified, and widespread, consumers will increasingly be able to use their wallets to reward climate leaders. Mandating climate impact disclosures for products and services can push this trend even further by identifying and pressuring climate laggards to improve. While consumers often have a misperception that more sustainable products are always more expensive, standardized climate labeling can show that climate-friendly choices often save money, whether it's choosing cheaper plant-based menu items at a restaurant or buying a heat-pump HVAC system that saves money over time.

10.5.6 Redirect fossil profits to climate infrastructure

Nearly all fossil fuel companies continue to invest in finding and extracting new fossil fuels and expanding their fossil infrastructure each year, despite the overwhelming scientific consensus that ending fossil fuel combustion is essential for solving our climate crisis. Yes, some fossil companies are also now investing in renewable energy and other climate solution technologies, but these investments are still a very small in comparison with the amounts that oil, gas, and coal companies invest in controlling and selling more fossil fuels. ExxonMobil invested only 0.22% of its capital expenditures in low-carbon projects between 2010 and 2018, while only 2.3% of BP's investments went into low-carbon projects during the same period [98]. The reason for this discrepancy in fossil versus non-fossil capital is that fossil investments are still often more profitable than investments in renewables, and fossil companies have technical expertise that makes additional fossil extraction easier than deploying renewables.

Politicians and investors often promote the idea that fossil companies should become diversified energy companies that shift from fossil to renewable energy, and fossil industry marketing often plays into this narrative. But this transition will not take place in a meaningful way until either (a) some combination of economic policies and market conditions make renewable energy investments more profitable than fossil extraction, or (b) fossil companies are mandated to invest most or all of their profits into renewables and other climate solutions.

Pricing climate pollution would help push fossil companies to transition much more quickly towards renewables. Another approach is for governments to simply mandate that fossil profits be invested in climate solution technologies instead of fossil infrastructure. While policymakers haven't yet taken this direct approach to reducing emissions from the fossil industry, some have started redirecting fossil profits in other ways. Windfall taxes on fossil profits have increasingly been proposed, particularly when fossil profits skyrocket during geopolitical supply disruptions, such as Russia's invasion of Ukraine. Windfall taxes can take away incentives for fossil companies to price gouge consumers while also preventing fossil companies from investing in new fossil infrastructure, and revenues from windfall taxes can also be redirected to subsidize climate solution technologies.

10.5.7 Re-allocate banking and investment portfolios

Historically, climate has not been a consideration for most investors. Over the past few decades many investors have increasingly moved assets into 'passive' index fund investments, which hold a diverse range of companies, and have generally offered cost and financial performance advantages over more actively managed financial strategies. This shift towards index funds means that most investors are invested in fossil fuels and other climate polluting companies, since conventional benchmark index funds invest in all parts of the economy, including dirty energy producers, agricultural polluters, and companies with links to deforestation [99].

The past few years have seen a substantial shift towards factoring climate and overall environmental, social, and governance (ESG) metrics into investment decision making. Many of the world's largest asset owners and asset managers have now joined climate and ESG investing pledges and organizations, such as the Glasgow Financial Alliance for Net Zero (GFANZ), Climate Action 100+ (CA100+), the Principles for Responsible Investment (PRI), Task Force on Climate-Related Financial Disclosures (TCFD), and Science Based Targets (SBTi), among others.

Unfortunately, most of these same investors are continuing to invest in fossil fuel companies and other climate polluters (both directly and through conventional index funds), and while they may be factoring more climate risks and opportunities into investment decision-making, even large investors with quantitative climate pledges are generally not planning to reach Net Zero until 2050. The largest asset managers are now offering a wide range of exchange traded funds (ETFs) and other products with climate and ESG labels, but these funds often still include fossil fuel companies and other climate polluters due to poorly designed conventional ESG ratings systems, leading to 'greenwashing' concerns for investors sincere about improving climate impacts [100].

The biggest debate in climate investing is whether investors should fully divest from fossil fuel companies and become 'fossil free.' Many large investors have justified maintaining holdings in fossil fuels and other large climate polluters by claiming they can use their shareholder power to push these companies to reduce emissions with shareholder resolutions. Proponents of fossil-free divestment, including many in the Rockefeller family whose Standard Oil empire became ExxonMobil and Chevron [101], argue that decades of shareholder engagement have done little to shift fossil companies away from climate pollution, and it is unlikely fossil companies will ever match climate rhetoric with business realities since they profit from continued climate inaction. Proponents of fossil divestment also make purely economic arguments, highlighting the fact that fossil companies on average underperformed global markets by over 50% from 2011 to 2021 [102], and fossil companies are likely to have substantial 'stranded assets' due to international climate targets and increased market competition from wind, solar, batteries, electric vehicles, and other clean technologies.

Moving investment portfolios toward zero-emissions holdings should start with a goal to make each portfolio net zero or climate positive as quickly as possible. While a portfolio doesn't need to be fully fossil free to reach climate targets, divesting from all coal, oil, and natural gas makes getting to net zero and climate positive much easier. The next step is a full Scope 1-3 climate accounting[7] to determine the financed GHG emissions connected to each individual investment, as well as the portfolio's total climate footprint. It is essential that this type of portfolio GHG analysis include 'downstream' Scope 3 climate impacts from when a company's products and services are used, since this is needed to accurately assess the climate footprint of fossil fuel holdings (where most GHGs come from the downstream burning of fuels produced). Downstream Scope 3 calculations also illuminate which companies are net climate positive, where the climate benefits from their products (such as renewable energy, EVs, and plant-based foods) outweigh the climate impacts from product production.

---

[7] https://ghgprotocol.org

Newer sustainable investing companies, such as Etho Capital (of which Ian was a co-founder), have shown how investors can improve both financial performance and climate sustainability by shifting to fully fossil-free strategies that eliminate emissions and are more authentically aligned with ESG goals. Figure 10.6 compares the performance of ETHO, Etho Capital's flagship Exchange Traded Fund, compared to the S&P 500 index from the date of ETHO's first full day of trading (20 November 2015). ETHO yielded about 19% more total return for the past six years compared to the S&P 500. That's about 3.3% per year additional return each year.

By shifting to sustainable index strategies that combine full Scope 1-3 climate data and deeper analysis of climate and ESG risks, investors can now get many of the same diversified performance advantages of conventional index investing while substantially improving climate and overall ESG portfolio impacts [103]. Analysis by Etho Capital has found that investors can already replace conventional global index strategies with ESG-aligned climate-positive alternatives while maintaining similar (or better) financial performance[8]. New green banks, like Atmos Financial and Ando Money, have also launched with missions that give individuals and businesses the ability to ensure that their savings finance clean technologies instead of fossil fuels.

Most investors already have the climate analysis tools and climate-friendly investment options to make their banking and investment portfolios climate positive long before 2050, it's just a matter of knowing where to look and how to go about the process. Some institutional investors still hide behind 'fiduciary duty' arguments against climate investing, but an honest assessment of projected climate policy and technology directions makes it clear that fiduciary duties require incorporating

Figure 10.6. Six-year performance of ETHO Climate Leadership US Index versus S&P 500 Index (ETHO inception on 11-19-2015 to 11-20-2021). Source: Yahoo finance using adjusted daily closing price, https://finance.yahoo.com.

---

[8] https://ethocapital.com/climate-positive-investing

climate risks and opportunities into all decisions [104]. As an example, some analysts predict that increasingly cheap energy storage could negatively affect one quarter of corporate debt within the next decade [105].

All of us should re-allocate our banking and investments as quickly as possible. Companies should ensure that corporate cash isn't invested in climate pollution and offer fossil-free and climate-positive funds for employee retirement plans. Financial regulators should push both asset owners and financial product creators towards full Scope 1-3 climate transparency, while also designing financial industry climate labeling standards and monitoring programs that culls greenwashing out of the market. Governments should enact laws that require public pensions to divest from fossil fuels and invest in climate solutions, and pension recipients, foundation employees, and university students and faculty should engage with their fund managers to demand fossil divestment.

Innovative investors are already blazing a trail to a climate-positive future. As climate regulations and clean technologies make fossil fuels unprofitable and obsolete, all investors will eventually follow, and smart financial policies can hasten this process.

10.5.8 Align central banking and monetary policy with climate targets

Central banks in large economies drive the creation of new money, which gives them a unique ability to combat climate change by linking monetary policy to climate criteria. Monetary policy can support climate infrastructure in several ways. First, central banks influence how commercial banks make lending decisions by controlling interest rates and dictating reserve requirements for commercial banking. Central banks therefore have the power to adjust interest rates and reserve requirements in ways that encourage commercial banks to lend to climate solution projects and discourage commercial banks from financing fossil fuels and other projects and companies tied to climate pollution.

Large central banks with reserve currencies also have the unique ability to influence economies through a process known as 'quantitative easing' (QE), where central banks effectively create new money to directly buy assets like corporate bonds and equity. Quantitative easing is generally deployed to help economies recover from economic downturns, such as the 2008 global financial crisis or the economic challenges posed by COVID-19. Some analysis has found that the trillions of dollars in recent QE from central banks has disproportionally gone to buy corporate debt and equity from fossil fuel companies, thereby likely making climate change worse [106] but QE could also be harnessed as a climate solution. Some researchers and central banks are now evaluating the potential to implement 'green quantitative easing' (or green QE), where central banks would preferentially buy corporate debt and equity linked to climate solutions, while also avoiding QE purchases linked to climate pollution [106].

10.5.9 Make new money for carbon removal + all climate solutions?

Central banks could play an even larger role in solving climate by making new money that is distributed directly to fund climate solutions. The concept of creating an entirely new climate currency has been gaining traction, and it is a particularly appealing concept to fund the scale up of carbon removal, since existing climate pollution in our atmosphere will only be removed once large new pools of capital are mobilized to pay for carbon removal services.

The most fully formed proposal for a new climate currency has been proposed by Global Carbon Reward, an international nonprofit effort that started in Australia[9]. The proposed new climate currency would be issued by a new Carbon Exchange Authority branch of the United Nations, which would ensure that climate currency is only issued to verified GHG mitigation or removal projects. The proposed Global Carbon Reward climate currency would be supported by the world's largest central banks, which would guarantee exchange rates with traditional currency.

Central bank backing would guarantee positive financial returns for the climate currency until global climate targets are achieved, providing incentives to participate in the carbon rewards market. Social and environmental factors beyond climate could also be assessed when the Carbon Exchange Authority determines how much climate currency to issue to each proposed climate solution project, providing additional financial incentives to scale climate solutions that maximize positive co-benefits while minimizing negative externalities.

While the Global Carbon Reward system is still at the proposal and piloting stage, it may be exactly the kind of big picture thinking that's needed to mobilize climate solution capital at the required scale. As we've discussed earlier, the scale climate pollution mitigation and removal necessary are only continuing to grow as the world delays in aligning climate action with what science says is necessary, and the more we delay funding climate action the more the costs of climate adaptation, resilience and reparations will also increase[10].

What's particularly appealing about the Global Carbon Reward concept is that this type of new climate currency would not come at any direct cost to taxpayers, and the climate currency system would also act as economic stimulant to a wide range of climate solution enterprises, potentially supporting millions of new jobs in the process while delivering a wide range of social and ecological co-benefits in the process. These factors potentially make a global climate currency approach more politically feasible than pollution pricing and more predictable than current cleantech subsidies, though these market-directing policies can coexist with the shared goal of re-engineering economies to scale climate solutions as quickly as possible.

It may make sense for a climate currency to first be limited to carbon removal projects, since CDR projects that remove pollution as their primary purpose avoid most of the additionality and leakage challenges in today's carbon offsets markets. CDR projects that are truly additional only exist because of the price paid for pollution removal but no government has yet provided a realistic plan for how this level of CDR will be funded. Thus far the biggest boosts to the CDR industry have come from a handful of technology companies that have voluntarily started to pay for pollution removal instead of carbon offsets [107]. If wisely implemented, funding CDR with a new climate currency could dramatically accelerate the pollution removal industry.

---

[9] https://globalcarbonreward.org

[10] https://globalcarbonreward.org/carbon-currency/pricing-theory/#pricing_cpm

10.6 Building your scenarios

Just as for non-financial incentives, the key to creating scenarios is to lay out your assumptions in detail about how financial incentives would need to change to enable a rapid transition to a climate-positive world. It's impossible to predict the effects of these changes with precision, but a structured process of creating a plausible and internally consistent set of such changes is essential to developing your intuition about constraints and opportunities.

10.6.1 Creating or adopting a business-as-usual scenario

Your business-as-usual (BAU) scenario should incorporate recent changes in financial incentives that are likely to result in measurable changes in emissions. If your chosen geography has recently adopted a price on emissions, for example, make sure that the BAU projections for utility sector and industrial emissions in particular reflect those policies.

10.6.2 Creating the climate-positive scenario

Koomey [115] is presents well-documented intervention scenarios for the buildings sector that you can use as a model when data permit. The details in that article are now old but it's a great example illustrating the appropriate level of detail. Just as for creating scenarios for technical options, the idea is to use historical experience as a guide to estimate potential effects of changes in financial incentives, quantifying as best you can. If your scenario is plausible, internally consistent, and well-documented, you'll be on safe ground.

10.7 Chapter conclusions

Moving to a climate-positive world means shifting trillions of dollars every year to zero-emissions and carbon removal technologies, as well as paying to alleviate and repair climate damages. We're already spending trillions to support our current high-pollution system, we just need to re-allocate these expenditures toward climate-positive alternatives. Fortunately, these shifts will also eliminate trillions of dollars every year in health costs, premature deaths, and a wide range of environmental harms. The political challenges of making these changes will not be trivial, but the overall story is clear: mobilizing money for a climate-positive planet will result in a much better world, with lower total costs and enhanced quality of life for virtually everyone.

Further reading

Buchner B *et al* 2021 *Global Landscape of Climate Finance 2021* (San Francisco, CA: Climate Policy Initiative) December https://www.climatepolicyinitiative.org/wp-content/uploads/2021/10/Full-report-Global-Landscape-of-Climate-Finance-2021.pdf. This report gives a clear picture of money flows of global climate finance.

Cullenward D and Victor D G 2020 *Making Climate Policy Work* (New York: Polity Press). This book gives a clear-eyed view of the real-world issues with pricing pollution.

Geddes A, Gerasimchuk I, Viswanathan B, Suharsono A, Corkal V, Roth J, Picciariello A, Tucker B, Doukas A and Gençsü I 2020 *Doubling Back and Doubling Down: G20 Scorecard on Fossil Fuel Funding* (Winnipeg: International Institute for Sustainable Development) 9 November https://www.iisd.org/publications/g20-scorecard. This report summarizes fossil fuel subsidies in detail.

Gore T 2020 *Confronting Carbon Inequality: Putting Climate Justice at the Heart of the COVID-19 Recovery* (Nairobi: Oxfam) 21 September https://www.oxfam.org/en/research/confronting-carbon-inequality. The wealthy contribute disproportionately to greenhouse gas emissions around the world, and this report documents their contributions.

IPCC 2022 *Climate Change 2022: Impacts, Adaptation and Vulnerability—The Working Group II Contribution to the Sixth Assessment Report* (Cambridge: Cambridge University Press) https://www.ipcc.ch/report/sixth-assessment-report-working-group-ii/. A compilation of findings and references on adaptation challenges.

IPCC 2022 *Climate Change 2022: Mitigation of Climate Change. Contribution of Working Group III to the Sixth Assessment Report of the Intergovernmental Panel on Climate Change* (Cambridge: Cambridge University Press) https://www.ipcc.ch/report/sixth-assessment-report-working-group-3/. Chapter 15 contains details and references on the challenge of financing the transition to a climate-positive world.

OECD 2021 *OECD Companion to the Inventory of Support Measures for Fossil Fuels 2021* (Paris: Organization for Economic Cooperating and Development) 30 March https://doi.org/10.1787/e670c620-en. Details on fossil fuel subsidies.

Parry I, Black S and Vernon N 2021 *Still Not Getting Energy Prices Right: A Global and Country Update of Fossil Fuel Subsidies* (Washington, DC: International Monetary Fund) Working Paper WP/21/236, 24 September https://www.imf.org/en/Publications/WP/Issues/2021/09/23/Still-Not-Getting-Energy-Prices-Right-A-Global-and-Country-Update-of-Fossil fuel-Subsidies-466004. A comprehensive analysis of fossil fuel subsidies and externalities.

UNEP 2021 *The Gathering Storm: Adapting to Climate Change in a Post-Pandemic World* (Nairobi: United Nations Environment Program) https://www.unep.org/resources/adaptation-gap-report-2021. This report is the most recent global analysis of adaptation challenges.

World Bank 2021 *State and Trends of Carbon Pricing 2021* (Washington, DC: World Bank) http://hdl.handle.net/10986/35620. This report gives the status on carbon pricing efforts around the world.

References

[1] IPCC 2022 *Climate Change 2022: Mitigation of Climate Change. Contribution of Working Group III to the Sixth Assessment Report of the Intergovernmental Panel on Climate Change*

(Cambridge: Cambridge University Press) https://ipcc.ch/report/sixth-assessment-report-working-group-3/

[2] Buchner B *et al* 2021 *Global Landscape of Climate Finance 2021* (San Francisco, CA: Climate Policy Initiative) https://climatepolicyinitiative.org/wp-content/uploads/2021/10/Full-report-Global-Landscape-of-Climate-Finance-2021.pdf

[3] UNEP 2021 *THE Gathering Storm: Adapting to Climate Change in a Post-Pandemic World* (Nairobi: United Nations Environment Program) https://unep.org/resources/adaptation-gap-report-2021

[4] Parry I, Black S and Vernon N 2021 Still not getting energy prices right: a global and country update of fossil fuel subsidies *Working paper* WP/21/236 International Monetary Fund, Washington, DC https://imf.org/en/Publications/WP/Issues/2021/09/23/Still-Not-Getting-Energy-Prices-Right-A-Global-and-Country-Update-of-Fossil-Fuel-Subsidies-466004

[5] Rennert K *et al* 2022 Comprehensive evidence implies a higher social cost of CO_2 *Nature* **610** 687–92

[6] Koomey J, Schmidt Z, Hausker K and Lashof D 2022 Exploring the black box: applying macro decomposition tools for scenario comparisons *Environ. Model. Softw.* **155** 105426

[7] Griffith S 2021 *Electrify: An Optimist's Playbook for Our Clean Energy Future* (Cambridge, MA: MIT Press)

[8] Griffith S 2020 *Solving Climate Change with a Loan* (San Francisco, CA: Otherlab) https://saulgriffith.com/blog/solving-climate-change-with-a-loan

[9] Osterwalder A and Pigneur Y 2010 *Business Model Generation: A Handbook for Visionaries, Game Changers, and Challengers* (Hoboken, NJ: Wiley)

[10] Griffith S 2021 *Electrify: An Optimist's Playbook for Our Clean Energy Future* (Cambridge, MA: MIT Press)

[11] Gore T 2020 *Confronting Carbon Inequality: Putting Climate Justice at the Heart of the COVID-19 Recovery* (Nairobi: Oxfam) https://oxfam.org/en/research/confronting-carbon-inequality

[12] Jones C M and Kammen D 2011 Quantifying carbon footprint reduction opportunities for US households and communities *Environ. Sci. Technol.* **45** 4088–95

[13] Kahn E 1988 *Electric Utility Planning and Regulation* (Washington, DC: American Council for an Energy-Efficient Economy)

[14] Lo H, Blumsack S, Hines P and Meyn S 2019 Electricity rates for the zero marginal cost grid *Electricity J.* **32** 39–43

[15] MIT 2016 *Utility of the Future: An MIT Energy Initiative Response to an Industry in Transition* (Cambridge, MA: Massachusetts Institute of Technology) http://energy.mit.edu/research/utility-future-study/

[16] Frew B, Milligan M, Brinkman G, Bloom A, Clark K and Denholm P 2016 Revenue sufficiency and reliability in a zero marginal cost future *Conf. paper* NREL/CP-6A20-66935 National Renewable Energy Lab (NREL), Golden, CO https://www.nrel.gov/docs/fy17osti/66935.pdf

[17] Bielen D, Burtraw D, Palmer K and Steinberg D 2017 The future of power markets in a low marginal cost world *Working paper* RFF WP 17-26 Resources for the Future, Washington, DC https://rff.org/publications/working-papers/the-future-of-power-markets-in-a-low-marginal-cost-world/

[18] Gramlich R and Lacey F 2020 *Who's the Buyer?: Retail Electric Market Structure Reforms in Support of Resource Adequacy and Clean Energy Deployment* (Washington, DC: Wind

Solar Alliance) https://windsolaralliance.org/wp-content/uploads/2020/03/WSA-Retail-Structure-Contracting-FINAL.pdf

[19] Aggarwal S, Corneli S, Gimon E, Gramlich R, Hogan M, Orvis R and Pierpont B 2019 *Wholesale Electricity Market Design for Rapid Decarbonization* (San Francisco, CA: Energy Innovation) https://energyinnovation.org/wp-content/uploads/2019/07/Wholesale-Electricity-Market-Design-For-Rapid-Decarbonization.pdf

[20] Corneli S 2020 A prism-based configuration market for rapid, low cost and reliable electric sector decarbonization *World Resources Institute Workshop on Market Design for the Clean Energy Transition: Advancing Long-Term Approaches (Washington, DC)* https://wri.org/events/2020/12/market-design-clean-energy-transition-advancing-long-term

[21] Gimon E 2020 Let's get organized! Long-term market design for a high penetration grid *World Resources Institute Workshop on Market Design for the Clean Energy Transition: Advancing Long-Term Approaches (Washington, DC, 16 December)* https://wri.org/events/2020/12/market-design-clean-energy-transition-advancing-long-term

[22] Tierney S F 2020 Wholesale power market design in a future low-carbon electric system: a proposal for consideration *White paper* 28 November Analysis Group, Boston, MA https://www.analysisgroup.com/Insights/publishing/wholesale-power-market-design-in-a-future-low-carbon-electric-system/

[23] Taylor J A, Dhople S V and Callaway D S 2016 Power systems without fuel *Renew. Sust. Energy Rev.* **57** 1322–36

[24] Pierpont B 2020 A market mechanism for long-term energy contracts to support electricity system decarbonization *World Resources Institute Workshop on Market Design for the Clean Energy Transition: Advancing Long-Term Approaches (Washington, DC)* https://wri.org/events/2020/12/market-design-clean-energy-transition-advancing-long-term

[25] WRI and RFF 2020 *Proc. of the Conf. on Market Design for the Clean Energy Transition: Advancing Long-Term Approaches* (Washington, DC: World Resources Institute and Resources for the Future) 16 and 17 December https://www.wri.org/events/2020/12/market-design-clean-energy-transition-advancing-long-term

[26] Brown T and Reichenberg L 2021 Decreasing market value of variable renewables can be avoided by policy action *Energy Econ.* **100** 105354

[27] Sepulveda N A, Jenkins J D, Edington A, Mallapragada D S and Lester R K 2021 The design space for long-duration energy storage in decarbonized power systems *Nat. Energy* **6** 506–16

[28] Mallapragada D S, Sepulveda N A and Jenkins J D 2020 Long-run system value of battery energy storage in future grids with increasing wind and solar generation *Appl. Energy* **275** 115390

[29] Das S, Hittinger E and Williams E 2020 Learning is not enough: diminishing marginal revenues and increasing abatement costs of wind and solar *Renew. Energy* **156** 634–44

[30] Johansson V, Thorson L, Goop J, Göransson L, Odenberger M, Reichenberg L, Taljegard M and Johnsson F 2017 Value of wind power—implications from specific power *Energy* **126** 352–60

[31] Cullenward D and Victor D G 2020 *Making Climate Policy Work* (New York: Polity Press)

[32] Weitzman M 1974 Prices versus quantities *Rev. Econ. Stud.* **61** 477–91

[33] World Bank 2021 *State and Trends of Carbon Pricing 2021* (Washington, DC: World Bank) http://hdl.handle.net/10986/35620

[34] Simpson R D 2022 How do we price an unknowable risk? *Issues in Science and Technology* Winter https://issues.org/social-cost-of-carbon-economics-simpson/

[35] IPCC 2018 *Global Warming of 1.5°C. An IPCC Special Report on the Impacts of Global Warming of 1.5°C Above Pre-industrial Levels and Related Global Greenhouse Gas Emission Pathways, in the Context of Strengthening the Global Response to the Threat of Climate Change, Sustainable Development, and Efforts to Eradicate Poverty* (Geneva: IPCC) https://ipcc.ch/sr15

[36] DeCanio S J 2003 *Economic Models of Climate Change: A Critique* (Basingstoke: Palgrave-Macmillan)

[37] Rennert K *et al* 2022 Comprehensive evidence implies a higher social cost of CO_2 *Nature* **610** 687–92

[38] Eilperin J and Brady D 2021 Biden is hiking the cost of carbon. It will change how the US tackles global warming *Washington Post* 26 February https://washingtonpost.com/climate-environment/2021/02/26/biden-cost-climate-change/

[39] Koomey J, Gordon D, Brandt A and Bergerson J 2016 *Getting Smart about Oil in a Warming World* (Washington, DC: Carnegie Endowment for International Peace) http://carnegieendowment.org/2016/10/04/getting-smart-about-oil-in-warming-world-pub-64784

[40] Gordon D 2022 *No Standard Oil: Managing Abundant Petroleum in a Warming World* (New York: Oxford University Press) https://nostandardoil.com

[41] Gordon D, Koomey J, Brandt A and Bergerson J 2022 *Know Your Oil and Gas: Generating Climate Intelligence to Cut Petroleum Industry Emissions* (Boulder, CO: Rocky Mountain Institute) https://rmi.org/insight/kyog/

[42] Brandt A R, Masnadi M S, Englander J G, Koomey J and Gordon D 2018 Climate-wise choices in a world of oil abundance *Environ. Res. Lett.* **13** 044027

[43] Jing L, El-Houjeiri H M, Monfort J-C, Brandt A R, Masnadi M S, Gordon D and Bergerson J A 2020 Carbon intensity of global crude oil refining and mitigation potential *Nat. Clim. Change* **10** 526–32

[44] Alvarez R A *et al* 2018 Assessment of methane emissions from the US oil and gas supply chain *Science* **361** 186–8

[45] Gordon D, Brandt A, Bergerson J and Koomey J 2015 *Know Your Oil: Creating a Global Oil-Climate Index* (Washington, DC: Carnegie Endowment for International Peace) http://goo.gl/Jly9Op

[46] Smith S J and Wigley M L 2000 Global warming potentials: 1. Climatic implications of emissions reductions *Clim. Change* **44** 445–57

[47] Smith S J and Wigley T M L 2000 Global warming potentials: 2. Accuracy *Clim. Change* **44** 459–69

[48] IPCC 2021 *Climate Change 2021: The Physical Science Basis. Contribution of Working Group I to the Sixth Assessment Report of the Intergovernmental Panel on Climate Change* ed V Masson-Delmotte *et al* (Cambridge: Cambridge University Press) https://ipcc.ch/report/sixth-assessment-report-working-group-i/

[49] Vohra K, Vodonos A, Schwartz J, Marais E A, Sulprizio M P and Mickley L J 2021 Global mortality from outdoor fine particle pollution generated by fossil fuel combustion: results from GEOS-Chem *Environ. Res.* **195** 110754

[50] Cohen A J *et al* 2017 Estimates and 25-year trends of the global burden of disease attributable to ambient air pollution: an analysis of data from the global burden of diseases study 2015 *Lancet* **389** 1907–18

[51] Muller N Z, Mendelsohn R and Nordhaus W 2011 Environmental accounting for pollution in the United States economy *Am. Econ. Rev.* **101** 1649–75

[52] Epstein P R *et al* 2011 Full cost accounting for the life cycle of coal *Ann. NY Acad. Sci.* **1219** 73–98

[53] Lee K and Greenstone M 2021 *Air Quality Life Index: Annual Update* (Chicago, IL: Energy Policy Institute, University of Chicago) https://aqli.epic.uchicago.edu/reports/

[54] Burnett R *et al* 2018 Global estimates of mortality associated with long-term exposure to outdoor fine particulate matter *Proc. Natl Acad. Sci.* **115** 9592

[55] Landrigan P J *et al* 2018 The Lancet commission on pollution and health *Lancet* **391** 462–512

[56] Schraufnagel D E *et al* 2019 Air pollution and noncommunicable diseases: a review by the Forum of International Respiratory Societies 2019; Environmental Committee, part 1: the damaging effects of air pollution *Chest* **155** 409–16

[57] Gao J *et al* 2018 Public health co-benefits of greenhouse gas emissions reduction: a systematic review *Sci. Total Environ.* **627** 388–402

[58] Choma E F, Evans J S, Hammitt J K, Gómez-Ibáñez J A and Spengler J D 2020 Assessing the health impacts of electric vehicles through air pollution in the United States *Environ. Int.* **144** 106015

[59] Dimanchev E G, Paltsev S, Yuan M, Rothenberg D, Tessum C W, Marshall J D and Noelle E S 2019 Health co-benefits of sub-national renewable energy policy in the US *Environ. Res. Lett.* **14** 085012

[60] WHO 2015 *Economic Cost of the Health Impact of Air Pollution in Europe: Clean Air, Health, and Wealth* (Copenhagen: World Health Organization Regional Office for Europe) https://euro.who.int/en/media-centre/sections/press-releases/2015/04/air-pollution-costs-european-economies-us$-1.6-trillion-a-year-in-diseases-and-deaths,-new-who-study-says

[61] Oudin A, Segersson D, Adolfsson R and Forsberg B 2018 Association between air pollution from residential wood burning and dementia incidence in a longitudinal study in Northern Sweden *PLoS One* **13** e0198283

[62] Ghosh R, Causey K, Burkart K, Wozniak S, Cohen A and Brauer M 2021 Ambient and household PM2.5 pollution and adverse perinatal outcomes: a meta-regression and analysis of attributable global burden for 204 countries and territories *PLoS Med.* **18** e1003718

[63] Taylor W L, Schuldt S J, Delorit J D, Chini C M, Postolache T T, Lowry C A, Brenner L A and Hoisington A J 2021 A framework for estimating the United States depression burden attributable to indoor fine particulate matter exposure *Sci. Total Environ.* **756** 143858

[64] Lobell D B, Tommaso S D and Burney J A 2022 Globally ubiquitous negative effects of nitrogen dioxide on crop growth *Sci. Adv.* **8** eabm9909

[65] Boucher J and Friot D 2017 *Primary Microplastics in the Oceans: A Global Evaluation of Sources and Sinks* (Gland: International Union for Conservation of Nature and Natural Resources)

[66] Leslie H A, van Velzen M J M, Brandsma S H, Vethaak A D, Garcia-Vallejo J J and Lamoree M H 2022 Discovery and quantification of plastic particle pollution in human blood *Environ. Int.* **163** 107199

[67] Parry I, Black S and Vernon N 2021 Still not getting energy prices right: a global and country update of fossil fuel subsidies *Working paper* WP/21/236 International Monetary Fund, Washington, DC https://imf.org/en/Publications/WP/Issues/2021/09/23/Still-Not-Getting-Energy-Prices-Right-A-Global-and-Country-Update-of-Fossil-Fuel-Subsidies-466004

[68] Erickson P, Down A, Lazarus M and Koplow D 2017 Effect of subsidies to fossil fuel companies on United States crude oil production *Nat. Energy* **2** 891–98

[69] Kotchen M J 2021 The producer benefits of implicit fossil fuel subsidies in the United States *Proc. Natl Acad. Sci.* **118** e2011969118

[70] Coady D, Parry I, Le N-P and Shang B 2019 *Global Fossil Fuel Subsidies Remain Large: An Update Based on Country-Level Estimates* (Washington, DC: International Monetary Fund) https://imf.org/en/Publications/WP/Issues/2019/05/02/Global-Fossil-Fuel-Subsidies-Remain-Large-An-Update-Based-on-Country-Level-Estimates-46509

[71] Geddes A *et al* 2020 *Doubling Back and Doubling Down: G20 Scorecard on Fossil Fuel Funding* (Winnipeg: International Institute for Sustainable Development) https://iisd.org/publications/g20-scorecard

[72] Krane J, Matar W and Monaldi F 2020 Fossil fuel subsidy reform since the Pittsburgh G20: a lost decade? *Research paper* Rice University's Baker Institute for Public Policy, Center for Energy Studies, Houston, TX

[73] UNEP and IISD 2019 *Measuring Fossil Fuel Subsidies in the Context of the Sustainable Development Goals* (Nairobi: United Nations Environment Program and International Institute for Sustainable Development) https://unep.org/resources/report/measuring-fossil-fuel-subsidies-context-sustainable-development-goals

[74] Bast E, Doukas A, Pickard S, van der Burg L and Shelagh W 2015 *Empty Promises: G20 Subsidies to Oil, Gas and Coal Production* (London: Overseas Development Institute and Oil Change International) http://priceofoil.org/2015/11/11/empty-promises-g20-subsidies-to-oil-gas-and-coal-production/

[75] Erickson P, van Asselt H, Koplow D, Lazarus M, Newell P, Oreskes N and Supran G 2020 Why fossil fuel producer subsidies matter *Nature* **578** E1–4

[76] Erickson P and Achakulwisut P 2021 *How Subsidies Aided the US Shale Oil and Gas Boom* (Stockholm: Stockholm Environment Institute) https://sei.org/publications/subsidies-shale-oil-and-gas/

[77] Bak-Coleman J B *et al* 2021 Stewardship of global collective behavior *Proc. Natl Acad. Sci.* **118** e2025764118

[78] OECD 2021 *OECD Companion to the Inventory of Support Measures for Fossil Fuels 2021* (Paris: Organization for Economic Cooperating and Development)

[79] Achakulwisut P, Erickson P and Koplow D 2021 Effect of subsidies and regulatory exemptions on 2020–2030 oil and gas production and profits in the United States *Environ. Res. Lett.* **16** 084023

[80] Prest B C and James H S 2021 *Climate Royalty Surcharges* (Washington, DC: National Bureau of Economic Research) https://nber.org/papers/w28564

[81] Ho J S, Shih J-S, Muehlenbachs L A, Munnings C and Krupnick A J 2018 Managing environmental liability: an evaluation of bonding requirements for oil and gas wells in the United States *Environ. Sci. Technol.* **52** 3908–16

[82] Raimi D, Nerurkar N and Bordoff J 2020 *Green Stimulus for Oil and Gas Workers: Considering a Major Federal Effort to Plug Orphaned and Abandoned Wells* (New York: Columbia University Center on Global Energy Policy and Resources for the Future) https://energypolicy.columbia.edu/research/report/green-stimulus-oil-and-gas-workers-considering-major-federal-effort-plug-orphaned-and-abandoned

[83] Davis L 2012 Modernizing bonding requirements for natural gas producers *Discussion paper* 2012-2 The Hamilton Project https://hamiltonproject.org/papers/modernizing_bonding_requirements_for_natural_gas_producers

[84] Weber J G, Ercoli T, Fitzgerald W, Nied P, Penderville M and Raabe E 2021 Identifying the end: minimum production thresholds for natural gas wells *Resour. Policy* **74** 102404

[85] Carbon Tracker 2022 Event horizon: a case study of holdback and the point of no return for decommissioning upstream oil and gas 'assets' *Carbon Tracker* July https://carbontracker. org/reports/event-horizon-a-case-study-of-holdback-analysis/

[86] Sadasivam N 2021 How bankruptcy lets oil and gas companies evade cleanup rules *Grist* 7 June https://grist.org/accountability/oil-gas-bankruptcy-fieldwood-energy-petroshare/

[87] Coady D, Parry I, Le N-P and Shang B 2019 *Global Fossil Fuel Subsidies Remain Large: An Update Based on Country-Level Estimates* (Washington, DC: International Monetary Fund) https://imf.org/en/Publications/WP/Issues/2019/05/02/Global-Fossil-Fuel-Subsidies-Remain-Large-An-Update-Based-on-Country-Level-Estimates-46509

[88] Desanlis H, Matsumae E, Roeyer H, Yazaki A, Ahmad M and Menon S 2021 *Funding Trends 2021: Climate Change Mitigation Philanthropy* (San Francisco, CA: ClimateWorks Foundation Global Intelligence) https://climateworks.org/report/funding-trends-2021-climate-change-mitigation-philanthropy/

[89] Brulle R J 2018 The climate lobby: a sectoral analysis of lobbying spending on climate change in the USA, 2000 to 2016 *Clim. Change* **149** 289–303

[90] Goldberg M H, Marlon J R, Wang X, van der Linden S and Leiserowitz A 2020 Oil and gas companies invest in legislators that vote against the environment *Proc. Natl Acad. Sci.* **117** 5111–2

[91] Dewan A 2021 Fossil fuel companies have over 500 people at COP26, more than any single country, report says *CNN* 8 November https://cnn.com/2021/11/08/world/cop26-climate-fossil-fuel-lobbying-intl/index.html

[92] InfluenceMap 2021 Corporate climate policy footprint: the 50 most influential companies and industry associations blocking climate policy action globally *Report* November InfluenceMap https://lobbymap.org/report/The-Carbon-Policy-Footprint-Report-2021-670f36863e7859e1ad7848ec601dda97

[93] Woellert L and Lefebvre B 2021 Washington's whipping boys *Politico* 6 July https://politico.com/newsletters/the-long-game/2021/07/06/washingtons-whipping-boys-493476

[94] Corporate Europe Observatory 2021 Stop the revolving door: fossil fuel policy influencers *Corporate Europe Observatory* 25 October https://corporateeurope.org/en/stop-revolving-door

[95] Bradham B, Tartar A and Warren H 2020 American politicians who vote against climate get more corporate cash *Bloomberg UK* 23 October https://bloomberg.com/graphics/2020-election-company-campaign-finance-climate-change/

[96] Toomey D 2017 How big money in politics blocked US action on climate change *Yale Environ. 360* 10 May https://e360.yale.edu/features/how-big-money-in-politics-blocked-u-s-action-on-climate-change

[97] House of Representatives, Committee on Oversight and Reform 2021 Analysis of the fossil fuel industry's legislative lobbying and capital expenditures related to climate change *Press Release* 28 October House of Representatives, Committee on Oversight and Reform, Washington, DC https://oversight.house.gov/news/press-releases/committee-analysis-of-fossil-fuel-industry-s-lobbying-reveals-public-praise-for

[98] Sunrise Project 2020 *The Passives Problem and Paris Goals: How Index Investing Trends Threaten Climate Action* (Sydney: The Sunrise Project) https://sunriseproject.org/wp-content/uploads/2020/01/Sunrise-Project-Report-The-Passives-Problem-and-Paris-Goals.pdf

[99] Quinson T 2022 Greenwashing is increasingly making ESG moot *Bloomberg* 16 March https://bloomberg.com/news/articles/2022-03-16/greenwashing-is-increasingly-making-esg-investing-moot-green-insight#xj4y7vzkg

[100] Egan M 2020 A $5 billion foundation literally founded on oil money is saying goodbye to fossil fuels *CNN* 18 December https://cnn.com/2020/12/18/investing/rockefeller-founda-tion-divest-fossil-fuels-oil/index.html

[101] Jeppesen H and Booth O 2021 A tale of two share issues: how fossil fuel equity offerings are losing investors billions *Carbon Tracker* https://carbontracker.org/reports/a-tale-of-two-share-issues/

[102] Bloomberg 2019 Climate ETF offers a scorecard on business environmental efficiency *Bloomberg* (video) https://bloomberg.com/news/videos/2019-09-11/climate-etf-offers-a-scorecard-on-business-environmental-efficiency-video

[103] Gordon K 2014 *Risky Business: The Economic Risks of Climate Change in the United States* (San Francisco, CA: Next Generation) http://riskybusiness.org/report/national/

[104] Parkin B 2016 Batteries may trip 'death spiral' in $3.4 trillion credit market *Bloomberg* 18 October https://bloomberg.com/news/articles/2016-10-18/batteries-may-trip-death-spiral-in-3-4-trillion-credit-market

[105] Matikainen S, Campiglio E and Zenghelis D 2017 *Policy Paper: The Climate Impact of Quantitative Easing* (London: Grantham Research Institute on ClimateChange and the Environment and Centre for Climate Change Economics and Policy) https://www.lse.ac.uk/granthaminstitute/publication/the-climate-impact-of-quantitative-easing/

[106] Pontecorvo E 2021 Meet the startup producing oil to fight climate change: how charm industrial became a go-to in big tech's mission to offset its carbon footprint *Grist* 18 May https://grist.org/climate-energy/lucky-charm/

[107] Koomey J G, Webber C A, Atkinson C S and Nicholls A 2001 Addressing energy-related challenges for the US buildings sector: results from the clean energy futures study *Energy Policy* **29** 1209–22

Chapter 11

Elevate truth

A lie can travel halfway around the world while the truth is putting on its shoes.
— Mark Twain

Chapter overview

- Because this fight is a political and moral one (in addition to having an economic dimension) elevating truth and fighting propaganda are critical to a successful transition to a climate-positive society.
- Reducing emissions means confronting the most powerful industry in human history and unwinding the many ways the economy has been tilted towards fossil fuels over more than a century.
- We need to stop treating the fossil fuel industry as legitimate participants in democratic debate and start treating them like the threat to the global climate that they've been for decades.
- We need to tell the truth about climate, root out fossil fuel funded corruption, mandate transparency and accurate disclosures, ban fossil fuel advertising, run mass-market public engagement campaigns, and hold the industry accountable for self-interested lies.

11.1 Introduction

Concern about climate change from increasing concentrations of greenhouse gases is based on some of the most well-established principles in physical science. The idea that greenhouse gases could warm the Earth is not a new one. We are now past the point where anyone needs to take seriously people who dispute what the US National Academy of Sciences [1] and virtually every other national scientific organization on the planet regard as 'settled facts': the Earth system is warming, humans are the cause, and continuously increasing greenhouse gas emissions will warm the Earth further [2].

Policy makers should understand that those who question those high-level findings and argue against immediate and rapid emissions reductions are no longer engaged in the scientific process, but in a propaganda exercise [3–5]. Of course, there will continue to be open questions about some details of climate science, and that's as it should be, but those details will not change the core findings of the science. The results have been confirmed in so many independent ways and by so many measurements that it is appropriate to consider the core findings as 'settled science', similar to gravity or relativity. Debate will continue on details, but the existence of continued scientific debate on those details is no reason to slow or stop efforts to reduce emissions to zero as quickly as we can [2].

Stoddard *et al* [6] explored why we've failed to reduce emissions decades after climate change was well understood to be a serious societal problem: 'a common thread that emerges across the reviewed literature is the central role of power, manifest in many forms'. The reality is that solving the climate problem means confronting and defeating the most powerful industry in human history. It is a power struggle to create the future, not simply a scientific debate [7, 8].

For too long, media outlets, government regulators, and scientists have taken objections to climate science as good-faith arguments, but that assumption has enabled the forces of denial and delay [3, 5]. Fighting against such tactics isn't easy, but the scientific community has learned a lot over the years. One key resource in that fight is the *Debunking Handbook 2020* [9], which outlines effective ways to combat disinformation and misinformation on climate and other topics. Others have started to apply public pressure to identify and publicize misleading fossil fuel advertising campaigns [10].

The broader fight to educate and train the public in the critical thinking skills needed to recognize and debunk misinformation is important for more than just climate change. These skills can be taught [11–13] and they are essential for preserving functioning democracies in the twenty-first century, but by themselves are no guarantee of an informed citizenry. It is incumbent on policy makers, aided by the scientific community and responsible companies, to continue to educate the public about the nature, scope, and urgency of the climate problem.

It is also important to develop new ways to combat misinformation, such as requiring social media and traditional media organizations to run climate public service announcements and prevent misleading content from running in the first place [10, 14]. Social media can do much better at limiting proliferation of climate misinformation, just as they now attempt to do for cigarettes, pornography, medical misinformation, human trafficking, and messages that promote terrorism. It's a complex and rapidly evolving area of law and policy, but without action against polluter-funded misinformation, achieving zero emissions will happen much more slowly than it should.

11.2 Public understanding about climate lags the science

The most comprehensive study of global views about climate change was conducted in 2021 by researchers from the Yale Program on Climate Change Communication, who used sophisticated demographic sampling techniques to survey over 76,000 people from 31 countries and territories [15]. This study illuminates several key facts about the current state of public understanding about climate.

First, the vast majority of people (79% to 94% in the countries sampled) understand that climate change is happening, and most people are concerned that climate disruptions will harm their own lives (57% to 92%) as well as future generations (59% to 91%). Second, a majority thinks climate action should be a high priority for governments (53% to 91%), and large majorities in most countries (with the exceptions of Egypt and Saudi Arabia) agree that governments should do more to address climate change.

While most people around the world are concerned about climate change, most people still don't understand that our current climate change is caused almost entirely by human activity, particularly the burning of fossil fuels [15]. As illustrated in figure 11.1, less than half of adults in most countries surveyed understand that climate change is caused 'mostly by human activities'. As shown in figure 11.2, even smaller percentages of people understand that solving climate change will require burning 'much less' oil, coal, and natural gas than we use today. We must bridge these gaps in understanding about the root causes of climate change if we're to implement climate solutions at the appropriate scale. We also need to study the rhetorical techniques used by the industry to confuse public understanding [16].

11.3 What we must do

Most people know that dangerous climate changes are happening, but they don't know why. We now need to help people understand what is causing climate change and what must be done to solve it. We also need to stop treating the fossil fuel industry and other large polluters as legitimate participants in policy debates and start treating them like the threat to the global climate that they've been for decades. That means telling the full truth about climate pollution, rooting out fossil fuel funded corruption, mandating transparency and accurate disclosures, banning fossil fuel advertising, running countervailing public engagement campaigns, and holding the industry accountable for self-interested lies.

11.3.1 Tell the truth about climate

We summarize the simple truths about climate change in chapter 1, following the framework from Nicholas [17]:

- It's warming.
- It's us.
- We're sure.
- It's bad.
- We can fix it, but we'd better hurry.

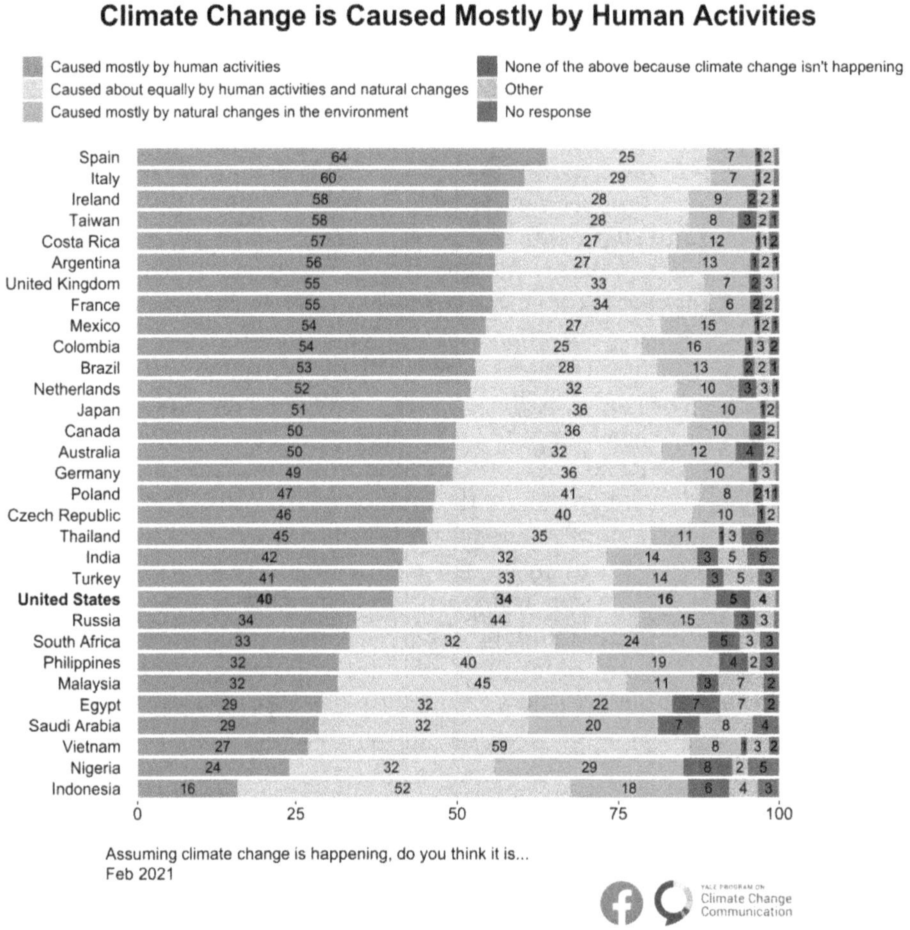

Climate Change is Caused Mostly by Human Activities

- Caused mostly by human activities
- Caused about equally by human activities and natural changes
- Caused mostly by natural changes in the environment
- None of the above because climate change isn't happening
- Other
- No response

Country	Caused mostly by human activities	Caused about equally by human activities and natural changes	Caused mostly by natural changes in the environment	None of the above because climate change isn't happening	Other	No response
Spain	64	25	7	1	2	1
Italy	60	29	7	1	2	1
Ireland	58	28	9	2	2	1
Taiwan	58	28	8	3	2	1
Costa Rica	57	27	12	1	1	2
Argentina	56	27	13	1	2	1
United Kingdom	55	33	7	2	3	1
France	55	34	6	2	2	1
Mexico	54	27	15	1	2	1
Colombia	54	25	16	1	3	2
Brazil	53	28	13	2	2	1
Netherlands	52	32	10	3	3	1
Japan	51	36	10	1	2	
Canada	50	36	10	3	2	
Australia	50	32	12	4	2	
Germany	49	36	10	1	3	1
Poland	47	41	8	2	1	1
Czech Republic	46	40	10	1	2	
Thailand	45	35	11	1	3	6
India	42	32	14	3	5	6
Turkey	41	33	14	3	5	3
United States	40	34	16	5	4	
Russia	34	44	15	3	3	
South Africa	33	32	24	5	3	3
Philippines	32	40	19	4	2	3
Malaysia	32	45	11	3	7	2
Egypt	29	32	22	7	7	2
Saudi Arabia	29	32	20	7	8	4
Vietnam	27	59	8	1	3	2
Nigeria	24	32	29	8	2	5
Indonesia	16	52	18	6	4	3

Assuming climate change is happening, do you think it is...
Feb 2021

YALE PROGRAM ON
Climate Change
Communication

Figure 11.1. Public opinion in major countries about the most important cause of climate change. Source: Reproduced with permission from *International Public Opinion on Climate Change* 2021 [15]. Yale University and Facebook Data for Good.

Solving this problem will create a better world, saving trillions of dollars and millions of lives every year. From society's perspective, the economic case for rapid climate action has never been more clear (and it gets clearer every day as we learn more and technology improves). While acting immediately would drive costs of future action down, every moment of delay makes cost of inaction larger, so delay is folly. We need immediate action that will reduce global emissions to zero as quickly as possible.

Climate solutions can always be made better, but existing technologies can get us most of the way to zero emissions, and most can scale immediately given changes in incentives and policies. Scale (which drives learning effects) is what will drive costs down for everyone, making climate mitigation an even better deal than it is today.

As we saw in global survey data above, most people understand that 'it's warming' and 'it's bad'. Fewer people understand that 'we're sure' that 'it's us'

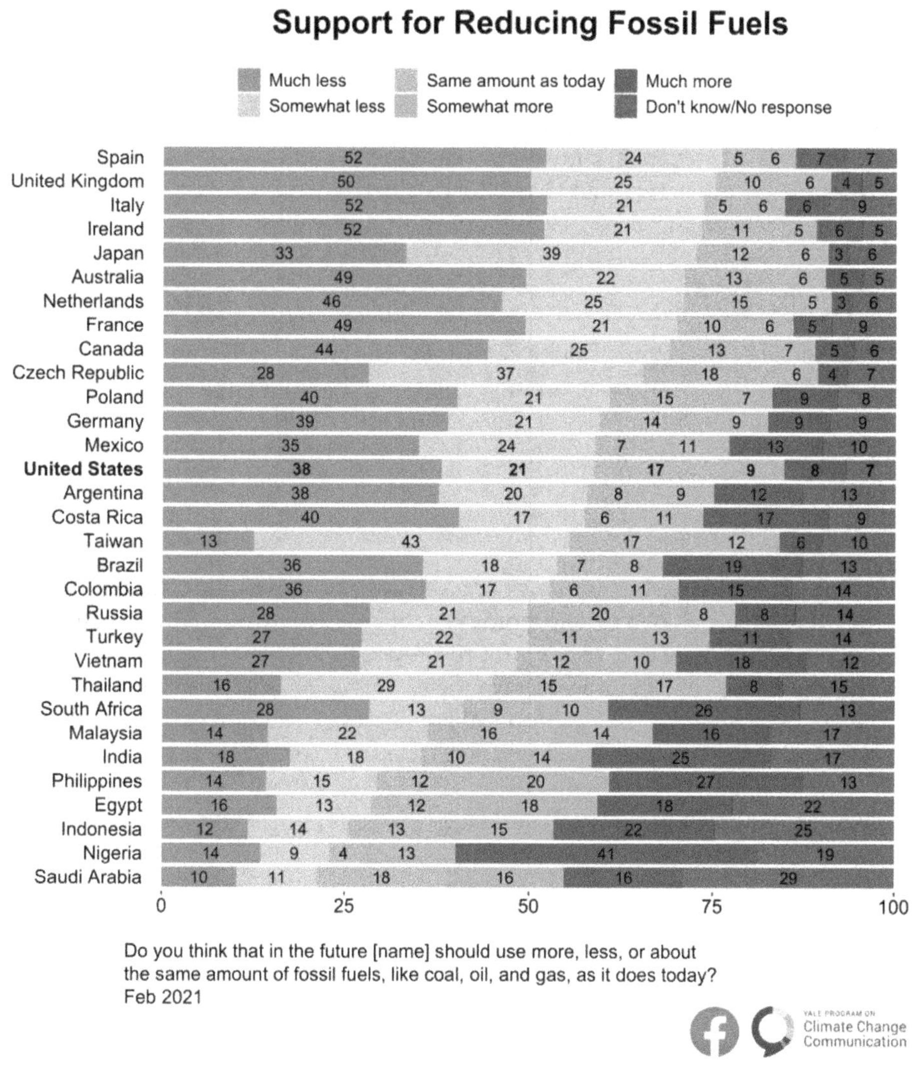

Support for Reducing Fossil Fuels

Legend: Much less | Same amount as today | Much more | Somewhat less | Somewhat more | Don't know/No response

Country	Much less	Somewhat less	Same amount as today	Somewhat more	Much more	Don't know/No response
Spain	52	24	5	6	7	7
United Kingdom	50	25	10	6	4	5
Italy	52	21	5	6	6	9
Ireland	52	21	11	5	6	5
Japan	33	39	12	6	3	6
Australia	49	22	13	6	5	5
Netherlands	46	25	15	5	3	6
France	49	21	10	6	5	9
Canada	44	25	13	7	5	6
Czech Republic	28	37	18	6	4	7
Poland	40	21	15	7	9	8
Germany	39	21	14	9	9	9
Mexico	35	24	7	11	13	10
United States	38	21	17	9	8	7
Argentina	38	20	8	9	12	13
Costa Rica	40	17	6	11	17	9
Taiwan	13	43	17	12	6	10
Brazil	36	18	7	8	19	13
Colombia	36	17	6	11	15	14
Russia	28	21	20	8	8	14
Turkey	27	22	11	13	11	14
Vietnam	27	21	12	10	18	12
Thailand	16	29	15	17	8	15
South Africa	28	13	9	10	26	13
Malaysia	14	22	16	14	16	17
India	18	18	10	14	25	17
Philippines	14	15	12	20	27	13
Egypt	16	13	12	18	18	22
Indonesia	12	14	13	15	22	25
Nigeria	14	9	4	13	41	19
Saudi Arabia	10	11	18	16	16	29

Do you think that in the future [name] should use more, less, or about the same amount of fossil fuels, like coal, oil, and gas, as it does today? Feb 2021

YALE PROGRAM ON Climate Change Communication

Figure 11.2. Public opinion in major countries about whether fossil fuel use should be reduced. Source: Reproduced with permission from *International Public Opinion on Climate Change* 2021 [15]. Yale University and Facebook Data for Good.

causing the warming. Even fewer understand how 'we can fix' our climate crisis. Challenging these misconceptions can help generate the political will necessary to enact needed change and encourage citizens to make more climate-friendly choices.

One of the most effective ways to convince people about the need for urgent action on climate is to explain that at least 97% of scientific articles on climate change support the consensus position that Earth is warming, humans are responsible, and that rapid emissions reductions are the solution [18–21]. Emphasizing that we're in a climate emergency, making climate risks personal, tying climate action to deeply held values, and explaining how we can fix the problem together (making a better world in the process) are key to communicating the threat and the opportunity of climate change in ways that will reach most audiences [22–26].

It's also essential to use the right climate messengers for each intended audience. As any marketing executive knows, people more readily accept information from trusted sources they respect who share their values. Who delivers a climate message and how it is delivered can be just as important as the content of the message.

11.3.2 Engage the public with climate truths through information campaigns

Considering the scope of our climate challenges, it's remarkable that massive climate public engagement campaigns aren't already commonplace. Effective climate communications should come from governments, traditional and online media, environmental nonprofits, and billions of individual conversations within trusted communities. We must share climate impacts and solutions in personal and actionable ways, but too often climate communications are vague and disempowering.

Effective climate communications should combine modern information technology with the same behavioral psychology techniques that marketing wizards have used for decades. There are now many resources to help make climate communications more effective, including ecoAmerica's *Communicating on Climate* 13 *Steps and Guiding Principles* [27], publications from the Yale Program on Climate Change Communications[1], and over a decade of research from the broader Behavior, Energy, and Climate Change (BECC) academic community[2]. Climate communicators at all levels should use these guides to personalize climate messages for everyone.

Governments should fund aggressive and frequent public service announcements to educate the public about the dangers of fossil fuels and the benefits of climate solutions. They should also mandate that traditional and online media platforms spread approved climate messages crafted with the guidance of science and marketing experts. We know that counter advertising worked to reduce tobacco use, there's no reason it wouldn't also work well for enhancing climate understanding and motivating climate action [28]. Governments should also fund nonprofits and mission-driven companies to invent new climate information technology tools that innovate new ways to engage the public in climate action.

Both traditional and online media should more effectively engage the public in climate action. Every time climate-related extreme weather is in the news, journalists should highlight the climate links, explain that fossil pollution is the main cause, and emphasize that solutions are ready to scale. Big technology companies also have a unique ability and responsibility to engage the public in climate action. Tech companies should be pushed to harness the same algorithms that they use to microtarget advertising to maximize public engagement on climate.

---

[1] https://climatecommunication.yale.edu
[2] https://beccconference.org/

Nonprofits and foundations can also be effective in spreading climate messages to wider audiences. Historically, large green groups have focused climate campaigns on reaching out to existing members to donate more and engage with elected leaders. While these 'preaching to the choir' campaigns likely have some benefits, these nonprofit-funded climate messages don't reach most people. Campaigns focused on partisan climate politics may also exacerbate political divides and make progress on policy more difficult.

Nonprofit climate messaging has also often deprioritized actions that individuals can take in their own lives, such as plant-based foods and electrifying homes and transportation. Calls for climate lifestyle changes are perceived as less popular and beneficial for nonprofit donations. While political engagement on climate is also important, donor-driven organizations should broaden the scope and reach of their climate messages, since many climate impacts come from individual choices. Both secular and religious nonprofit organizations can also spread powerful calls to climate action that emphasize the moral and social justice dimensions of our climate crisis, with some leaders like Pope Francis already highlighting the moral imperative of climate action to millions of followers.

Last but certainly not least, we all have a role to play in communicating climate truths within our communities. Many people best absorb information from trusted friends, family members, neighbors, colleagues, and community leaders. We can all be effective climate communicators for those who trust us, and climate messages that resonate with each of us as individuals will likely also resonate with friends and family. Online media technology tools also can empower us to amplify our most powerful climate truths to a much wider audience. But even for those with no desire to communicate more widely, every conversation is an opportunity to spread climate truths and advocate for action.

11.3.3 Require climate education

Public climate engagement is much easier when everyone starts from common understanding. Climate literacy is now as vital as basic math and science, since climate change will affect many aspects of our lives, from lifestyle choices to political decisions to our careers, to our personal safety. Mandating climate education can ensure better public understanding of key facts about our climate challenges and solutions, including how local communities should prepare for climate disruptions while moving to zero emissions as quickly as possible.

Effective climate curriculum should start with educating the educators. Most teachers devote little time to climate science in their classrooms, and educators who do teach about climate change often feel underqualified or even teach inaccurate information. A 2016 *Science* article found that a majority of middle- and high-school teachers in the US include only an hour or two of instruction about climate change over the course of an entire academic year, and 30% of teachers devoted less than an hour [29]. Moreover, despite 79% of Americans supporting climate science education, 70% of middle school science teachers and 55% of high-school science teachers in the US do not understand the scientific consensus that climate change is caused primarily by fossil fuels and other human activities [30], and 40% of teachers who include climate into their science curriculum teach it inaccurately [31]. The climate education situation is likely better in some countries, but in many countries it is likely worse, given global polling on levels of climate understanding.

Italy has become the first country to mandate climate science in public education [32], making knowledge of climate fundamentals as essential as reading, writing, and arithmetic. Other countries are now starting to follow Italy's lead, including Finland[3] and Mexico[4], and education regulators in cities, counties, provinces and states can set climate education standards when national governments fail to act. There are many curriculum resources and standards on climate already freely available from nonprofit, government, and academic groups, including the Next Generation Science Standards, the National Center for Science Education (NCSE), Stanford University's Climate Change Education Project, and ACE (formerly the Alliance for Climate Education, now renamed Action for the Climate Emergency).

Basic knowledge of climate science and solutions can also be added as requirements for college and university degrees, as well as professional certifications, including teaching credentials. Climate solutions and disruptions will affect many careers, so professionals should a working knowledge of relevant climate information, from electricians preparing homes for a future free of oil and gas, to architects designing for floods and wildfires, to investment advisors shifting clients to climate-positive investing strategies. Companies of all sizes can incorporate climate solutions into their employee training programs, whether it's educating chefs how to cook with induction electric ranges or instructing product designers how to make more climate-friendly design decisions. Climate education can and should come from many sources, and there are many available resources to do it better.

11.3.4 Root out fossil-funded corruption

Just stating the truth isn't enough to win this battle. It's a political fight, and that means confronting powerful interests directly and working toward an orderly shutdown of the fossil fuel industry. It also means rooting out polluter-funded corruption everywhere it exists, because such corruption allows climate polluters to elevate their mistruths and exercise power effectively.

As discussed in earlier chapters, many of the largest climate polluters are also the largest political donors and lobbyists, and there's often a revolving door between government and lucrative industry lobbying jobs. This is legalized corruption, and it can be found almost everywhere, at all levels of government.

One of many case studies in such corruption is Pacific Gas and Electric (PG&E), the California utility whose poorly maintained infrastructure has caused over 1500 fires in the past decade [33], resulting in convictions for dozens of felony crimes [34], over 250,000 claims from victims of wildfires and natural gas explosions, and over

---

[3] https://thegeep.org/learn/countries/finland
[4] https://www.weforum.org/agenda/2020/07/mexico-fighting-climate-change-classroom

400 deaths[5]. PG&E continues to be one of the largest political donors in California, paying millions of dollars every year to political lobbyists and Democratic and Republican politicians alike. About 80% of politicians in California have taken money from PG&E[6], and 69% of PG&E political lobbyists have previously held government jobs[7].

There are also many examples of illegal direct bribes paid by climate polluters to regulators or politicians, and that blatant corruption is clearly important and pernicious. According to researchers at Stanford Law School, the oil and gas industry has been the most frequent violator of the Foreign Corrupt Practices Act[8], and large climate polluters have been caught in the act of bribing US officials as well. An outrageous recent example is in the US state of Ohio, where a major state utility bribed state and local politicians so they would pass legislation favorable to that company [35].

It's also important to understand that 'soft corruption' is more widespread and can be even more insidious than direct political donations, lobbying, and illegal bribery. Funding fake experts, subtly influencing university research through major donations, and trumpeting misleading research to foster predatory delay are common tactics of this powerful industry [3], but often these tactics fly under the radar. They all need to be exposed and fought.

Systemic corruption is enabled by money in politics. The US has allowed unlimited corporate spending to influence elections and equated political donations with free speech, enshrining *Citizens United* as the law of the land[9]. This approach is counterproductive to the goal of rapid emissions reductions and the US (and other countries moving down this path) need to reverse it to enable more rapid progress on climate and many other important issues. Unlimited corporate money makes it much easier for the fossil fuel industry and other polluters to impede climate action.

Corporations are not people and money is not speech. In a true democracy, those with more money should not have a louder voice than those with less. In that sense, the fight for strengthening democracy is also a fight for climate action and justice.

11.3.5 Mandate transparency and accurate climate disclosures

Taking action against climate change requires accurate information. For governments, deciding which companies to regulate or cajole depends on metrics that reflect climate reality. For companies, having accurate metrics for climate success is needed to drive institutional change and marketing zero emissions products. For investors, knowing which companies are truly facing the climate challenge is

---

[5] https://www.firevictimtrust.com/Docs/Letter_from_the_Trustee_6-21-22.pdf
[6] https://www.abc10.com/article/news/local/abc10-originals/fire-power-money-california-wildfires-investigation-pge/103-c273fb35-1c43-4d9a-9bdc-3d7971e5540b
[7] https://www.opensecrets.org/orgs/pg-e-corp/summary?topnumcycle=2022&toprecipcycle=2022&contribcycle=2022&outspendcycle=2022&id=D000000290&lobcycle=2022
[8] https://fcpa.stanford.edu/industry?industry=Oil+%26+Gas
[9] https://www.fec.gov/legal-resources/court-cases/citizens-united-v-fec/

essential for making informed investment choices. For individuals, knowing what works and what doesn't is critical for deciding how best to reduce emissions.

In every case, disclosure of accurate climate risks, costs, benefits, and full life-cycle emissions is needed for informed decision making. Government has a critical role to play in developing such metrics and mandating companies to accept and use them, just as they now do for financial instruments and food nutrient labels. Investors and donors can help by supporting research on such metrics.

France has set an example by passing the Climate and Resilience Law in 2021, which contains a sweeping set of climate-related measures, including mandates for climate labeling on many consumer products and requirements that products must prove climate-related claims with full disclosure of calculation methodology [36][10].

11.3.6 Ban advertising and sponsorships for fossil fuel technologies

When the US finally decided that smoking was a threat to public health, a ban on tobacco advertising followed in 1971 [37]. Subsequently, the World Health Organization developed the framework convention for tobacco bans on advertising, which had been adopted by countries representing only about 12% of the world's population by 2014[11]. The evidence for the effectiveness of partial bans on tobacco advertising is mixed, but comprehensive bans have generally been found to be more effective at reducing tobacco consumption [28, 38–42].

In France, the Climate and Resilience Law already bans ads for fossil fuel companies, and the advertising ban will extend to polluting vehicles by 2028, while local governments have been given authority to further regulate ads for fossil advertising and sponsorships [43]. Many other governments, nonprofits, media outlets, advertising professionals, and citizen groups are now also exploring other bans on fossil technology advertising, including diverse efforts in over a dozen countries[12], and a European Citizens' Initiative pushing to ban advertising and sponsorship from climate polluting industries throughout Europe[13]. Advertising for fossil fuels and fossil-burning technologies can also be forced to include warning labels related to climate and human health damages, similar to the health warnings on tobacco products in many countries, or climate-polluter ads can be required to promote climate-friendly alternatives. In addition to advertising bans, France has also required that vehicle ads must now promote walking, cycling, or public transport instead of driving [44], alongside a hashtag urging climate-friendly transportation.

It's worth exploring whether a ban on fossil fuel advertising would speed up the transition to zero emissions. The fossil fuel industry, automakers, airlines, and other large polluters clearly believe advertising works because they keep doing it. Perhaps

---

[10] https://www.climate-laws.org/geographies/france/laws/law-no-2021-1104-on-the-fight-on-climate-change-and-resilience
[11] https://www.who.int/europe/health-topics/tobacco/banning-tobacco-advertising-sponsorship-and-promotion
[12] https://verbiedfossielereclame.nl/only-words/
[13] https://banfossilfuelads.org/about-us/

it's time to start clamping down on that advertising, just as some important countries have done for tobacco.

11.3.7 Deplatform climate misinformation and disinformation

To elevate climate truths we also need to curb the bombardment of climate lies that come through both traditional and online media. Climate disinformation is often funded by fossil fuel interests with the goal of causing public doubts about climate change's true severity, causes, and solutions, following a similar playbook that the tobacco industry deployed for over half a century [3, 4]. Political TV news stations and talk radio have been spreading climate disinformation for decades, but online media and social media silos have expanded the spread of climate mistruths, both deliberate (disinformation) and unintentional (misinformation).

The largest tech companies have now also become the largest spreaders of climate misinformation and disinformation. The core of the problem is the 'attention economy' business model that maximizes advertising revenue by feeding users content that maximizes engagement. Searching for climate content on social media or YouTube can quickly lead to content by climate denialists, and as people engage with this disinformation then tech platform algorithms feed them similar disinformation. This algorithmic amplification of climate lies is happening everywhere, and while some of the large tech companies have taken tepid steps to combat the problem, they have financial incentives to not do very much, since deplatforming popular sources of disinformation can anger users and reduce revenues.

It doesn't have to be this way. News media once did much more to elevate facts of deliberately distorted fictions, and technology companies have all the tools they need to filter out climate lies. But substantial government regulation will be needed to root harmful disinformation out of traditional and online media. Given what's at stake, we should treat climate disinformation the same way we treat harmful content that's already subject to mandatory editorial filters, such as child pornography, propaganda from terrorists, and others advocating violence and hate. The European Digital Services Act[14] should be a big step in this direction, with the new regulations expected to force tech giants to remove a wide range of harmful disinformation from their platforms [45], including climate disinformation [46], though much will depend on how the new laws are enforced [47]. Other governments can now follow Europe's lead, and both new and old media companies should be pushed to do much more to stem the flow of mistruths.

11.3.8 Hold the fossil fuel industry and other bad actors accountable for climate lies

The fossil fuel industry has known about climate risks for decades yet has repeatedly lied and misled the public about this issue [3, 4, 48–53]. It is critical for government, individuals, companies, and others who support climate action to publicize and correct these lies as loudly as they can. It is also important for government to explore

---

[14] https://ec.europa.eu/commission/presscorner/detail/en/ip_22_2545

the use of anti-fraud and other legal measures to bring consequences to bear on individuals and companies who knowingly spread lies to advance their own interests.

Large repeat distributors and funders of climate disinformation could even face criminal penalties beyond fines, including prison time and suspensions of rights to do business. The landmark French Climate and Resilience Law[15] codified the new crime of 'ecocide', with large illegal polluters soon subject to criminal penalties. Given that climate disinformation likely leads to more pollution, it is logical to expand ecocide-type laws to include similar criminal liabilities for deliberate mass distribution of climate lies. Unrestricted climate change will bring horrific consequences for billions of people around the world, and the people causing it should see consequences for lying about that risk for private gain.

11.4 Chapter conclusions

The fight to stabilize the climate is primarily a political and moral fight, and we need to treat it that way. For too long, the scientific community has argued with fossil fuel funded fake experts as if they were good-faith participants in public debate about technical issues. Reframing these actors as the self-interested agents of predatory delay makes it clear that a different approach to this misinformation is needed if we are to reduce emissions rapidly enough.

It is also important to focus on climate change as the emergency that it is. People and institutions can move quickly when there is an emergency, but many lack that sense of 'emergency urgency' for climate change. The extended time horizons for consequences make this problem challenging but treating it as an emergency requiring an immediate response is an entirely proportional to the real climate risks we face.

Human-induced climate change is a scientific reality, and its decisive mitigation is a moral and religious imperative for humanity… We are not faced with two separate crises, one environmental and the other social, but rather one complex crisis which is both social and environmental.

—Pope Francis

Further reading

Franta B 2018 Early oil industry knowledge of CO_2 and global warming *Nat. Clim. Change* **8** 1024–5 https://doi.org/10.1038/s41558-018-0349-9. Franta shows that the oil industry knew climate change was a real danger to society as early as the 1950s.

Franta B 2021 Weaponizing economics: Big Oil, economic consultants, and climate policy delay *Environ. Politics* **31** 555–75 https://doi.org/10.1080/09644016.2021.1947636. Franta documents how the oil industry used economic modelers and consultants to foster predatory delay starting as early as the 1980s.

---

[15] https://www.iea.org/policies/12874-climate-and-resilience-law

Franta B 2021 Early oil industry disinformation on global warming *Environ. Politics* **30** 663–8 https://doi.org/10.1080/09644016.2020.1863703. A history of oil industry-funded propaganda in the 1980s.

Lewandowsky S *et al* 2020 *The Debunking Handbook 2020* (St Lucia, Australia: George Mason University Center for Climate Change Communication) https://sks.to/db2020. The definitive primer on how to fight misinformation and disinformation, with applications to climate change.

Mann M E 2021 *The New Climate War: The Fight to Take Back Our Planet* (New York: Public Affairs). A clear-eyed recent view of the political and scientific struggles about climate change.

Markowitz E, Hodge C, Harp G, St John C, Marx S M, Speiser M, Zaval L and Perkowitz R 2014 *Connecting on Climate: A Guide to Effective Climate Change Communication* (New York: Center for Research on Environmental Decisions (CRED) at Columbia University and ecoAmerica) https://doi.org/10.7916/d8-pjjm-vb57. A terrific guide to effective climate communications.

Oreskes N and Conway E M 2010 *Merchants of Doubt: How a Handful of Scientists Obscured the Truth on Issues from Tobacco Smoke to Global Warming* (New York: Bloomsbury). This eye-opening book traces common threads among those arguing against government action to protect the public good.

Oreskes N 2015 The fact of uncertainty, the uncertainty of facts and the cultural resonance of doubt *Phil. Trans. R. Soc.* A **373** 20140455 https://doi.org/10.1098/rsta.2014.0455. A more recent and academic treatment covering some similar ground to *Merchants of Doubt*.

Schneider S H 2009 *Science as a Contact Sport: Inside the Battle to Save Earth's Climate* (Washington, DC: National Geographic). The politics and science of climate change are front and center in this classic book from Schneider.

Stoddard I *et al* 2021 Three decades of climate mitigation: why haven't we bent the global emissions curve? *Annu. Rev. Environ. Resources* **46** 653–89 https://doi.org/10.1146/annurev-environ-012220-011104. This article is the definitive social science analysis of the reasons we've failed to truly face the climate challenge.

References

[1] NAS 2010 *Advancing the Science of Climate Change* (Washington, DC: National Academy of Sciences) http://www.nap.edu/catalog.php?record_id=12782

[2] IPCC 2021 *Climate Change 2021: The Physical Science Basis. Contribution of Working Group I to the Sixth Assessment Report of the Intergovernmental Panel on Climate Change* ed V Masson-Delmotte *et al* (Cambridge: Cambridge University Press) https://www.ipcc.ch/report/sixth-assessment-report-working-group-i/

[3] Oreskes N and Conway E M 2010 *Merchants of Doubt: How a Handful of Scientists Obscured the Truth on Issues from Tobacco Smoke to Global Warming* (New York: Bloomsbury)

[4] Oreskes N 2015 The fact of uncertainty, the uncertainty of facts and the cultural resonance of doubt *Phil. Trans. R. Soc.* A **373** 20140455

[5] Lamb W F, Mattioli G, Levi S, Roberts J T, Capstick S, Creutzig F, Minx J C, Müller-Hansen F, Culhane T and Steinberger J K 2020 Discourses of climate delay *Global Sustainab.* **3** e17

[6] Stoddard I *et al* 2021 Three decades of climate mitigation: why haven't we bent the global emissions curve? *Annu. Rev. Env. Resour.* **46** 653–89

[7] Mann M E 2021 *The New Climate War: The Fight to Take Back Our Planet* (New York: Public Affairs)

[8] Schneider S H 2009 *Science as a Contact Sport: Inside the Battle to Save* Earth's Climate (Washington, DC: National Geographic)

[9] Lewandowsky S *et al* 2020 *The Debunking Handbook 2020* (St Lucia, Australia: George Mason University Center for Climate Change Communication) https://sks.to/db2020

[10] Timperley J 2021 Advertising with a conscience *Lancet Planet. Health* **5** e118–9

[11] Koomey J 2017 *Turning Numbers into Knowledge: Mastering the Art of Problem Solving* 3rd edn (El Dorado Hills, CA: Analytics)

[12] Mayfield M 2013 *Thinking for Yourself: Developing Critical Thinking Skills Through Reading and Writing* 9th edn (Belmont, CA: Wadsworth)

[13] Hughes W, Lavery J and Doran K 2014 *Critical Thinking: An Introduction to the Basic Skills* 7th edn (Peterborough, ON: Broadview)

[14] Satariano A 2022 EU takes aim at social media's harms with landmark new law *The New York Times* 22 April https://www.nytimes.com/2022/04/22/technology/european-union-social-media-law.html

[15] Leiserowitz A, Carman J, Buttermore N, Wang X, Rosenthal S, Marlon J and Mulcahy K 2021 *International Public Opinion on Climate Change* (New Haven, CT: Yale Program on Climate Change Communication and Facebook Data for Good) https://climatecommunication.yale.edu/publications/international-public-opinion-on-climate-change

[16] Supran G and Oreskes N 2021 Rhetoric and frame analysis of ExxonMobil's climate change communications *One Earth* **4** 696–719

[17] Nicholas K 2021 *Under The Sky We Make: How To Be Human In A Warming World* (New York, NY: G.P. Putnam's Sons) https://www.penguinrandomhouse.com/books/665274/under-the-sky-we-make-by-kimberly-nicholas-PhD/

[18] Cook J *et al* 2016 Consensus on consensus: a synthesis of consensus estimates on human-caused global warming *Environ. Res. Lett.* **11** 048002

[19] Oreskes N 2004 The scientific consensus on climate change *Science* **306** 1686

[20] Doran P T and Zimmerman M K 2009 Examining the scientific consensus on climate change *Eos Trans. AGU* **90** 22–3

[21] van der Linden S, Leiserowitz A, Rosenthal S and Maibach E 2017 Inoculating the public against misinformation about climate change *Glob Chall.* **1** 1600008

[22] Fischetti M 2021 We are living in a climate emergency, and we're going to say so *Sci. Am.* 12 April https://www.scientificamerican.com/article/we-are-living-in-a-climate-emergency-and-were-going-to-say-so/

[23] Hickel J 2021 What would it look like if we treated climate change as an actual emergency? *Current Affairs* 15 November https://www.currentaffairs.org/2021/11/what-would-it-look-like-if-we-treated-climate-change-as-an-actual-emergency

[24] Markowitz E *et al* 2014 *Connecting On Climate: A Guide To Effective Climate Change Communication* (New York: Center for Research on Environmental Decisions (CRED) at Columbia University and ecoAmerica)

[25] Spratt D and Morton J 2018 *How To Communicate A Climate Emergency: Messaging For Safe Climate Mobilisation* (Sydney: Breakthrough Center for Climate Restoration) https://www.breakthroughonline.org.au/guides

[26] Renkl M 2022 How to talk about 'extreme weather' with your angry uncle *The New York Times* 25 July https://www.nytimes.com/2022/07/25/opinion/climate-change-conservatives.html?referringSource=articleShare

[27] EcoAmerica 2013 *Communicating On Climate: 13 Steps And Guiding Principles* (Washington, DC: EcoAmerica) http://ecoamerica.org/wp-content/uploads/2013/11/Communicating-on-Climate-13-steps_ecoAmerica.pdf

[28] Saffer H 2004 *The Effect of Advertising on Tobacco and Alcohol Consumption* (Cambridge, MA: National Bureau of Economic Research)

[29] Plutzer E, McCaffrey M, Hannah A L, Rosenau J, Berbeco M and Reid A H 2016 Climate confusion among US teachers *Science* **351** 664–5

[30] Cheskis A, Marlon J, Wang X and Leiserowitz A 2018 *Americans Support Teaching Children about Global Warming* (New Haven, CT: Yale Program on Climate Change Communication) https://climatecommunication.yale.edu/publications/americans-support-teaching-children-global-warming/

[31] NCSE, Plutzer E and Hannah A L 2016 *Mixed Messages: How Climate Change is Taught in America's Public Schools* (Oakland, CA: National Center for Climate Science Education) https://ncse.ngo/mixed-messages-how-climate-change-taught-americas-public-schools

[32] Bentson C 2019 Italy makes climate change education compulsory: Italy will be the first country to mandate climate change studies in schools *ABC News* 7 November https://abcnews.go.com/International/italy-makes-climate-change-education-compulsory/story?id=66817133

[33] Russell G, Blunt K and Smith R 2019 PG&E sparked at least 1,500 California fires. Now the utility faces collapse *Wall Street J.* 13 January https://www.wsj.com/articles/pg-e-sparked-at-least-1-500-california-fires-now-the-utility-faces-collapse-11547410768

[34] Rodriguez O R and Liedtke M 2022 PG&E reaches $55 million deal to avoid criminal charges in counties ravaged by recent wildfires *KQED* 11 April https://www.kqed.org/news/11910835/pge-reaches-55-million-deal-to-avoid-criminal-prosecution-in-counties-ravaged-by-recent-wildfires

[35] USA Today Network Ohio Bureau 2021 Selling out in the statehouse: a timeline deep dive inside HB 6 and the biggest government scandal in Ohio state history *The Columbus Dispatch* 3 June https://www.dispatch.com/in-depth/news/politics/2021/06/03/ohio-corruption-house-bill-6-bribery-timeline-larry-householder/5248218001/

[36] Aurelien B 2021 France passes climate law, but critics say it falls short *The New York Times* https://www.nytimes.com/2021/07/20/world/europe/france-climate-law.html

[37] Brandt A M 2007 *The Cigarette Century: The Rise, Fall, and Deadly Persistence of the Product that Defined America* (New York: Basic)

[38] Saffer H and Chaloupka F 2000 The effect of tobacco advertising bans on tobacco consumption *J. Health Econ.* **19** 1117–37

[39] Quentin W, Neubauer S, Leidl R and König H H 2007 Advertising bans as a means of tobacco control policy: a systematic literature review of time-series analyses *Int. J. Public Health* **52** 295–307

[40] Blecher E 2008 The impact of tobacco advertising bans on consumption in developing countries *J. Health Econ.* **27** 930–42

[41] Lange T, Hoefges M and Ribisl K M 2015 Regulating tobacco product advertising and promotions in the retail environment: a roadmap for states and localities *J. Law, Med. Ethics* **43** 878–96

[42] Kasza K A, Hyland A J, Brown A, Siahpush M, Yong H-H, McNeill A D, Lin L and Cummings K M 2011 The effectiveness of tobacco marketing regulations on reducing smokers' exposure to advertising and promotion: findings from the International Tobacco Control (ITC) Four Country Survey *Int. J. Env. Res. Public Health* **8** 321–40

[43] RFI 2022 France clamps down on 'zero carbon' advertising to avoid greenwashing *RFI* 16 April https://www.rfi.fr/en/france/20220416-france-clamps-down-on-zero-carbon-advertising-to-avoid-greenwashing

[44] Ormesher E 2022 France mandates car ads must urge viewers to walk, cycle or take public transport instead *The Drum* 6 January https://www.thedrum.com/news/2022/01/06/france-mandates-car-ads-must-urge-viewers-walk-cycle-or-take-public-transport

[45] Tidey A 2022 EU strikes deal to force tech giants to tackle disinformation *Euro News* 25 April https://www.euronews.com/my-europe/2022/04/22/eu-on-cusp-of-deal-to-force-tech-giants-to-tackle-disinformation

[46] Hanley S 2022 EU Digital Services Act is a big deal for climate advocates *Clean Technica* 25 April https://cleantechnica.com/2022/04/25/eu-digital-services-act-is-a-big-deal-for-climate-advocates/

[47] Satariano A 2022 EU takes aim at social media's harms with landmark new law: the Digital Services Act would force Meta, Google and others to combat misinformation and restrict certain online ads. How European officials will wield it remains to be seen *The New York Times* 22 April https://www.nytimes.com/2022/04/22/technology/european-union-social-media-law.html

[48] Benjamin L, Bhargava A, Franta B, Toral K M, Setzer J and Tandon A 2022 *Policy Briefing: Climate-Washing Litigation–Legal Liability for Misleading Climate Communications* (London: The Climate Social Science Network) https://www.lse.ac.uk/granthaminstitute/publication/climate-washing-litigation-legal-liability-for-misleading-climate-communications/

[49] Franta B 2018 Early oil industry knowledge of CO_2 and global warming *Nat. Climate Change* **8** 1024–5

[50] Franta B 2021 Weaponizing economics: big oil, economic consultants, and climate policy delay *Environ. Politics* **31** 555–75

[51] Franta B 2021 Early oil industry disinformation on global warming *Environ. Politics* **30** 663–8

[52] Bonneuil C, Choquet P-L and Franta B 2021 Early warnings and emerging accountability: Total's responses to global warming, 1971–2021 *Global Environ. Change* **71** 102386

[53] McMullen J 2022 The audacious PR plot that seeded doubt about climate change *BBC News* 23 July https://www.bbc.com/news/science-environment-62225696

Solving Climate Change
A guide for learners and leaders
Jonathan Koomey and Ian Monroe

Chapter 12

Bringing it all together

In preparing for battle, I have always found that plans are useless, but planning is indispensable.

—General Dwight D Eisenhower

Chapter overview

- The process of analyzing scenarios for a climate-positive world is akin to telling good stories about the world you want to create.
- The end-result of your climate-positive plan will be summarized in a graph showing major contributors to emissions reductions relative to a 'current trends continued' scenario.
- The best approach to creating your plan is to use simple spreadsheet models. Complex models can be useful for estimating key parameters as input to the spreadsheet model, but they often limit the flexibility of your thinking and require lots of supporting data and analysis. Those constraints mean they are not the right tool for creating your climate-positive scenario.
- Solving climate is a team sport, requiring a range of skills and knowledge as well as flexibility in responding appropriately as reality evolves.
- Key cross-cutting issues include balancing costs, risks, and tradeoffs, addressing interactions between measures, confronting issues of equity and justice, treating population, health, and education with care, incorporating adaptation and resilience, and truly making governance work.
- Identify emissions reduction options with co-benefits such as reduced air and water pollution, labor savings, or other productivity improvements. These options are likely to be the cheapest for society so they should feature prominently in your climate-positive plan.
- Pinpoint the biggest and fastest growing sources of emissions and address those first, moving down the list until you've identified how to eliminate emissions for every important task in the economy.

12.1 Introduction

In previous chapters we described the building blocks of credible emission reduction plans. This chapter describes the process of pulling those data together in a coherent way.

For each major task (such as water heating, cooking, or mobility), scenario planning exercises project key drivers of energy and emissions into the future assuming current trends continue, then project an alternative path for each task that moves society to become climate positive by 2040. The purpose of the exercise is to develop intuition and quantitative understanding around how hard that transition will be.

No one can predict the future of human systems with accuracy. The purpose of this exercise is to use knowledge of current trends, technology characteristics, human and institutional behavior, and physical principles to create a simple model and use it to conduct systematic 'thought experiments' [1]. The futurist John Platt called this 'mapping the future', which is a way to explore a range of possibilities to help people create the future they prefer. Another name for such work is scenario analysis.

12.2 Telling a good story[1]

The purpose of scenario analysis, as explained by Peter Schwartz in his now classic book *The Art of the Long View,* is to explore several possible futures in a systematic way [2]. No matter how things turn out, the analyst will have on the shelf a scenario (or story) that resembles that future and will have thought through in advance the likely consequences of that scenario on the decisions at hand. Such a story 'resonates in some ways with what [people] already know and leads them from that resonance to reperceive the world.'

The steps for successful scenario development are as follows:

- *Define the question:* Scenarios should be developed in advance of some major strategic decision so that when the time to decide arrives, the options are laid out in detail and the implications of each decision path are made manifest. Such a decision might be 'How can we most rapidly reduce our company's direct emissions to zero?'
- *Determine the key factors affecting the system:* The next step is to identify the key factors determining whether the focal decision will succeed or fail. These factors might include such things as customer acceptance of a new technology or competitor response to the new strategy. They could also include demographics, inflation, commodity prices, technological change, public policies, or political instability. Some of them are beyond the control of any individual or institution ('predetermined elements'), and others are subject to human influence.

---

[1] This section is adapted from Koomey [1]. Used with permission.

- *Evaluate factors by importance and uncertainty:* Of the trends and driving forces identified in the previous step, a small number will be both highly important and highly uncertain (see also Morgan and Henrion's book *Uncertainty* [3]). These are the issues upon which differences in the scenarios will be based. Predetermined elements don't differ among scenarios.
- *Choose the scenario set:* The most important and uncertain factors should be combined in various ways to define the set of scenarios to be considered. These factors are the dimensions of uncertainty that the scenarios will explore. For example, the scenario set for an analysis of the costs of reducing carbon emissions might include variations in fuel prices and costs of different technologies, which are two of the key sources of uncertainty. One scenario (a worst case for carbon reductions) might combine low fossil fuel prices with pessimistic technology costs for non-fossil sources. Another scenario might combine high fossil fuel prices with optimistic technology costs. The set of scenarios chosen also frames the story lines that will emerge from the scenarios.
- *Add detail to the scenarios:* Each of the key driving forces and factors from the second step should be assessed for each of the main scenarios. Usually, these factors can be assigned relative rankings (e.g. 'Inflation should be higher than average in scenario 1 and lower than average in scenario 2'), which can then be translated into actual numbers. The exact numbers are not so important as long as they reflect in a credible way people's expectations about how the future might unfold in a particular scenario. The scenario must then be woven into a plausible story describing a course of events that would result in the scenario's final outcome. Each scenario must be named in a way that vividly reminds people of its essence.
- *Determine the implications*: Revisit the decision identified in step 1, and assess how that decision would fare in each of the different scenarios. If the decision (e.g. rapidly reducing emissions) would lead to success (as defined in step 2) for all or most of the scenarios considered, that decision is probably a wise one. If it is likely to be successful in only one scenario, the decision is 'a high-risk gamble…especially if the company has little control over the likelihood of the required scenario coming to pass.'
- *Define metrics for success:* Once the scenarios have been developed, it is crucial to identify a few key indicators by which decision-makers can monitor the evolution of events. These indicators are used to determine which of the different scenarios best matches up with actual events and can be used to trigger an institutional response (e.g. 'if we miss our emissions reduction target by more than 5% in any year, then do X'). If you can't measure it, you can't manage it, so defining appropriate indicators is crucial to success.

Schwartz believes that scenarios are successful when 'they are both plausible and surprising; when they have the power to break old stereotypes; and when the makers assume ownership of them and put them to work'. The institutional process of creating scenarios (what Schwartz calls 'holding a strategic conversation') can also reveal new opportunities for an organization and promote team building among its employees.

Another name for scenario analysis is the commonly used scientific technique of creating a thought experiment. All scientists use this technique, in which they imagine certain real-world complexities to be absent and then think through the consequences for the situation they are investigating. For example, Galileo imagined a world in which the complicating effects of friction did not exist and, by doing so, was able to make great strides in what would come to be called the field of kinetics (the study of moving objects). Such thought experiments are as valuable in business as in science. Instead of limiting your imagination by current circumstances and constraints, imagine a world in which some key constraints don't exist, and ask yourself how you or your institution might respond.

Most of Schwartz's examples of scenario analysis have only a small quantitative component, but many other futurists are obsessed with numbers and computer models. Most explorations of the future with which we are directly familiar err by focusing too much on the mechanics of forecasting and quantitative analysis (e.g. on modeling tools) and far too little on careful scenario development based on the procedures described above. Quantitative analysis can lend coherence and credence to scenario exercises by elaborating on consequences of future events, but modeling tools should support that process rather than drive it, as is so often the case.

The relationship between data, anecdotes, models, and scenarios is summarized in figure 12.1, adapted from Ghanadan and Koomey [4]. Raw data are difficult both to analyze and interpret, so people rely more on anecdotes (which illustrate concepts well but have little quantitative rigor) or models (the results of which often don't tell a coherent story but are at least easy to evaluate in a quantitative way). Good

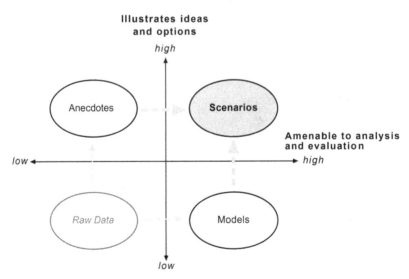

Figure 12.1. How scenarios relate to models, stories, and data. Source: Koomey [1], adapted from Ghanadan and Koomey [4]. Reproduced with permission.

scenarios combine the narrative power of anecdotes with the additional quantitative rigor that comes from the use of models (even simple ones).

In our years of involvement with development of energy and climate policy, we have been most struck by how few resources are devoted to sensible scenario development and associated data and how many are devoted to the development of different modeling tools to assess such policies. Computer tools are sexy and appealing (at least to the funding agencies). Data and scenario analysis, upon which the results generally hinge, are virtually always given short shrift.

Tens of millions of dollars are spent every year on models whose capabilities are redundant with others, usually because an agency with money wants its own in-house modeling capability and is unwilling, because of institutional rivalry or personal biases, to adopt a pre-existing framework. Policy makers fail to realize that models are ALL unable to predict the future in an accurate way and that small improvements in modeling methodology are made irrelevant by inadequate scenario development.

12.3 The end of the journey

The end goal is to create with a graph that looks like figure 12.2 (which is the same as figure 3.2), taken from the scenarios detailed in [5], using decomposition methods from [6, 7]. This graph uses 100 year global warming potentials (GWPs). We supplement this graph with one that shows cumulative savings by category from 2020 to 2050 for both 20 year and 100 year GWPs, as in figure 12.3.

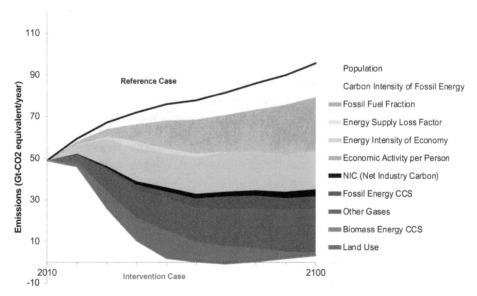

Figure 12.2. Sources of emissions reductions over time from a widely cited zero-emissions scenario. Source: Scenario results from [5], decomposed using methods from [6, 7]. There are no savings from biomass CCS and a tiny increase in emissions from economic activity per person (not shown) in this scenario.

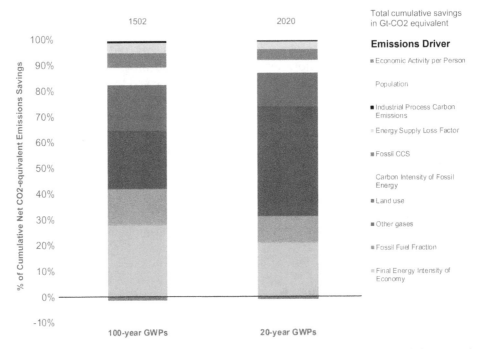

Figure 12.3. Cumulative emissions reductions from 2020 to 2050 from a widely cited zero-emissions scenario. Source: Scenario results from [5], decomposed using methods from [6, 7]. There are no savings from biomass CCS and a tiny increase in emissions from economic activity per person (not shown) in this scenario.

The reference case envisions a world in which current trends continue. The intervention case reaches net-zero emissions by 2040. As with all such analyses, there are quirks to investigate and explain.

This scenario is one of the few in which changes in population growth affect emissions. Lower population growth results in this scenario from efforts to promote education and empower women [5].

Net emissions hit zero but then rise. Why should that be? The existence of carbon capture in this scenario allows the continued operation of fossil gas plants that would otherwise be retired, and the improvements in the energy intensity of the economy that are rapid in the first few decades of the scenario slow down dramatically mid-century. These driving factors, which are embodied in a large and complex model, lead to an increase in primary energy demand and consequent increase in fossil fuel use towards the end of the twenty-first century.

These are the kinds of questions that systematically analyzing zero-emissions scenarios should raise, which is one reason conducting such exercises has value. A world that has achieved net-zero emissions shouldn't (and probably wouldn't) allow backsliding in total emissions, and it shows why using carbon capture could lead to a perverse result. Models are not reality, but they can help highlight issues to consider and pitfalls to avoid.

Your climate-positive scenario will likely have categories of emissions reductions that differ from those in figures 12.2 and 12.3. The availability of data and local circumstances will dictate which options you explore.

12.4 Big models or simpler spreadsheets?

The most widely used integrated assessment models use large and complicated models to capture the many interactions that affect the feasibility and speed of emissions reductions. While these interactions are sometimes important, the use of big models (the care and feeding of which require significant resources) also has disadvantages.

Making key assumptions transparent is much easier with simpler models. Big models also often contain buried assumptions in their data and structure [1]. A recent example is in [8], which showed that a common assumption (that climate damages and climate investments should be treated equally by rational actors in integrated assessment models) has significant implications for the results. Weighting losses more heavily than gains, as many decision-makers do, implies much faster mitigation than is typically shown in benefit-cost analyses for climate.

The data and structure of these models often lags recent developments because reality changes fast and the effort to update the models is significant. For example, a systematic review of treatment of wind and solar technologies in emissions reduction scenarios found 'the trend of rapid cost declines has been structurally under-estimated in virtually all future energy scenario analyses and suggest that even the most recent studies refer to obsolete or very conservative values' [9]. There's even a name for that effect, 'assumption drag', which was coined by William Ascher in his classic book on forecasting from late 1970s [10].

The structure of models also sometimes limits the questions analysts ask. For example, most models have difficulty implementing accelerated capital retirement scenarios, and many scenarios simply ignore it, but as we've discussed in previous chapters, it's a critical part of sensible and suitably rapid emissions reductions plans [11, 12].

It is for the reasons above that we encourage students to build simple spreadsheet models for their climate-positive scenarios. This approach teaches the students about data, key drivers, and underlying interdependencies more reliably than just running a big model would. It also allows students to incorporate many kinds of data (including results from big modeling exercises) in a way that allows for more comprehensive analysis. The time constraints inherent in a one quarter class also dictate a simpler approach, but the main substantive reasons for this choice are the ones listed above.

12.5 Solving climate is a team sport

As we discuss in chapter 2, climate change is the ultimate adaptive challenge, because the rate and scope of the changes needed to solve the problem will stretch us to the limit. In addition, the solutions must involve changes in behavior and institutional structure, not just technology, because the problem is so pressing. It will also require us to learn how to work together in new ways, and to grow and evolve rapidly.

Ertel and Solomon write

It's nearly impossible for any one senior executive–or small leadership team–to solve adaptive challenges alone. They require observations and insights from a wide range of people who see the world and your organization's problems differently. And they require combining those divergent perspectives in a way that creates new ideas and possibilities that no individual would think up on his or her own. [13: p 10]

As Ertel and Solomon argue, tackling adaptive challenges requires 'strategic conversations' that help define the problem and generate innovative ideas for solving it. That's why we think of solving the climate problem as a 'team sport'.

That means choosing a team with different perspectives and skills will make solving the problem easier. It means being open to ideas from outside your group. And it means using a simpler more inclusive spreadsheet model that makes it easier to incorporate 'outside the box' ideas rather than a more rigid and complex modeling tool. Remember, there's no such thing as accuracy in 2050, just do the best you can (and document everything!)

12.6 Cross-cutting issues

12.6.1 Balancing costs, benefits, and tradeoffs

Adaptive challenges are complicated and solving climate change will involve shifting money flows of tens of trillions of dollars over the next few decades [14]. A real emissions reduction plan needs to consider the costs and benefits of each possible action, and while we don't require students to fully assess scenario economics we do encourage them to consider costs and benefits in choosing the options for their plan.

That means reading literature on costs for different options and comparing results across studies. It also means focusing on mass produced technologies where they make sense, because those are most likely to benefit from learning-by-doing related cost reductions.

12.6.1.1 Background
Many economic modelers try to find the 'optimal' solution for the climate problem. They make estimates of the costs of reducing emissions and the avoided damages from increased emissions (as in figure 12.4) and characterize the point where the two

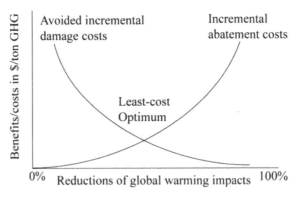

Figure 12.4. Standard application of benefit-cost analysis to the climate problem. Source: Reproduced with permission from Koomey [23].

curves cross as the optimal one [15]. That point is where the marginal cost of reducing emissions is equal to the marginal benefits from reducing them.

In this view, reducing emissions beyond that point would imply that we would have paid too much for emissions reductions—the costs for incremental emissions reductions would exceed the benefits. This particular model has worked reasonably well for short-term analyses of other kinds of pollution, such as sulfur dioxide, which is why economists have tried to apply it to the climate problem.

The most prominent example is that of Nobel prize winning professor William Nordhaus, the father of benefit-cost analysis for climate. In his 2018 Nobel acceptance speech he said:

> ...one of the most amazing results of IAMs is the ability to calculate the optimal carbon price...This concept represents the economic cost caused by an additional ton of carbon dioxide emissions (or more succinctly carbon) or its equivalent...In an optimized climate policy (abstracting away from various distortions), the social cost of carbon will equal the carbon price or the carbon tax.

Nordhaus discusses optimality (with associated 'uncertainty') based on his tacit belief that it is possible to characterize accurately the real risk to the climate of our continued emissions of greenhouse gases in terms of costs and benefits decades and centuries hence.

The McKinsey cost curve [16, 17] shown in figure 3.3 is an attempt to replicate the stylized curve of incremental abatement costs in figure 12.3. The literature on 'conservation supply curves' contains many other examples [18–22]. While this way of representing costs has heuristic value, this mode of analysis ultimately fails for climate.

First, the idea of accurately calculating costs and benefits decades and centuries hence strains credulity. The literature on forecasting accuracy for energy systems and many other technical areas should give pause to anyone who holds this belief [10, 24–32]. What can 'optimal' mean when the marginal costs and benefits can't be estimated with precision?

Another reason why this simple model doesn't work is the presence of so many societal co-benefits of climate action [33], with these co-benefits dispersed throughout the society. If society invests in energy efficiency, for example, it makes a single capital investment that reduces greenhouse gas emissions, energy use, and other pollutants. While there have been many attempts to add co-benefits to the supply and demand curve framework, they always suffer from the inability to precisely allocate parts of the capital investment to each of the co-benefits, and when the costs and benefits are experienced by different economic actors, things get even more complicated. The idea of marginal costs and benefits governing societal decision making in this context seems to us to be problematic.

The simple model also fails because it assumes certain things about how the economy works, such as no increasing returns to scale, perfect information, and perfect rationality, but those mostly don't apply to real economies. It's a highly stylized model that can yield insights in simple situations, but the climate problem is far too complex. The underlying assumption flows from the view that there IS a unique optimal path, based on outputs of models that assume constant or decreasing returns to scale and no other sources of path dependence.

Once increasing returns are introduced to models (and these effects are pervasive throughout the real economy) then the idea of a single unique optimal path no longer holds [34, 35]. For example, the classic analysis in [36] introduced stylized learning effects in a model of future climate scenarios and found that strong path dependence was the result. The model results indicated two clusters of scenarios with essentially similar costs but vastly different emissions, indicating that there isn't one unique optimal path, just lots of possible paths, and we get to choose which one we prefer.

The economist Brian Arthur, in [34], explains the intellectual history of increasing returns in economic modeling. The problem of path dependence from increasing returns has been well known for decades and many economists acknowledge it as an issue, but their stylized models almost always fall back on standard production functions such as Cobb–Douglas (which assumes constant returns to scale). As Arthur argues, this choice has historically been justified by economists preferring models that yield a single unique equilibrium, for computational convenience as well as elegance, but that's not the way the real world works.

12.6.1.2 Focus on total societal costs and benefits

For the reasons described above we think the simplified marginal costs and benefits approach doesn't work for climate. Instead, we advocate comparing *total net societal costs* (*after adjusting for co-benefits*) for getting to zero emissions for multiple scenarios. It's fine to use more micro studies to give guidance on which lower emissions resources might be cheaper than others, but it's ultimately total net societal costs that should inform and govern our strategic choices.

This realization also points towards the importance of whole system integrated design as the path forward, as discussed in chapter 3 and in [37–39]. Co-benefits from climate action come in part from integrated design strategies that consider the whole system and count costs and benefits for accomplishing the tasks at hand.

For example, a zero-energy home with a super-tight shell and heat recovery ventilation can often get by with a tiny heat pump for heating and cooling, and sometimes no heating system is needed at all, even in cold climates [40]. Similarly, moving to a full electric vehicle allows for cost savings and design opportunities that don't exist for plug-in hybrids that include both fossil fuel and electric power trains.

Long-time energy researcher Amory Lovins [41] calls this approach 'tunneling through the cost barrier' to identify whole system options that would be overlooked when analysis is focused strictly on marginal costs. Redesigning the system from scratch to accomplish the task at hand can ultimately result in lower total system costs even if there are localized increases in marginal costs along the way.

12.6.1.3 Opportunity costs

The idea of opportunity costs comes from economics. Investing in one emissions reduction option means you can't spend that money on something else. Getting to zero emissions will be hard and cost trillions, but we'd be spending most of that money anyway to replace the current energy system over time. This time we get to spend it on energy sources that cost less in societal terms than the fuels they displace, and society as a whole will be better off than it would otherwise be (in part because the co-benefits of climate action, in terms of reduced pollution of all kinds, are so large [42–49]).

Opportunity costs come into play in two major ways:

1. Fast emissions reductions require effective deployment of resources. Choosing the options that are already the cheapest or will soon likely be cheapest in direct cost terms if we expand production means we get more bang for the buck.
2. The opportunity costs of moving too slowly are big but avoidable. Keeping fossil resources online too long, expanding fossil infrastructure that will soon become stranded assets, and not replacing the worst polluting fuels as quickly as we can all have deleterious climate and other environmental effects. If we move more quickly we can avoid these opportunity costs.

The danger, in our view, is not in spending too much on climate mitigation as conventional economic thinking would indicate, the danger is in spending too little and moving too slowly. We've dithered for so long that little about the transition to zero emissions will be optimal and now we just need to move as quickly as we can. We're in a climate emergency, and that should change our calculus about opportunity costs.

12.6.1.4 Political tradeoffs

One of the biggest issues in designing a zero-emissions world is that individual and institutional interests often diverge from society's interests. This conclusion is particularly true for fossil fuel investors, but it applies more broadly, and it points towards the importance of power and politics in designing a successful transition [50]. We are not political scientists, sociologists, or politicians, and we view our role as explaining what is necessary to preserve a livable world so that others can design strategies to move the politics in the right direction.

The primacy of power in managing the transition comes to the fore both within and among countries. As we have argued, solving the climate problem means ending fossil fuels, and that means powerful interests will lose money and influence. There's no happy ending for the fossil fuel industry, only the management of a just transition in which their influence and power will be vastly diminished.

There may be ways to mollify fossil fuel companies in the near term, and there may also be opportunities for them to diversify into other energy sources. For example, the oil and gas industry's technical and logistical skills would be applicable to tapping geothermal energy and building offshore wind turbines, but the history of large companies managing big societal shifts does not make us optimistic that many of these companies will survive the transition.

12.6.2 Interactions between measures

Some emissions reductions interact. For example, large-scale use of biomass energy or reforestation can affect food production, and reductions in the carbon intensity of electricity supply can reduce the emissions savings from fuel switching. When creating a climate stabilization plan it is critical to ensure that the most important of these interactions are treated explicitly, and that the assumptions about those interactions are well documented. Otherwise, there is a real danger of double counting savings.

12.6.3 Equity and climate justice

Because the costs and benefits of climate action are distributed unequally across time, space, income, race, nationality, and age, decisions on balancing climate mitigation with adaptation and suffering are at their core *moral choices*, not solely questions of economics, a discipline predominantly focused on economic efficiency [51–57].

The faster and more deeply we reduce emissions, the more rapidly we'll reduce damages from this problem, the more quickly we'll address long-standing questions about equity and justice, and the more completely we'll correct disparities between the wealthy people who are predominantly the cause of unrestricted climate change (and who have the means to alleviate the problem), and disadvantaged people, poorer people, or people yet unborn who bear most of the costs but have few (or no) means to cope with those changes.

The IPCC's Working Group III report [42] discusses the concept of a 'just transition' towards sustainability, with the goal of not imposing 'hardship on already marginalized populations within and between countries', 'just transition':

....refers to a set of principles, processes and practices aimed at ensuring that no people, workers, places, sectors, countries or regions are left behind in the move from a high-carbon to a low-carbon economy. It includes respect and dignity for vulnerable groups; creation of decent jobs; social protection; employment rights; fairness in energy access and use and social dialogue and democratic consultation with relevant stakeholders.

The emphasis on fairness, justice, and equity highlights the moral dimensions of the climate challenge.

12.6.4 Energy security, food/water security, and geopolitics

The importance of energy and climate to geopolitics was made clear again in 2022 when Russia invaded Ukraine. Natural gas and oil prices spiked around the world, a reflection of the globalized trade in these energy products.

Dependence on volatile fuels has been a concern since the 1970s [58–61]. The focus of policy makers has traditionally been on energy *imports* and on *energy independence*, but this focus is misguided. When fuels are traded on global markets it is dependence on the fuel itself that exposes countries to price volatility and economic shocks [62].

Fortunately, climate action and geopolitical stability can go hand in hand. Getting to zero emissions means ending fossil fuels, which are the source of fuel price volatility. Dependence on these fuels creates the risk of extortion from fossil fuel exporting nations (such as Russia, on whose fossil gas Europe traditionally depends, and Saudi Arabia, which is still one of the key swing producers for the world's oil markets).

Fossil fuels also affect in-country politics for most nations. In the US the power of fossil fuel producers makes passing climate legislation difficult or impossible in some states and at the Federal level. Similar issues have arisen in Australia, the UK, Canada, and in many other fossil fuel producing nations.

Dependence on Russia's natural gas has made a comprehensive response to Russia's invasion of Ukraine difficult, but the invasion and subsequent Russian threats have made clear the costs of dependence on fossil fuels supplied by a hostile power. Breaking Russia's grip on Europe is one of several reasons why phasing out fossil fuels as rapidly as we can is in the best interests of society at large.

Similar issues arise for conflict between nations driven by water scarcity [63] and food supplies [64], with complex interactions between fossil fuels, water use, and agricultural production. Fossil fuels use significant water compared to alternatives [65, 66], making it clear that ending fossil fuels will have benefits for water use and agriculture as well. Some carbon capture technologies also use significant water [67], raising similar issues.

The structure of property rights for water matters greatly, and inequities in who is allowed to exploit water resources (which flow from the structure of property rights in each country) will only intensify as the climate warms. Those inequities will inevitably lead to intra- and inter-national conflicts. The intense drought in Chile in recent years is one of many examples [68].

12.6.5 Population, health, and education

We discuss population as a key driver of emissions in chapter 3. It's the first term in the Kaya identity and is often incorrectly identified as the most important driver of emissions growth. Economic activity per person is usually more important over time [6], at least in recent decades, and in almost all future scenarios conducted recently, global population is projected to level off in the mid to late twenty-first century [69].

Previous research has documented that population growth affects emissions, but that changes in population growth are affected by many complex ethical and human factors related more broadly to societal development. O'Neill [70] cautions that 'the fact that a particular phenomenon is a quantitatively significant driver of emissions does not mean that it is also an important policy lever'.

Choices that affect societal development can also affect population (and thus emissions) but most scenarios avoid discussing such choices as explicitly driven by the goal of emissions reductions [71]. That said, understanding the complex dynamics of population and economic development is critically important for creating more useful emissions scenarios [70–72].

The van Vuuren scenario [5] shown in figure 12.2 above is one of the few to explicitly treat the effects of changes in population growth on emissions. Those researchers are careful to treat the issue as one of societal development, female education, female empowerment, and availability of family planning services, not driven by a goal to reduce emissions. Such efforts would likely have that happy side effect, however. Project Drawdown also calls out availability of reproductive health services for women and equal access to education for girls as a key option that would result in a more just, lower emissions world[2].

12.6.6 Adaptation and resilience

Human systems were designed to suit a relatively narrow temperature range, but climate change (even at 1.5 °C above pre-industrial levels) will require us to adapt to changing conditions. The first and most important priority is reducing emissions to zero as quickly as we can, which is why we wrote this book. But even if we're successful at achieving that goal, we'll also need to build in more resilience to society's key systems.

We encourage students to add discussion of ways to increase societal resilience at the end of their climate solutions plan. For example, increased resilience means hardening electricity grids against more intense heat, installing air conditioning more widely, raising the height of key infrastructure (such as airports and sewage treatment) to avoid flooding, upgrading building standards, redrawing flood maps, and reforming assessments of household and business risk for insurance purposes.

12.6.7 Making governance work

Effective institutional governance is critical to achieving a climate-positive world. The IPCC's most recent mitigation report [42] states (in the Technical Summary): 'Achieving the global transition to a low-carbon, climate-resilient and sustainable world requires purposeful and increasingly coordinated planning and decisions at many scales of governance including local, subnational, national and global levels (*high confidence*).'

Sometimes only government can do what needs to be done, and we need an honest discussion about what kind of government we want and what we want it to do for us. Sometimes we'll want more government, such as when we find lead in children's toys, salmonella in peanut butter, poison in medicines, an unsustainable health care system, or fraudulent assets and a lack of transparency in the financial world. We know from experience that only government can fix those things. Sometimes we'll want less government, such as when old and conflicting regulations and incentives get in the way of phasing out fossil fuels. Only government can fix that too (although the private sector has some lessons to teach on that score). And sometimes we'll want the same government, just delivered more efficiently, like many governments have done in recent years by applying information technology to speed up the delivery of services.

When it comes to government, more is not better. Less is not better. Only better is better. And better is the goal for which we as a society should strive.

---

[2] https://drawdown.org/solutions/health-and-education/technical-summary

One common theme for those opposed to action on climate is a deep concern about government. It is so deep, in fact, that these folks appear unable or unwilling to recognize the reality of the climate problem. This is exactly backwards—once you accept that only government can do certain things about the climate problem, we move that discussion to where it should be, focusing on the question 'what kind of government do we want, and how can we make it work best?' Government is us, it is not an alien force, and we will, as the old proverb says, get the government we deserve. If we don't learn better ways to govern ourselves, we're going to be in big trouble, given the scope and nature of the climate problem.

The late Stephen Schneider asked 'can democracy survive complexity?', which is exactly the right question [73]. We'd better hope the answer is yes.

The challenges we face, whether climate change or financial meltdowns, have in common the failure of governing institutions to align private incentives with the public good. Now we face new realities, with technological and financial power beyond the imagination of the people of two centuries ago, and new environmental challenges that require new ways of working together [74]. We must design institutions that recognize those realities and use our new capabilities to align private interests with broader societal goals [75].

Private enterprise is the best means yet devised for driving down costs and spreading the use of technology, but we need *well-regulated capitalism*. Otherwise, we end up with lead in children's toys, testing of drugs on unsuspecting patients, fraud and theft by corporate cronies, and rivers that catch on fire. The challenge is to create the right kind of check on corporate power, keeping the spirit of innovation alive while curtailing corporate excesses.

12.7 Focus on what matters most

Speed is of the essence. Focus first on the biggest fastest growing emissions segments and get them to zero as quickly as you can. Then move down the list, with the goal of eliminating emissions from every human activity throughout the economy.

Use findings from other studies to rank order your emissions reduction and carbon removal options, addressing local conditions, resource constraints, and interactions between measures. We don't ask for a complete tally of costs and benefits for our class but do expect students to use their judgment when choosing options, focusing on the most cost-effective options first.

Often the best way to identify the cheapest options is to find those with multiple benefits in addition to emissions reductions. There are many such options with co-benefits, and they almost always can be captured at negative net cost for society.

12.8 Key pieces of the puzzle

Figure 12.5, which is identical to figure 2.3, summarizes the overarching goals of climate action, the eight pillars of climate action, and the four key actors essential to making climate action a success. In chapters 4–11 we listed key action items for the four key actors. In this section we boil down those action items into five high-level recommendations:

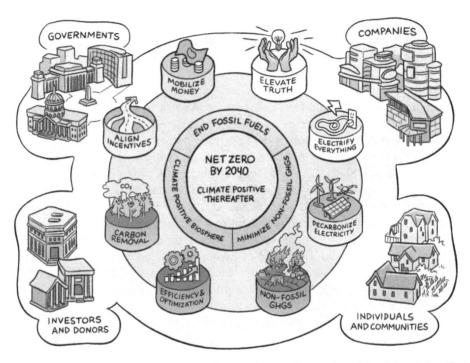

Figure 12.5. Summary of climate action goals, pillars, and actors. Source: Copyright Violet Kitchen 2022.

- *Fix rules and norms*: The fossil fuel industry has had more than a century to rig the economy in its favor. Changing these rules of the game is essential for reaching a climate-positive world as quickly as we can. That means eliminating fossil fuel subsidies, both direct and indirect, everywhere we find them. It also means changing property rights, lending practices, measurement standards, efficiency standards, environmental laws, regulatory processes, and many other institutional practices to favor zero emissions.
- *Smooth the transition*: Financial incentives are critical for accelerating the transition to a climate-positive world. We'll need to abandon a significant fraction of existing fossil capital before its accounting life ends, and that means paying off some of the owners of those assets if they retire that infrastructure early. We also need to pay people to electrify more quickly than they would otherwise be inclined to do. Pricing emissions is also potentially helpful (especially for industry and the power sector), but it's only one of many supporting policies that are needed.
- *Put purchasing power to work*: One of society's biggest levers is the immense purchasing power of governments, companies, and individuals, which totals in the tens of trillions of dollars every year globally. Channeling even a fraction of those expenditures toward zero-emissions options in the near term would allow a massive scale-up of these options, accompanied by significant cost reductions from learning effects, network externalities, and economies of scale.

- *Use information technology to minimize, optimize, and iterate*: Information technology is our 'ace in the hole'. It can help us better optimize our processes, better match our energy and material flows to the tasks at hand, and reduce our environmental impacts. It is also well suited to an adaptive challenge like climate change that will require us to iterate and change direction as developments dictate.
- *Invest in the future, sensibly*: One of the most important functions of government is to support long term research, development, demonstration, and deployment of promising zero-emissions technologies, but we need more. The private sector can be good at 'picking winners' that maximize private wealth, but only government can pick winners that put us on the best path for society, and that's what we need to truly face the climate challenge.

A portfolio approach for R&D investments can be beneficial, but technologies that fail to scale and achieve cost reductions after many attempts must be abandoned, because we can't afford to waste talent, time, money, and effort on things that don't work. 'All of the above' for zero-emissions options doesn't mean continuing to invest in failed technologies, it means trying everything and determining what path is quickest and most efficacious, discarding what isn't working.

12.9 Chapter conclusions: creating a climate-positive world

To avoid the worst effects of climate change we will need to re-assess and re-design every task in the economy with the twin goals of eliminating emissions and pulling carbon from the atmosphere. Your emissions reduction plan should prioritize options with significant co-benefits first because these are likely to be the cheapest from society's perspective. A climate-positive world will be a better world in every way that matters, but we need to envision that world, tell compelling analysis-based stories about it, and then help make it a reality.

No battle plan survives contact with the enemy.
—Helmuth von Moltke the Elder

Further reading

Asheim G B, Fæhn T, Nyborg K, Greaker M, Hagem C, Harstad B, Hoel M O, Lund D and Rosendahl K E 2019 The case for a supply-side climate treaty *Science* **365** 325 https://doi.org/10.1126/science.aax5011. The end-game for fossil fuel production is a supply-side climate treaty. It's the only way to avoid incentives for individual countries to keep producing fossil fuels as others phase them out.

Erickson P, Lazarus M and Piggot G 2018 Limiting fossil fuel production as the next big step in climate policy *Nat. Clim. Change.* **8** 1037–43 https://doi.org/10.1038/s41558-018-0337-0. One of several recent analyses pointing toward increased effort on restricting fossil fuels on the supply side.

Green F and Denniss R 2018 Cutting with both arms of the scissors: the economic and political case for restrictive supply-side climate policies *Clim. Change* **150** 73–87 https://doi.org/10.1007/s10584-018-2162-x. Another analysis pointing toward increased effort to restricting fossil fuels on the supply side.

Moore F C, Lacasse K, Mach K J, Shin Y A, Gross L J and Beckage B 2022 Determinants of emissions pathways in the coupled climate–social system *Nature* **603** 103–11 https://doi.org/10.1038/s41586-022-04423-8. A state-of-the-art review of drivers of emissions reductions, emphasizing both social and technical factors.

Morgan M G and Henrion M 1992 *Uncertainty: A Guide to Dealing with Uncertainty in Quantitative Risk and Policy Analysis* (New York: Cambridge University Press). The classic text on uncertainty in policy analysis.

Newell P and Simms A 2019 Towards a fossil fuel non-proliferation treaty *Clim. Policy.* **20** 1043–54 https://doi.org/10.1080/14693062.2019.1636759. Another article arguing for a treaty to phase out fossil fuel production.

Shi L and Moser S 2021 Transformative climate adaptation in the United States: trends and prospects *Science* **327** eabc8054 https://doi.org/10.1126/science.abc8054. A high-level review article on adaptation in the US.

Schwartz P 1996 *The Art of the Long View: Planning for the Future in an Uncertain World* (New York: Doubleday). The definitive book on creating compelling scenarios.

Sovacool B K, Burke M, Baker L, Kotikalapudi C K and Wlokas H 2017 New frontiers and conceptual frameworks for energy justice *Energy Policy* **105** 677–91 https://doi.org/10.1016/j.enpol.2017.03.005. Nice review of issues around energy justice.

References

[1] Koomey J 2017 *Turning Numbers into Knowledge: Mastering the Art of Problem Solving* 3rd edn (El Dorado Hills, CA: Analytics)

[2] Schwartz P 1996 *The Art of the Long View: Planning for the Future in an Uncertain World* (New York: Doubleday)

[3] Morgan M G and Henrion M 1992 *Uncertainty: A Guide to Dealing with Uncertainty in Quantitative Risk and Policy Analysis* (New York: Cambridge University Press)

[4] Ghanadan R and Koomey J 2005 Using energy scenarios to explore alternative energy pathways in California *Energy Policy* **33** 1117–42

[5] van Vuuren D P *et al* 2018 Alternative pathways to the 1.5 °C target reduce the need for negative emission technologies *Nat. Clim. Change* **8** 391–7

[6] Koomey J, Schmidt Z, Hummel H and Weyant J 2019 Inside the black box: understanding key drivers of global emission scenarios *Environ. Model. Softw.* **111** 268–81

[7] Koomey J, Schmidt Z, Hausker K and Lashof D 2022 Exploring the black box: applying macro decomposition tools for scenario comparisons *Environ. Model. Softw.* **155** 105426

[8] Kulkarni S, Hof A, van der Wijst K and van Vuuren D 2022 Disutility of climate change damages warrants much stricter climate targets *Research Square* doi: 10.21203/rs.3.rs-1788130/ 8 August 2022

[9] Xiao M, Junne T, Haas J and Klein M 2021 Plummeting costs of renewables—are energy scenarios lagging? *Energy Strategy Rev.* **35** 100636

[10] Ascher W 1978 *Forecasting: An Appraisal for Policy Makers and Planners* (Baltimore, MD: Johns Hopkins University Press)

[11] Cui R Y *et al* 2019 Quantifying operational lifetimes for coal power plants under the Paris goals *Nat. Commun.* **10** 4759

[12] Grubert E 2020 Fossil electricity retirement deadlines for a just transition *Science* **370** 1171

[13] Ertel C and Solomon L K 2014 *Moments of Impact: How to Design Strategic Conversations that Accelerate Change* (New York: Simon and Schuster)

[14] IEA 2021 *Net Zero by 2050: A Roadmap for the Global Energy Sector* (Paris: International Energy Agency) https://www.iea.org/reports/net-zero-by-2050

[15] Ackerman F, DeCanio S J, Howarth R B and Sheeran K 2009 Limitations of integrated assessment models of climate change *Climatic Change* **95** 297–315

[16] Enkvist P-A, Nauclér T and Rosander J 2007 A cost curve for greenhouse gas reduction *McKinsey Quarterly* 1 February https://www.mckinsey.com/business-functions/sustainability/our-insights/a-cost-curve-for-greenhouse-gas-reduction

[17] Enkvist P-A, Dinkel J and Lin C 2010 Impact of the financial crisis on carbon economics: version 2.1 of the global greenhouse gas abatement cost curve *McKinsey Stustainability* 1 January https://www.mckinsey.com/business-functions/sustainability/our-insights/impact-of-the-financial-crisis-on-carbon-economics-version-21

[18] Jackson T 1991 Least-cost greenhouse planning: supply curves for global warming abatement *Energy Policy* **19** 35–46

[19] Meier A 1982 Supply curves of conserved energy *PhD Thesis* Energy and Resources Group, University of California, Berkeley http://escholarship.org/uc/item/20b1j10d

[20] Rosenfeld A, Atkinson C, Koomey J G, Meier A, Mowris R and Price L 1993 Conserved energy supply curves *Contemp. Policy Issues* **11** 45–68

[21] Krause F and Koomey J G 1989 Unit costs of carbon savings from urban trees, rural trees, and electricity conservation: a utility cost perspective *Conf. on Urban Heat Islands (Berkeley, CA, 23–24 February)* LBL-27311

[22] Koomey J G, Atkinson C, Meier A, McMahon J E, Boghosian S, Atkinson B, Turiel I, Levine M D, Nordman B and Chan P 1991 The potential for electricity efficiency improvements in the US residential sector *LBNL Report* LBL–30477 Lawrence Berkeley National Laboratory, Berkley, CA https://eta.lbl.gov/publications/potential-electricity-efficiency

[23] Koomey J G 2012 *Cold Cash, Cool Climate: Science-Based Advice for Ecological Entrepreneurs* (El Dorado Hills, CA: Analytics)

[24] DeCanio S J 2003 *Economic Models of Climate Change: A Critique* (Basingstoke: Palgrave-Macmillan)

[25] Craig P, Gadgil A and Koomey J 2002 What can history teach us? A retrospective analysis of long-term energy forecasts for the US *Annual Review of Energy and the Environment 2002* ed R H Socolow, D Anderson and J Harte (Palo Alto, CA: Annual Reviews) pp 83–118 LBNL-50498

[26] Koomey J G, Craig P, Gadgil A and Lorenzetti D 2003 Improving long-range energy modeling: a plea for historical retrospectives *Energy J* **24** 75–92

[27] Scher I and Koomey J G 2011 Is accurate forecasting of economic systems possible? *Clim. Change* **104** 473–9

[28] Laitner J A S, DeCanio S J, Koomey J G and Sanstad A H 2002 Room for improvement: increasing the value of energy modeling for policy analysis *Proc. 2002 ACEEE Summer*

Study on Energy Efficiency in Buildings (Asilomar, CA, August) (Washington, DC: American Council for an Energy Efficient Economy)

[29] Hultman N E, Koomey J G and Kammen D M 2007 What history can teach us about the future costs of US nuclear power *Environ. Sci. Technol.* **41** 2088–93

[30] Huss W R 1985 Comparative analysis of company forecasts and advanced time series techniques using annual electric utility energy sales data *Int. J. Forecasting* **1** 217–39

[31] Huss W R 1985 What makes a good load forecast? *Public Utilities Fortnightly* 28 November pp 3–11

[32] Huss W R 1985 Can electric utilities improve their forecast accuracy? The historical perspective *Public Utilities Fortnightly* 26 December pp 3–8

[33] Köberle A C, Vandyck T, Guivarch C, Macaluso N, Bosetti V, Gambhir A, Tavoni M and Rogelj J 2021 The cost of mitigation revisited *Nat. Clim. Change* **11** 1035–45

[34] Arthur W B 1990 Positive feedbacks in the economy *Sci. Am.* February pp 92–9

[35] Arthur W B 1994 *Increasing Returns and Path Dependence in the Economy* (Ann Arbor, MI: The University of Michigan Press)

[36] Gritsevskyi A and Nakicenovic N 2000 Modeling uncertainty of induced technological change *Energy Policy* **28** 907–21

[37] Stansinoupolos P, Smith M H, Hargroves K and Desha C 2008 *Whole System Design: An Integrated Approach to Sustainable Engineering* (New York: Routledge)

[38] Lovins A, Bendewald M, Kinsley M, Bony L, Hutchinson H, Pradhan A, Sheikh I and Acher Z 2010 *Factor Ten Engineering Design Principles* (Old Snowmass, CO: Rocky Mountain Institute) https://www.rmi.org/our-work/areas-of-innovation/office-chief-scientist/10xe-factor-ten-engineering/

[39] Lovins A 2010 *Integrative Design: A Disruptive Source of Expanding Returns to Investments in Energy Efficiency* (Old Snowmass, CO: Rocky Mountain Institute) http://www.rmi.org/rmi/Library/2010-09_IntegrativeDesign

[40] Harvey L D 2015 *A Handbook on Low-Energy Buildings and District-Energy Systems: Fundamentals, Techniques and Examples* (New York: Routledge)

[41] Lovins A B 2005 *Energy End-Use Efficiency* (Amsterdam: Rocky Mountain Institute for InterAcademy Council) https://rmi.org/insight/energy-end-use-efficiency/

[42] IPCC 2022 *Climate Change 2022: Mitigation of Climate Change. Contribution of Working Group III to the Sixth Assessment Report of the Intergovernmental Panel on Climate Change* (Cambridge: Cambridge University Press) https://www.ipcc.ch/report/sixth-assessment-report-working-group-3/

[43] Vohra K, Vodonos A, Schwartz J, Marais E A, Sulprizio M P and Mickley L J 2021 Global mortality from outdoor fine particle pollution generated by fossil fuel combustion: results from GEOS-Chem *Environ. Res.* **195** 110754

[44] Roberts D 2020 Air pollution is much worse than we thought: ditching fossil fuels would pay for itself through clean air alone *Vox* 12 August https://www.vox.com/energy-and-environment/2020/8/12/21361498/climate-change-air-pollution-us-india-china-deaths

[45] Burnett R *et al* 2018 Global estimates of mortality associated with long-term exposure to outdoor fine particulate matter *Proc. Natl Acad. Sci.* **115** 9592

[46] Schraufnagel D E *et al* 2019 Air pollution and noncommunicable diseases: a review by the Forum of International Respiratory Societies 2019; Environmental Committee, part 1: the damaging effects of air pollution *Chest* **155** 409–16

[47] Schraufnagel D E *et al* 2019 Air pollution and noncommunicable diseases: a review by the Forum of International Respiratory Societies 2019; Environmental Committee, part 2: air pollution and organ systems *Chest* **155** 417–26

[48] Gao J *et al* 2018 Public health co-benefits of greenhouse gas emissions reduction: a systematic review *Sci. Total Environ.* **627** 388–402

[49] Choma E F *et al* 2020 Assessing the health impacts of electric vehicles through air pollution in the United States *Environ. Int.* **144** 106015

[50] Stoddard I *et al* 2021 Three decades of climate mitigation: why haven't we bent the global emissions curve? *Annu. Rev. Environ. Resour.* **46** 653–89

[51] Bullard R D 2008 *Dumping in Dixie: Race, Class, And Environmental Quality* 3rd edn (Boulder, CO: Westview)

[52] Bullard R D (ed) 1999 *Confronting Environmental Racism: Voices from the Grassroots* (Boston, MA: South End)

[53] Howarth R B 2011 Intergenerational justice *The Oxford Handbook of Climate Change and Society* ed J S Dryzek, R B Norgaard and D Schlosberg (Oxford: Oxford University Press) ch 23 pp 338–52

[54] Rezai A, Duncan K F and Taylor L 2012 Global warming and economic externalities *Econ. Theory* **49** 329–51

[55] Howarth R B 1996 Climate change and overlapping generations *Contemp. Econ. Policy* **14** 100–11

[56] Rosen R A and Guenther E 2015 The economics of mitigating climate change: what can we know? *Technol. Forecast. Soc. Change* **91** 93–106

[57] DeLong J B 2022 *Slouching Towards Utopia: An Economic History of the Twentieth Century* (New York: Basic Books)

[58] Freeman S *et al* 1974 *A Time to Choose: America's Energy Future (Final Report of the Energy Policy Project of the Ford Foundation)* (Cambridge, MA: Ballinger)

[59] Schurr S H, Darmstadter J, Perry H, Ramsay W and Russell M 1979 *Energy in America's Future: The Choices Before Us* (Baltimore, MD: Johns Hopkins University Press for Resources for the Future)

[60] Stobough R *et al* 1979 *Energy Future* (New York: Ballantine)

[61] Bohi D and Montgomery W D 1982 *Oil Prices, Energy Security, and Import Policy* (Washington, DC: Resources for the Future)

[62] Lovins A B *et al* 2004 *Winning the Oil Endgame: Innovation for Profits, Jobs, and Security* (Old Snowmass, CO: Rocky Mountain Institute) https://www.rmi.org/insights/knowledge-center/winning-oil-endgame/

[63] Gleick P H 1993 Water and conflict: fresh water resources and international security *Int. Security* **18** 79–112

[64] Hendrix C and Brinkman H-J 2013 Food insecurity and conflict dynamics: causal linkages and complex feedbacks *Stability* **2** 26

[65] Grubert E and Kelly T S 2018 Water use in the United States energy system: a national assessment and unit process inventory of water consumption and withdrawals *Environ. Sci. Technol.* **52** 6695–703

[66] Peer R A M, Grubert E and Sanders K T 2019 A regional assessment of the water embedded in the US electricity system *Environ. Res. Lett.* **14** 084014

[67] Fuhrman J, McJeon H, Patel P, Doney S C, Shobe W M and Clarens A F 2020 Food–energy–water implications of negative emissions technologies in a +1.5 °C future *Nat. Clim. Change* **10** 920–7

[68] Bartlett J 2022 'Consequences will be dire': Chile's water crisis is reaching breaking point *The Guardian* 1 June https://www.theguardian.com/world/2022/jun/01/chiles-water-crisis-mega-drought-reaching-breaking-point

[69] Tebaldi C *et al* 2021 Climate model projections from the scenario model intercomparison project (scenarioMIP) of CMIP6 *Earth Syst. Dynam.* **12** 253–93

[70] O'Neill B C, Dalton M, Fuchs R, Jiang L, Pachauri S and Zigova K 2010 Global demographic trends and future carbon emissions *Proc. Natl Acad. Sci.* **107** 17521

[71] Bongaarts J and O'Neill B C 2018 Global warming policy: is population left out in the cold? *Science* **361** 650

[72] O'Neill B C, Liddle B, Jiang L, Smith K R, Pachauri S, Dalton M and Fuchs R 2012 Demographic change and carbon dioxide emissions *Lancet* **380** 157–64

[73] Schneider S H 2009 *Science as a Contact Sport: Inside the Battle to Save Earth's Climate* (Washington, DC: National Geographic)

[74] Orr D W 2009 *Down to the Wire: Confronting Climate Collapse* (Oxford: Oxford University Press)

[75] Klein E 2022 What America needs is a liberalism that builds *The New York Times* 29 May https://www.nytimes.com/2022/05/29/opinion/biden-liberalism-infrastructure-building.html

IOP Publishing

Solving Climate Change
A guide for learners and leaders
Jonathan Koomey and Ian Monroe

Chapter 13

Our climate-positive future

If we do not address climate change in a timely manner, it does not really matter what we do on human rights, on education or on health, because destruction on the planet will be so severe, everything else will fall by the wayside.
—Christiana Figueres[1]

Chapter overview

- We need immediate action on climate. That means now, not in a few years, or decades hence.
- All that matters is rapid emissions reductions, getting to zero greenhouse gas emissions as quickly as we can, then pulling carbon from the atmosphere by making the biosphere 'climate positive'.
- Defeating the forces of predatory delay will take new ways of working together, new institutional arrangements, and new ways of thinking about the world.
- We have all the technology we need to get to 80%–90% emissions reductions in coming decades, and we'll sort out that last 10%–20% soon enough, assuming we ramp up investments in research, development, demonstration, and deployment as we should.
- A climate-positive world will be a better one than the one it replaces, but only if we act quickly.

In his 'I have a dream' speech in 1963, Dr Martin Luther King Jr, spoke of the 'fierce urgency of now' and warned about 'the tranquilizing drug of gradualism'[2].

---

[1] https://www.newstatesman.com/the-environment-interview/2022/01/christiana-figueres-if-we-fail-on-climate-we-fail-on-human-rights-education-and-health
[2] https://www.ihaveadreamspeech.us

He was talking about the immediate need to address racial and social justice in the face of ongoing strife, but his words carry weight today when speaking about climate change.

If you take only one lesson on climate from this book, it's that we have no more time to waste [1]. When we started working on this issue years ago there was still time to implement modest changes to solve the problem, but no longer. We've squandered thirty years and now need to reduce emissions to zero as quickly as we can.

There are no longer gradual, moderate solutions on climate, only rapid deep cuts to greenhouse gas emissions starting now. Immediate absolute emissions cuts are the only relevant metric for success. Long-term targets are only as useful as the supporting action plans to make them a reality in the near term. And defeating those arguing for what Alex Steffen calls 'predatory delay' (our era's version of 'tranquilizing gradualism') is the only way we'll get through this period of societal evolution without serious disruptions of modern society.

For governments, that means stopping all new fossil infrastructure while rapidly approving non-combustion generation technologies, transmission lines, and energy storage. It also means setting aggressive timelines for phasing out sales of fossil technologies, retiring existing facilities and equipment, and phasing out myriad fossil fuel subsidies and regulatory advantages. It means aggressive efforts to promote efficient electricity use. And it means a halt to industrial-scale deforestation everywhere, *starting now*.

For consumers, that means every purchase of autos, appliances, and homes from now on needs to be electric, and they must never again vote for politicians who fight climate action, *starting now*.

For companies, that means no more contributions to climate denying politicians, no more investments in fossil infrastructure, no more selling products that lock-in greenhouse gas emissions for decades, *starting now*.

For investors and donors, that means eliminating all support for fossil fuels and funding research, development, demonstration, and deployment of a broad portfolio of zero-emissions options and innovations, *starting now*.

Whether our society will be up to facing the climate challenge is an open question. After studying this topic for decades, we are convinced that society *can* do it, the question is whether society *will* do it. To pass successfully through this stage of societal evolution will require us to develop new ways of working together, new institutional arrangements, and new worldviews[3]. It won't be easy.

The fossil fuel industry says calls for urgent action are 'alarmist' and that modest changes will be sufficient, but *it is not alarmist to be alarmed about something that is genuinely alarming*. As we've described throughout this book, climate change threatens the continued orderly development of human civilization, and that should be plenty alarming to all sensible folks.

---

[3] https://alexsteffen.substack.com/p/old-thinking-will-break-your-brain?s=r

The facts that give us the most hope about solving the climate problem are these:

1. We have everything we need to begin deep and rapid emissions reductions NOW. We just need to decide to do it. Getting to 80%–90% reductions in most places is feasible using technology that is now commercial or near commercial. The last 10%–20% will be harder, but technology is improving rapidly, and the more we do now, the more possibilities open up for the future.
2. Many mass-produced products fall in price at predictable rates with expansion of production. Solar panels are now 90+% cheaper, batteries are 80+% cheaper, and wind generation is 1/3 cheaper than a decade ago. If we build more, costs come down, and the more we build, the cheaper zero-emissions technology will get. Most models ignore or downplay these effects, but they are real and powerful, and will make climate action much cheaper from a *direct costs* perspective than most people think.
3. From a *societal* perspective, climate action is already much cheaper than preserving the status quo and has been for a long time. If we accurately count avoided pollution costs from moving away from fossil fuels (as we should) the transition will look a whole lot cheaper for society, and once you factor in cost reductions from scaling technology (from point 2) we are convinced that total societal costs will be much less than just carrying on business as usual.
4. There is a growing realization that we need to *retire* existing fossil capital on rapid and predictable schedules, to allow new technology to take hold. Past energy transitions happened organically and were slower than they needed to be because few actors retired capital early. We need to do that for climate, and if we do, things will happen much more rapidly than economic models predict (because most of those ignore or downplay forced retirements).
5. Information technology is our ace in the hole. In past energy transitions we didn't have computers and sensors to aid the transition but this time we do, and they will allow us to (a) move bits instead of atoms; (b) substitute smarts for parts; (c) dynamically control energy supply and demand; (d) collect high value data; and (e) help us design better gadgets and systems—all of these in ways that we never could before.

We need to get started now [1]. It's a climate emergency [2]. Let's get to work [3]!

It is not the critic who counts, not the man who points out how the strong man stumbled, or where the doer of deeds could have done them better. The credit belongs to the man who is actually in the arena; whose face is marred by dust and sweat and blood; who strives valiantly; who errs and comes short again and again; who knows the great enthusiasms, the great devotions, and spends himself in a worthy cause; who, at the best, knows in the end the triumph of high achievement; and who, at worst, if he fails, at least fails while daring greatly, so that his place shall never be with those cold and timid souls who know neither victory nor defeat.

—Theodore Roosevelt

References

[1] UN 2022 *UN Climate Report: It's 'Now Or Never' To Limit Global Warming To 1.5 Degrees* (Geneva: United Nations) https://news.un.org/en/story/2022/04/1115452

[2] Fischetti M 2021 We are living in a climate emergency, and we're going to say so *Sci. Am.* 12 April https://www.scientificamerican.com/article/we-are-living-in-a-climate-emergency-and-were-going-to-say-so/

[3] Hickel J 2021 What would it look like if we treated climate change as an actual emergency *Current Affairs* 15 November https://www.currentaffairs.org/2021/11/what-would-it-look-like-if-we-treated-climate-change-as-an-actual-emergency

Solving Climate Change
A guide for learners and leaders
Jonathan Koomey and Ian Monroe

Appendix A

Introduction to the climate problem (long form)

Appendix overview

- Following Nicholas [3], we summarize the current state of knowledge about the climate problem in a few phrases:
  - It's warming.
  - It's us.
  - We're sure.
  - It's bad.
  - We can fix it (but we'd better hurry).
- These conclusions are based on some of the most well-established principles in physical science, corroborated by the work of thousands of scientists and many lines of independent evidence, measurements, and analysis. Every reputable academy of science, the United Nations, and the vast majority of the world's national governments and multinational corporations agree with these core climate truths.
- Keeping Earth's average temperature from rising much more than 1.5 °C from pre-industrial levels will require immediate, rapid, and sustained greenhouse gas emissions reductions. It will also almost certainly require removal of carbon pollution from the atmosphere.
- Society should aim to hit 'net zero' greenhouse gas emissions globally by 2040 at the latest, sooner if possible, and every year thereafter should be 'net climate positive' until enough climate pollution is removed to reverse climate damages.
- Governments, companies, investors, and communities that can move faster should do so, and their early action will drive deployment-related cost reductions associated with learning-by-doing and economies of scale that will benefit the entire world.

Climate change is a complex issue, but the basic outlines of the problem are well known. In her book *Under the Sky We Make*, sustainability scientist and author, Kimberly Nicholas summarizes the climate issue in a few simple phrases [3]:

- It's warming.
- It's us.
- We're sure.
- It's bad.
- We can fix it.
- Michael E Webber, a professor of energy resources at the University of Texas at Austin, rightly adds: 'But we'd better hurry'[1].

This appendix explains why aggressive climate action is urgent, using Nicholas's framing (with Webber's addition) to structure the discussion. Chapter 1 contains a more compact version of this argument.

A.1 It's warming

Scientists have used many methods to estimate global average temperatures. Reliable and comprehensive direct measurements began around 1850, but proxy methods, using tree rings, ratios of isotopes, and other techniques can yield estimates going back thousands or even millions of years. Of course, the further back we look, the more complicated it is to do accurate assessments, but even the proxy estimates are based on well-established principles of physical science.

Figure A.1 shows a summary of the instrumental temperature record since 1850, created by researchers at the University of Oxford [4] and continually updated. These data combine four well accepted time-series of temperatures into what they call a **climate change index**. Such temperature data are expressed relative to a base year (or an average over a set of years).

This graph is expressed as a change in temperature relative to a 1850–1900 baseline, which is one way to characterize what we call a **pre-industrial baseline.** The industrial revolution, which was the beginning of large-scale exploitation of fossil fuels, started in the late 1700s, but really didn't gather steam until the late 1800s. That fact combined with widespread instrumental temperature data only extending back to about 1850 have led scientists to use the 1850–1900 baseline in the most recent research.

Temperature changes in the 1800s were mostly driven by volcanoes and other natural forces. The effect of Krakatoa in 1883 was particularly dramatic, with volcanic ash cooling the Earth by more than 0.1 °C over a year or two. Temperatures in the 1900s were also affected by such events but increased by more than 1 °C by 2020 due mainly to increasing concentrations of greenhouse gases during this period (see below). Each decade since the 1970s has been hotter than the

---

[1] https://twitter.com/MichaelEWebber/status/1335950979334893572?s=20

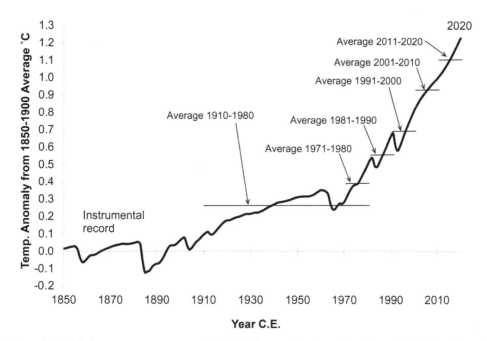

Figure A.1. Global average temperatures, 1850–2020. Source: Based on data through 2020 from https://globalwarmingindex.org, first assessed and presented in [4].

last, and almost all of the hottest years ever recorded have occurred in the past two decades[2].

Warming since pre-industrial times has pushed the Earth out of the comfortable and stable temperature range in which human civilization developed [5, 6]. The *rate* of temperature change in the last century (as well as the rate of change in the underlying drivers of temperature change) is also much more rapid than humanity and the Earth have experienced in thousands of years, which is another reason for concern, as we discuss below.

We can plot the two components of the instrumental record, the man-made ('anthropogenic') and the natural forcings (such as volcanoes and solar activity) plus the total (which corresponds to the instrumental record in figure A.1). Figure A.2 shows those results.

The anthropogenic forcings proceed smoothly because they are driven mainly by increasing concentrations of greenhouse gases over time, while the natural forcings are highly variable. On balance, the natural forcings contribute no-net warming over the 1850 to 2019 period. To first approximation, warming to date is 100% driven by human activities.

Kaufman *et al* created a temperature record going back 12 000 years in 2020 [5, 6] as shown in figure A.3. We also include the instrumental record from figure A.1 tacked on at the end of the period.

---

[2] https://climate.copernicus.eu/copernicus-globally-seven-hottest-years-record-were-last-seven

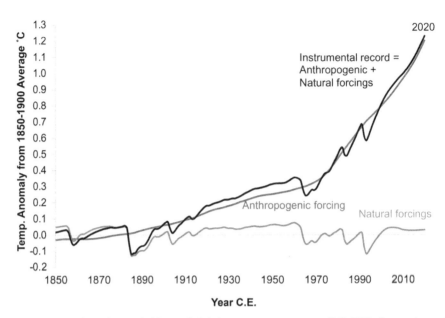

Figure A.2. Man-made and natural drivers of global average temperatures, 1850–2020. Source: latest data: https://globalwarmingindex.org; reference: [4].

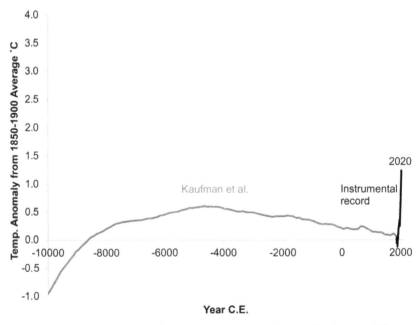

Figure A.3. Global average temperatures for the past 12 000 years. Source: Kaufman *et al* for past 12 000 years [2] and https://globalwarmingindex.org for the instrumental record [4] as shown in figure A.1.

This graph tells an important story about the Holocene, the period over the past 10 000 years when civilization developed. Global temperatures in this period were remarkably stable, with a modest rise in temperature to levels about 0.5 °C above pre-industrial levels 6–7000 years ago (−4000 to −5000 common era or CE). Temperatures started declining gradually around that time, and at the beginning of the industrial revolution (late 1700s/early 1800s), global temperatures reached roughly zero on the y-axis of our graph. Temperatures oscillated in the 1800s but started to increase rapidly in the 1900s to current levels, which are higher than any temperatures the globe has seen since before civilization began millennia ago. The increase in temperatures between −10 000 and −8000 CE is related to a naturally occurring increase in carbon dioxide (CO_2) occurring before that time.

The *rate* of temperature change in the last century (as well as the rate of change in the underlying drivers of temperature change) is also much more rapid than humanity and the Earth have experienced in thousands of years, which is another reason for concern, as we discuss below.

This warming is reflected in other indicators. Northern Hemisphere summer sea-ice extent has reached historically low levels in recent years, while Antarctic sea ice saw significant declines in the early to mid-twentieth century [7] with Antarctic sea-ice extent falling below long-term averages from 2015 onwards[3]. Sea levels keep increasing as land-based glaciers melt [8] and a warmer ocean expands, as shown in figures A.4 and A.5 [9, 10]. Sea level rise has also been accelerating in recent years

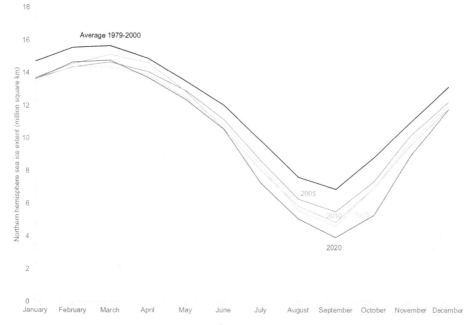

Figure A.4. Arctic summer sea-ice extent (million km²). Source: National Snow and Ice Data Center https://nsidc.org.

---

[3] https://nsidc.org/data/seaice_index/

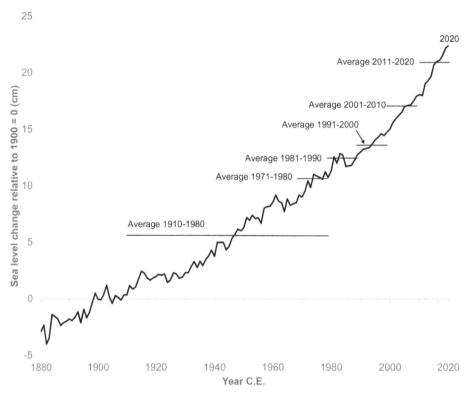

Figure A.5. Global average sea levels relative to 1900. Source: The 2 Degrees Institute https://www.sealevels. org.

[11, 12], at the same time as global average upper ocean heat content has increased rapidly, as shown in figure A.6 [13]. Finally, satellite measurements of the Earth's energy balance have confirmed that greenhouse gases are trapping heat, just as scientists expected [14].

It's important to understand that **global warming** does *not* mean that everywhere is breaking temperature records all the time. Natural cycles and fluctuations still exist, and places can still experience record cold temperatures on a given day. Many places are experiencing extremes in both directions from climate change, particularly the **weakening of the jet stream** in the Northern Hemisphere linked to melting Arctic sea ice [15], which can allow hot air to move further north and cold air to move farther south (contributing to localized **polar vortex** cold snaps). However, despite localized variability, average annual and daily high temperatures are increasing almost everywhere we look.

We have even longer-term data for historical temperatures going back millions of years, as presented in Burke *et al* [16]. The uncertainties grow as we move back in time, but it is clear Earth was a lot cooler (−3 °C to −5 °C from our 1850 to 1900 baseline) for most of the 300 000 years before the Holocene, with kilometers-deep ice sheets covering significant fractions of Earth's land masses.

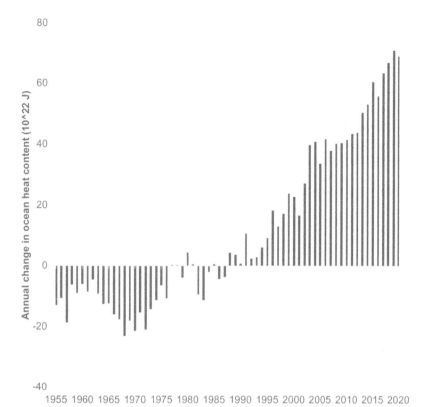

Figure A.6. Global annual change in upper ocean heat content (10^{22} J). Source: Upper ocean defined as the top 700 m. Change measured relative to a 1955 to 2006 baseline. https://www.climate.gov/news-features/understanding-climate/climate-change-ocean-heat-content, https://www.ncei.noaa.gov/access/global-ocean-heat-content/.

Temperatures about 3 million years ago were about at about current levels, and before that (50–60 million years ago) temperatures were much hotter (10 °C to 15 °C above 1850–1900). Sea levels in that warmer period were more than one hundred meters higher than they are today [17].

Multiple independent lines of evidence are consistent with a rapidly warming Earth. The world's foremost climate science authority, the **United Nations Intergovernmental Panel on Climate Change** (IPCC) concluded in 2021 [18] that 'global surface temperature has increased faster since 1970 than in any other 50 year period over at least the last 2000 years'.

A.2 It's us

We know why average temperature have increased over the past century: emissions of **greenhouse gases** (GHGs) related to human activity have increased substantially since pre-industrial times, leading to increasing **concentrations** of these gases in the atmosphere. GHGs like **carbon dioxide** (CO_2) trap energy from the Sun that otherwise would radiate to space, acting like a blanket that has been getting thicker for more than two centuries. To first approximation, warming to date is 100% driven by human activities.

Global temperature changes have historically corresponded very closely with atmospheric GHG concentrations [18]. The basic science behind how greenhouse gases warm the Earth through the **greenhouse effect** has been understood since the 1800s, and pre-industrial levels of atmospheric GHGs have been instrumental in keeping Earth warm enough for life to evolve. But while GHG levels have naturally fluctuated over millennia, the rapid rise in greenhouse gases since 1900, driven by human activities, is unprecedented in Earth's history.

Figure A.7 shows man-made emissions of all greenhouse gases from 1850 to 2020, which have increased rapidly and exponentially. There are four major categories of greenhouse gases: carbon dioxide, methane (CH_4), nitrous oxide (N_2O), and other

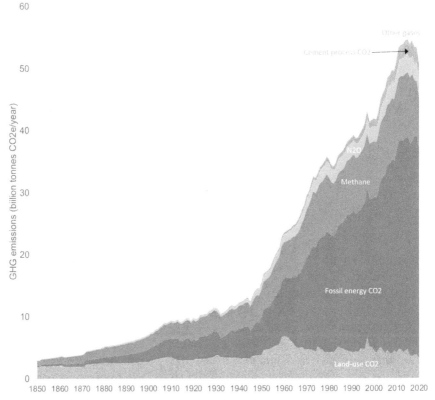

Figure A.7. Greenhouse gas emissions expressed as CO_2 equivalent 1850 to 2020. Source: Fossil, cement, and net land-use CO_2 emissions 1850 to 1958 from CDIAC archives: https://cdiac.ess-dive.lbl.gov/trends/emis/tre_glob_2013.html. Fossil, cement, and net land-use CO_2 emissions 1959 to 2020 from [19]. Fossil CO_2 emissions include combustion from flaring, solids, liquids, and gases. Methane, N_2O, and F-gas emissions from 1850 to 2019 from PIK, taken from https://www.climatewatchdata.org/data-explorer/historical-emissions. Global warming potential values for methane and N_2O adjusted to reflect AR6 100 year values in table 2.1 (PIK data uses AR4 values as described in [20]). Emissions of methane, N_2O, and F-gases for 2020 estimated assuming emissions stay at 2019 levels (just like they did during the 2009 recession).

gases (mostly what are called 'F-gases' that contain fluorine). The non-CO_2 gases are converted to what's called **CO_2 equivalent** (CO_2e), which is the equivalent amount of CO_2 that would result in the same warming effect as emissions of these other gases over a 100 year period. This technique allows us to approximate the total warming effect for all gases over time.

Total greenhouse gas equivalent emissions have increased at a rate of 2.2%/year from 1850 to 2020, with the emissions from fossil energy use growing at a 3.1%/year rate over that period. There have been periods of modest declines, but the overall upward trend has been inexorable.

The most important warming gas is carbon dioxide, which comes mainly from combustion of fossil fuels for energy, but also from deforestation and production processes for cement, steel, aluminum and other materials. Figure A.8 shows sources and sinks by decade for CO_2 over time [206].

Figure A.8 shows that burning of fossil fuels and land-use changes are the main sources adding CO_2 to the biosphere. The ocean and land have historically taken up about half of these emissions, but what remains goes into the atmosphere and stays there for centuries. As these emissions continue over time, the concentrations of CO_2 in the atmosphere go up, **climate feedback loops** (like wildfires and melting permafrost) can further accelerate GHG emissions and associated warming.

Scientists have different methods to assess past concentrations of trace gases over time. One way is to drill for ice cores and extract and analyze air samples from bubbles in the ice. For more recent times (since 1959) we have detailed direct measurements of atmospheric concentrations from the observatory on Mauna Loa in Hawaii and other locations.

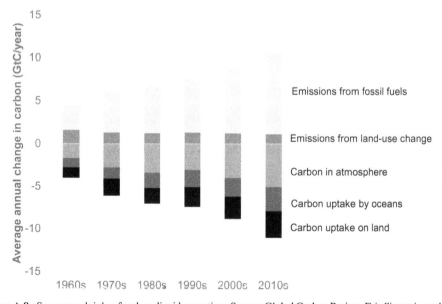

Figure A.8. Sources and sinks of carbon dioxide over time. Source: Global Carbon Project, Friedlingstein *et al* [19].

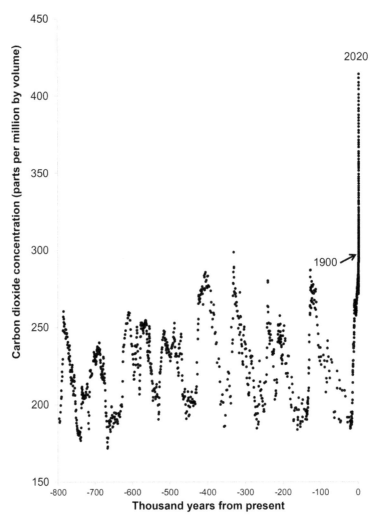

Figure A.9. CO$_2$ concentrations in the atmosphere for the past 800 000 years. Source: Law Dome ice cores [212]. Moana Loa data (https://gml.noaa.gov/ccgg/trends/).

Figure A.9 shows that over the last 800 000 years, Earth's atmosphere never held more than about 300 **parts per million** of CO$_2$ [21] corresponding to significantly cooler temperatures than in the Holocene (as shown in Burke *et al* [16]). In 2020 we hit 414 parts per million, and in May of 2022 we hit 421 parts per million. The increase over historical levels has occurred mostly in the span of about one and a half centuries. When it comes to CO$_2$ concentrations, we've moved rapidly into uncharted territory [22], driven by the increasing CO$_2$ emissions shown in figure A.7 above.

Recently, scientists have estimated the *rate of change* in CO$_2$ concentrations over this same period, to compare to historical trends, creating what they call 'the carbon skyscraper', as shown in figure A.10. This graph shows the rate of change in CO$_2$

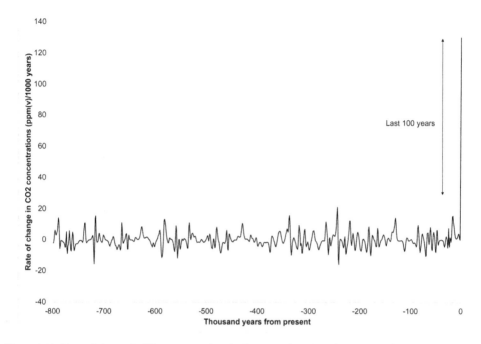

Figure A.10. Rate of change in CO_2 concentrations in the atmosphere (ppm/1000 years). Source: Strauss [213]. https://www.climatecentral.org/report/the-carbon-skyscraper.

concentrations in parts per million per thousand years, and the rate of change (driven by changes in the last century or so) is many times higher than anything in the historical record. That means the warming effect of increased carbon dioxide concentrations is increasing much more rapidly than any time in the past eight hundred thousand years, a trend that is consistent with the temperature data presented above.

The story is similar for methane (CH_4). Figure A.11 shows methane concentrations over time for the past two thousand years. By 2020, methane concentrations were 2.3 times higher than in 1850.

The data for nitrous oxide concentrations shows the same shape as for methane, but the increase since 1850 (only 22%) is less dramatic than for CO_2 and CH_4. Figure A.12 shows those data.

Growth in other gases is far more dramatic, particularly after 1960, as shown in figure A.13.

Figure A.14 shows CO_2 equivalent concentrations (including all warming agents) from 1850 to 2020, which reached just over 500 parts per million (ppm) in 2020[4]. For comparison, concentrations in 1850 were about 305 ppm. That means current concentrations are now about 1.6 times pre-industrial levels.

_____

[4] https://gml.noaa.gov/aggi/aggi.html

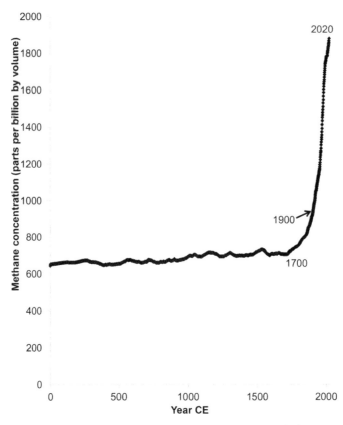

Figure A.11. CH$_4$ concentrations in the atmosphere since year 0 CE. Source: Meinshausen *et al* [214] for years 0 to 2014 CE. NOAA concentrations for 2015 to 2020.

We know from models and measurements that those increases in concentrations are consistent with long-term temperature increases of the magnitude our instruments measure. You can verify this for yourself, once you know (based on multiple lines of independent evidence) that every doubling of greenhouse gas concentrations implies an increase in temperature in coming centuries of about 3 °C [215].

An increase in greenhouse gas concentrations to 1.6 times pre-industrial levels implies an ultimate warming effect of 3 °C × 0.6 °C or 1.8 °C, assuming those concentrations remain constant for the next few centuries. Some of that warming has been masked by aerosols and other effects (see below) and time lags in the system ensure that the full warming effect won't be realized for centuries, but this is a reasonable order of magnitude estimate of the ultimate warming to which current levels of greenhouse gas concentrations commit the world.

One subtlety is that methane and black carbon, unlike CO$_2$, nitrous oxide, and many other warming agents, have relatively short average lifetimes in the atmosphere. That means that actions to reduce these pollutants could reduce concentrations relatively soon, and thus mitigate at least some of the warming effect implied by figure A.14 for 2020.

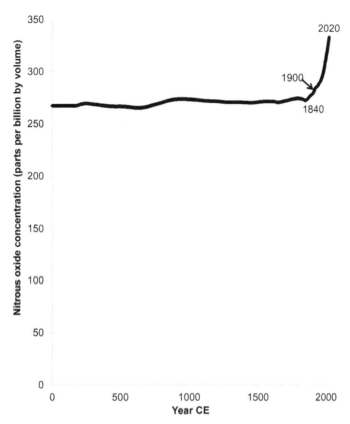

Figure A.12. N_2O concentrations in the atmosphere since year 0 CE. Source: Meinshausen *et al* [214] for years 0 to 2014 CE. NOAA concentrations for 2015 to 2020.

Another subtlety is that we know at least one way to reduce CO_2 concentrations on human time scales, and that's to increase uptake of CO_2 by biomass (planting trees, in colloquial parlance). Some scientists are also investigating ways to extract CO_2 from the atmosphere using 'direct-air capture' technology and storing it underground, but that's more speculative [207–210] and will likely exacerbate conflicts over energy, water, food, and land use [211]. Many recent mitigation scenarios include some form of reforestation or direct-air carbon capture because we've dithered over the past few decades while emissions increased, and we're out of time.

The Intergovernmental Panel on Climate Change (IPCC) in 2021 tallied the main drivers of warming to the 2010 to 2019 period, relative to the 1850 to 1900 period [18]. Those results are shown in figure A.15. Higher positive numbers mean more warming effect, and negative numbers mean that factor causes cooling.

The most important driver of increases in global surface temperatures since pre-industrial times has been increasing concentrations of carbon dioxide, mostly from combustion of fossil fuels, with a significant additional contribution from changes in land-use patterns (such as deforestation). Next is methane, driven by land-use

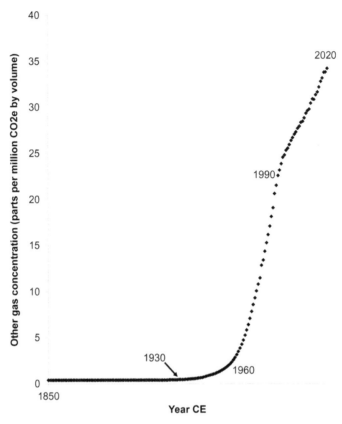

Figure A.13. Other greenhouse gas concentrations in the atmosphere since 1850. Source: Meinshausen *et al* [214] for years 1850 to 2014 CE. NOAA concentrations for 2015 to 2020.

changes, energy production, and agriculture, followed by non-methane volatile organic compounds and carbon monoxide, halogenated gases (such as the chlorofluorocarbons being phased out under the Montreal Protocol for ozone depleting chemicals, plus some others), nitrous oxides (also mainly from agriculture), black carbon, and airplane contrails. The cooling agents are sulfur dioxide aerosols (small particles that reflect sunlight), nitrogen oxides, land-use reflectance and irrigation, organic carbon, and ammonia.

The cooling effects just about cancel out the effect of all factors but CO_2 and half of the warming associated with methane, but that doesn't mean that we can ignore the other warming agents. As we start to phase out fossil fuels, the cooling effects from the aerosols will also be reduced, so we'll need to compensate by also rapidly minimizing the shorter-lived warming agents such as methane, some fluorinated gases, and black carbon.

A.3 We're sure

Uncertainties always exist in science, but when we confirm scientific findings by multiple, independent lines of evidence, we call them 'facts'. There may still be

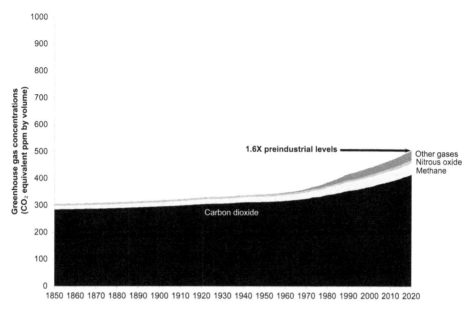

Figure A.14. CO_2 equivalent concentrations of greenhouse gases in the atmosphere, 1850 to 2020. Source: NOAA, Meinshausen *et al* [214], and Koomey calculations.

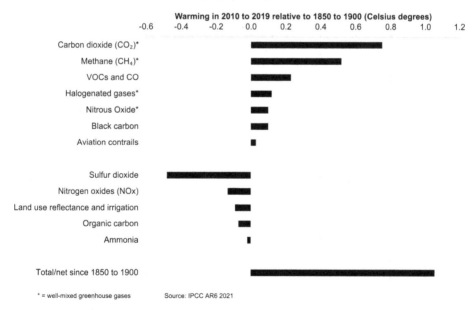

Figure A.15. Contributors to warming in 2010 to 2019 compared to 1850 to 1900. Source: ICPP Working Group I, Summary for Policy Makers, Sixth Assessment Report [23].

uncertainty about some details, but the preponderance of the evidence points towards these facts accurately describing how the physical world operates. Every credible scientific organization now agrees that burning fossil fuels and other human activities are almost entirely responsible for current climate change, based on decades of research from thousands of scientists. Human-caused climate change is now as much of a settled scientific fact as gravity.

The US National Academy of Sciences [24], which is not known for its wild speculation, concluded in 2010:

> A strong, credible body of scientific evidence shows that climate change is occurring, is caused largely by human activities, and poses significant risks for a broad range of human and natural systems....
>
> Some scientific conclusions or theories have been so thoroughly examined and tested, and supported by so many independent observations and results, that their likelihood of subsequently being found to be wrong is vanishingly small. Such conclusions and theories are then regarded as settled facts. This is the case for the conclusions that the Earth system is warming and that much of this warming is very likely due to human activities.

The academies of science for 80 other countries (including China, India, Russia, Germany, Japan, Brazil, and the UK) have released comparable statements about the science of climate [25]. The Intergovernmental Panel on Climate Change, the global scientific body charged with investigating this issue, stated with uncharacteristic bluntness in 2007, that 'warming of the climate system is unequivocal, as is now evident from observations of increases in global average air and ocean temperatures, widespread melting of snow and ice and rising global average sea level'. [26] IPCC reports in 2013 [27] and 2021 [18] were even more emphatic, as was the World Meteorological Association in 2021 [28].

That greenhouse gases warm the Earth is a finding based on some of the most well-established principles in physical science, as well as extensive measurements. The largest historical source of warming is carbon dioxide, and we know for a fact that the increase in carbon dioxide concentrations after the industrial revolution (particularly after the mid-1800s) was fueled predominantly by human activity.

One reason that we know that fossil fuels and land-use changes are the cause of the measured increases in CO_2 concentrations is because the total carbon emitted since the dawn of the industrial age is about twice as large as the total amount that remains in the atmosphere nowadays (the other half was absorbed by the oceans), and there are no other known sources of carbon that could account for such an increase in the atmosphere's CO_2 content.

In addition, scientists can measure the prevalence of different isotopes of carbon in the atmosphere to understand the sources of CO_2. Carbon from recent biological activity (such as deforestation) contains a different mix of carbon isotopes than carbon from fossil fuels and other geological sources, and scientists have measured changes in the concentrations of those isotopes in the atmosphere that are consistent with the additional carbon coming from human-linked sources [29]. So it's virtually certain that humans are the cause of elevated CO_2, as well as causing similar increases in the past two centuries in concentrations of methane, nitrous oxides, and other GHGs [18].

These changes are warming the planet, as shown in figures A.1–A.3, and as predicted with surprising accuracy as early as 1975 in *Science* [30] and in 1982 by scientists at Exxon [31]. The best current estimates are that global average surface temperatures have increased 1.1 °C–1.2 °C since 1900.

The physical science results supporting these conclusions go back at least a century to the Swedish scientist Svante Arrhenius, who in 1896 calculated the first climate sensitivity of 4 °C to 5 °C for a doubling of carbon dioxide concentrations. Arrhenius's analysis was supported by earlier measurements made by Eunice Foote and John Tyndall that demonstrated the heat trapping abilities of CO_2 [32]. The idea that greenhouse gases could warm the Earth is not a new one, and in fact the first informed speculation about this topic was by the mathematician Joseph Fourier in the 1820s [33].

These concerns are also validated by actual measurements of the climate system that corroborate theoretical predictions [34]. Satellite measurements show, for example, that as greenhouse gases have built up in the atmosphere, the heat emitted from the Earth has been declining at wavelengths that exactly correspond to those absorbed by various GHGs [35]. Similar measurements show that thermal radiation back to Earth's surface from the atmosphere, which we'd expect to increase if greenhouse gases trap heat, has been increasing exactly as we thought it would [36]. And the amount of heat stored in the oceans over the past few decades has been rising rapidly, which is consistent with a warming planet [23, 37, 38][5].

We can also examine data on some key indicators of warming, which by most accounts are changing at rates equaling or exceeding our worst-case predictions of just a few years ago [18, 39]. These measurements are one of the main reasons why scientists are so alarmed about humanity's effect on the planet's temperature.

In summarizing these and other measurements, the IPCC [26] concluded in 2007 that 'observational evidence from all continents and most oceans shows that many natural systems are being affected by regional climate changes, particularly temperature increases'. By 2021, the IPCC [23] was even more explicit: 'Human-induced climate change is already affecting many weather and climate extremes in every region across the globe. Evidence of observed changes in extremes such as heatwaves, heavy precipitation, droughts, and tropical cyclones, and in particular, their attribution to human influence, has strengthened since AR5 [in 2013]'.

The observations cited above are powerful evidence for a warming world, but they are not the only ones [18]. Glaciers are melting [8, 10], growing seasons are lengthening, and insects and birds are changing their long-established patterns [41]. Something's clearly happening to the climate, and these indicators are consistent with what the last one and a half centuries of science has been saying about this problem all along.

---

[5] Also see https://www.climate.gov/news-features/understanding-climate/climate-change-ocean-heat-content.

A.4 It's bad

If we continue on our current trajectory, we're committing to a doubling or tripling of greenhouse gas concentrations compared to pre-industrial times. Figure A.16 shows the result from one widely cited 'current trends continued' case (also known as SSP-2) created around 2015 or so, from Fricko *et al* [44].

A tripling of greenhouse gas concentrations means, using our simple math from above and the knowledge that every doubling results in about 3 °C of ultimate warming, we're committing the Earth to 1.5 doublings or 4.5 °C. As above, the ultimate warming level assumes that concentrations in 2100 remain constant for centuries as temperatures equilibrate, and the actual realized warming by 2100 will be lower than that (because equilibration takes time)

Figure A.17 shows the implications of that emissions trajectory for global temperatures, resulting in an increase of about 3.6 °C above pre-industrial levels by 2100. This result is typical for assessments of a 'current trends continued' path circa 2015, which fall in the range of 3 °C to 4 °C above pre-industrial times. In the past few years, with significant climate action in some major countries, the 'current trends continued' case has improved even more [45–48], to more like 2.5 °C to 3.5 °C.[6]

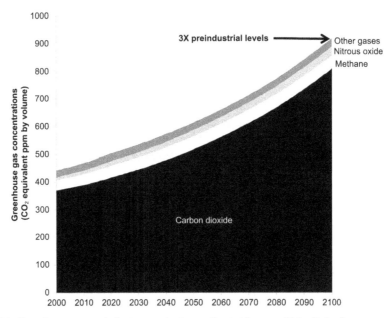

Figure A.16. Greenhouse gas equivalent concentrations estimated from a widely cited reference case published in 2017, based on SSP-2. Source: Fricko *et al* [44], Grubler *et al* [53], Koomey calculations.

---

[6] Also see https://thebreakthrough.org/issues/energy/3c-world.

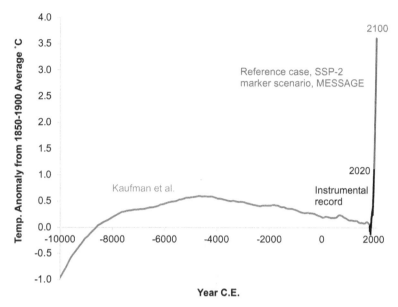

Figure A.17. Historical temperatures contrasted with the reference case projection from a prominent 2017 study. Source: Kaufman *et al* [5] for the past 12 000 years [5], https://globalwarmingindex.org for the instrumental record [4], and Fricko *et al* [44] for the reference case, SSP-2 marker scenario [53].

In 2012, when one of us (Koomey) wrote a book assessing options for reducing emissions [49], the current path at that time implied an increase of about 5 °C by 2100. Since 2012 we've made great progress in bringing down the cost of clean technology, we've shut down many coal plants, we've implemented and strengthened policies to reduce emissions, and we've realized that some of the more dire projections circa 2010 were based on overestimates of exploitable reserves of coal [50]. Of course, there is great uncertainty in any projection into the future, but we need a benchmark against to measure progress, and 2.5 °C–3.5 °C by 2100 is as good an estimate as any.

That improvement is small comfort, however. Even our current trajectory would be a disaster for the Earth and most other species [43, 51]. It would raise Earth's average temperature to a level not seen for about four million years (see Burke *et al* [16]), and to do so with unprecedented speed.

The most important direct effect of increasing temperatures is stress on natural and human systems [43]. Greenhouse gases keep more energy in the climate system, and that energy must go somewhere. Where it goes is into extreme rainfall and temperature events, which will become increasingly difficult or impossible for humans to manage, and ever more devastating for natural systems (whose ability to adapt is even more limited) [52].

Figure A.18 illustrates how a warming climate 'loads the dice' and makes extreme temperature events more likely and the extremes more extreme, just as more moisture in the air makes high precipitation events more likely (and oddly enough, makes droughts in some places more likely and more intense as well) [23, 42, 54–58].

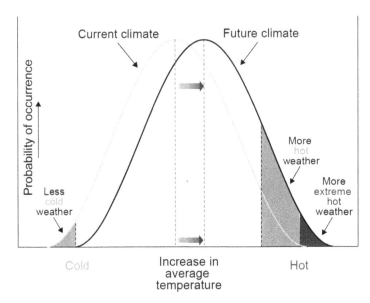

Figure A. 18. Increasing temperatures load the dice and make extreme temperatures more likely (and the extremes more extreme). Source: Reproduced with permission from [49]. Adapted from a graph made by the University of Arizona, Southwest Climate Change Network http://www.southwestclimatechange.org.

This concern is not a theoretical one. In a NASA report published in 2012, James Hansen and his colleagues examined distributions of temperatures for the period 1951 to 1980, comparing them to those from the 1980s, the 1990s, and the 2000s [59]. In each succeeding decade, the shifting of the distribution to towards higher temperatures became more pronounced as the climate warmed.

Warmer oceans mean more evaporation and an overall increase in moisture content in the air. That means extreme rainfall becomes more common, and the data for the US show that effect (as do broader measurements for the Northern Hemisphere [216]). Figure A.19 shows the percent of US land area subject to extreme one-day rainfall totals, and that percentage has clearly increased since the 1910 to 1980 period, a fact that will not be a surprise to US residents facing unprecedented flooding in recent years. Similar patterns for extreme precipitation events have been observed all over the world in recent years [18].

As we showed in figure A.17 above, the current global environment is accustomed to a relatively narrow temperature range, one that has prevailed for thousands of years. Ecosystems can sometimes migrate slowly (over many millennia, not decades or centuries), but that migration is limited by geography and other constraints. For example, a forest biome can gradually move up the mountainside as the climate warms (soil and geography permitting), but once it reaches the peak there's nowhere else to go, and extinction is the result. On our current path, thousands of plant and animal species that have existed for eons will be driven to extinction in the span of a century or so [60–62].

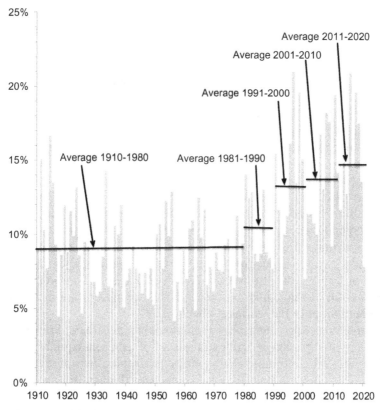

Figure A.19. Extreme precipitation events in the US. Source: https://www.ncdc.noaa.gov/extremes/cei/graph/us/01-12/4.

Humans and their support systems are also vulnerable to a warming climate. Heatwaves will become ever more frequent, and people in locations without air conditioning will either add it (which will worsen climate change), suffer, or even die (as tens of thousands of Europeans did during the heat wave of 2003). Our wastewater treatment and water supply systems are designed to handle current conditions but will be difficult and expensive to adapt to a warming world's rapidly rising sea level and increasingly intense rainstorms. Wildfires will become more frequent and more intense [40]. Pollen and its associated respiratory effects will worsen [64]. Climate change will also increase risks of cross-species viral transmission [65, 66].

With even small increases in sea level, low lying coastal areas will become increasingly vulnerable to storm surges, putting millions of lives at risk, particularly in the developing world. Those areas will also suffer from increased saltwater intrusion into groundwater supplies. A 'current trends continued' path implies more than 0.5 m rise in sea level by 2100 [23, 67–69], which would represent significant challenges to human society, and sea level rise will continue for centuries, barring substantial carbon removal from the atmosphere.

There are also indirect effects. One of the most important is an increase in the acidity of the oceans, caused by more dissolved CO_2 (which creates carbonic acid). This development will make life increasingly difficult for many types of aquatic life, with rates of acidification proceeding more rapidly than at any time in the past 65 million years [70]. The acidification effect (plus the increase in ocean temperatures) means that coral reefs will likely be a thing of the past by the end of the twenty-first century, and will pose increasing challenges to marine life of all types [71, 72]. It also is one reason why schemes such as those proposed to inject particles into the atmosphere to cool the Earth are ultimately chimerical—as long as more CO_2 dissolves into the oceans, the acidification effect will intensify, and just reflecting more sunlight won't fix it.

Remember also that the climate sensitivity measures the *average* temperature change for a doubling of greenhouse gas concentrations. Changes at Earth's poles have been [73] and will be much larger (that's just how the system works). The most likely case for climate sensitivity combined with the reference case emissions forecast would ultimately lead to an ice-free planet Earth and sea level rises much bigger than even recent projections. It also means that large releases of carbon trapped in the permafrost and in methane hydrates beneath the ocean floor are much more likely, and that would amplify the warming effect.

Perhaps the most worrying aspect of climate change is the unknowable but non-zero probability that pushing Earth's climate out of its recent equilibrium might lead to 'tipping points', discontinuous change, and catastrophic disruptions of weather and climate [74, 75]. One example that has preoccupied scientists for decades is the possibility that melting ice could disrupt the Gulf Stream in the North Atlantic [77], and there has been evidence of a weakening of those currents in recent decades [78–80]. The rapid release of carbon and methane from melting permafrost and methane from warming oceans are two more. The probability and consequences from catastrophic events are inherently *unknowable*, and that reality disrupts the standard benefit–cost model for assessing the economics of climate mitigation [76, 81–84].

A.5 We can fix it (but we'd better hurry)

Let's start by defining what we mean by 'fixing' climate change. Solving climate change starts with stabilizing global surface temperatures at the lowest possible level above pre-industrial times, and that means *reducing emissions to zero as soon as possible* [85].

Many climate action plans now focus on getting climate pollution to **net zero**, the point where GHG emissions are reduced far enough that any remaining remissions are counterbalanced by GHG emission removals from natural and engineered systems. Net zero by 2050 (or sooner) targets have now been adopted by many governments, as well as many of the world's largest companies and investors, catalyzed by goals of the United Nation's 2015 **Paris Agreement**.

Unfortunately, net zero is not enough. Even if Earth gets to net zero GHG emissions tomorrow, without additional action we're still locked into at decades of additional warming, and a new climate equilibrium that leaves the world worse off than it is today. Droughts, floods, heatwaves, wildfires, rising seas, crop failures, and other climate-linked disasters will continue to get worse. Many more lives and livelihoods will be lost, and some once thriving places may become unlivable.

To truly solve climate change, we must view net zero as a transition point rather than an end goal. We should aim instead to get our planet to net **climate positive**, sometimes also called **carbon negative**, where there is a net removal of carbon from the atmosphere every year. Our collective goal should be to make Earth climate positive until we have returned the climate to a state that's as close as possible to the pre-industrial range for which human civilization and the ecosystems around us have evolved. We should also aim to repair the damages caused by climate change as much as we can and compensate those who have suffered when damages can't be repaired. Fixing climate change requires our best efforts to not just stop the bleeding, but also heal the wounds.

This doesn't mean that we shouldn't aim to reach (and surpass) net zero as soon as possible, but net zero by 2050 is not soon enough, as we discuss below. Because of the long residence time of the most important greenhouse gases in the atmosphere, warming is to first approximation proportional to *cumulative* emissions [87, 88]. That means every molecule emitted matters, and that if we stop emitting greenhouse gases, warming will eventually stop [89, 90]. This makes the climate problem different from other types of air pollution, which generally stay in the atmosphere for a much shorter time than do emissions of carbon dioxide, nitrous oxide, and many of the other gases.

The importance of cumulative emissions means we have no time to lose in reducing our emissions, a fact that is not widely enough appreciated. We've already dithered for more than three decades as emissions kept increasing, and every day we delay getting to climate positive makes the situation worse [85].

The good news is that we already have almost all the climate technologies we need, and these solution technologies are improving every day. The same technologies that can get us to net zero can propel us beyond to climate positive and a regenerative climate future. The biggest climate challenges are sociological, political, and economic rather than technological. Society has plenty of money to pay for the transition but shifting this money out of climate pollution and into solutions requires navigating powerful entrenched interests, perverse incentives, bureaucratic barriers, and outdated ways of thinking. Solving climate change will also require much better communications about our climate impacts and solutions, engaging all levels of society in shifting to a climate-positive economy. We can do it, but we need to decide to do it.

A.5.1 What is a warming limit?

The warming limit approach has its origins in the realization that stabilizing the climate at a certain temperature to minimize climate risks (e.g. a warming limit of 1.5 °C or 2 °C above pre-industrial times) implies a particular emissions budget, which represents the total cumulative greenhouse gas emissions compatible with that temperature goal [91]. That budget also implies a set of emissions pathways that are well defined and tightly constrained (particularly now that we've squandered the past three decades by not reducing emissions). This risk-minimization approach, which can also be described as 'working toward a goal', also involves assessing the cost effectiveness of different paths for meeting the normatively determined target [49].

A warming limit is more than just a number (or a goal to be agreed on in international negotiations). It embodies a way of thinking about the climate problem that yields real insights [92]. A warming limit is also a value choice that is informed by science. It should not be presented as solely a scientific 'finding', but as a value judgment that reflects our assessment of societal risks and our preferences for addressing them.

The warming limit approach was first suggested for externalities more generally by Baumol [93] and was explored (then quickly dismissed) by Nordhaus [94, 95]. It had its first fully developed incarnation in 1989 in Krause *et al* [1] (which was subsequently republished by Wiley in 1992 [2]). It was developed further in Caldeira *et al* [96] and Meinshausen *et al* [97], and served as the basis for the International Energy Agency's analysis of climate options in 2010, 2011, 2012, and 2020 [98–101].

This way of thinking was developed as a counterpoint to the prevailing 'benefit–cost' approach favored by the economics community, and it has many advantages [92]. It encapsulates our knowledge from the latest climate models on how cumulative emissions affect global temperatures, placing the focus squarely on how to stabilize those temperatures. It places the most important value judgment up-front, embodied in the normatively determined warming limit, instead of burying key value judgments in economic model parameters or in ostensibly scientifically chosen concepts such as the discount rate. It gives clear guidance for the rate of emissions reductions required to meet the chosen warming limit, thus allowing us to determine if we're 'on track' for meeting the goal and allowing us to adjust course if we're not hitting those near-term targets.

The warming limit approach also allows us to estimate the costs of delaying action or excluding certain mitigation options and provides an analytical basis for discussions about equitably allocating the emissions budget. Finally, instead of pretending that we can calculate an 'optimal' technology path based on guesses at mitigation and damage cost curves decades hence, it relegates economic analysis to the important but less grandiose role of comparing the cost effectiveness of currently available options for meeting near-term emissions goals [92].

The warming limit approach shows that delaying action is costly, required emissions reductions are rapid, and most proved reserves of fossil fuels will need to stay in the ground, and large amounts of carbon will need to be removed from the atmosphere if we're to stabilize the climate and repair the damage. These ideas are familiar to some, but many still don't realize that they follow directly from the warming limit framing.

- Delaying emissions reductions forecloses options and makes achieving climate stabilization much more difficult [102]. 'Wait and see' for the climate problem is foolish and irresponsible, which is obvious when considering cumulative emissions under a warming limit. The more fossil infrastructure we build now, the faster we'll have to reduce emissions later. If energy technologies changed as fast as computers there could be justification for 'wait and see' in some circumstances, but they don't, so it's a moot point.

- Absolute global emissions will need to turn down immediately and approach climate positive in the next few decades [103] if we're to have a good chance to keep global temperatures 'well below 2 °C', as many in the scientific community advocate [104]. The emissions pathways given the current carbon budgets are tightly constrained. Even if the climate sensitivity is at the lowest end of the range included in IPCC reports (1.5 °C), that only buys us another decade in the time of emissions peak [105], which indicates that the findings on emissions pathways are robust, even in the face of uncertainties in climate sensitivity.
- The rate of emissions reductions, which is a number that can be measured, is one way to assess whether the world is on track to meet the requirements of a particular warming limit. We know what we need to be doing to succeed, and if we don't meet the tight time constraints imposed by that cumulative emissions budget in one year, we need to do more the next year, and the next, and the next. It's a way of holding policy makers' proverbial feet to the fire.
- The concept of 'stranded fossil fuel assets' that can't be burned, popularized by Bill McKibben [106] and Al Gore [107], follows directly from the warming limit framing. In fact, Krause *et al*'s 1989 book, *Energy Policy in the Greenhouse* [1] had a chapter titled 'How much fossil fuel can still be burned?', so the idea of stranded assets is not a new insight (but it is a profound one).

A.5.2 An evolution in thinking

The warming limit approach continues to be helpful in building the case for urgent action on climate [108–110] but its original policy usefulness rested on the overarching assumption that there was still time left to address the crisis. As time has passed and climate damages continue to mount it has become clear to us that the world is running out of time [111], and given uncertainties in estimating the carbon budget [112, 113], it is in our judgment better to focus on concrete emissions reductions goals instead of worrying too much about how much carbon budget is left.

The Intergovernmental Panel on Climate Change completed its report on scenarios based on a 1.5 °C warming limit in 2018 [104]. That study found, after reviewing many studies that meet the 1.5 °C warming limit, that cutting global greenhouse gas emissions at least in half from 2020 to 2030 (with subsequent reductions to follow at a similar pace) is a critically important milestone for success. This rough rule of thumb, which was enshrined as a 'carbon law' by Johan Rockstrom and his colleagues in a seminal article in *Science* in 2017 [103], describes the minimum level of effort needed to hit a 1.5 °C warming limit. Rockstrom's 'law' implies halving of absolute emissions in each decade starting in 2020, reaching close to net zero emissions by 2050.

When talking about rapid emissions reductions it is customary to refer to rates of change as a percentage of base year emissions, instead of often-used exponential rates of decline [49]. This convention is used because rates of decline reach astronomical levels in percentage terms as emissions approach zero, and percentages of a base year maintain an intuitive physical meaning throughout the analysis period. With this convention, a 5%/year decline in emissions relative to 2020 would result in a halving of annual emissions by 2030 (5%/year × 10 years) and complete elimination of emissions by 2040 (5%/year × 20 years).

As an example, consider a prominent emissions reduction case, called the low energy demand (LED) scenario, that would keep warming below the 1.5 °C limit [53]. It is unusual in that it does not rely on technological carbon removal options (such as carbon capture and storage) and it includes more aggressive improvements in energy intensity over time. It also includes significant increases in carbon removal from reforestation and land-use changes over time.

The LED scenario shows total greenhouse gas emissions reductions relative to 2020 of about 5.5%/year (expressed as above as a percentage of 2020 emissions levels). This pace of reductions implies that total global emissions reach 45% of 2020 levels in 2030.

Figure A.20 shows projected emissions from 2020 to 2100 from the LED scenario, using the same categories as shown in figure A.7 above. By 2030, the land-use sector moves from emitting CO_2 to absorbing it because the scenario

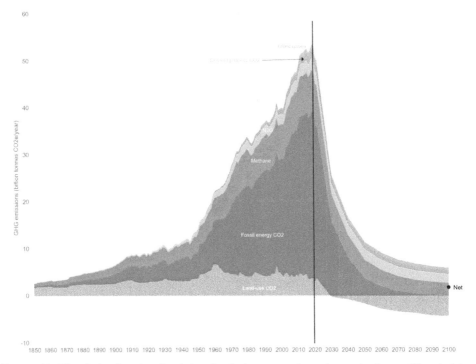

Figure A.20. Historical and projected emissions for the low energy demand scenario. Source: Historical data from figure A.7 and LED scenario from Grubler *et al* [53].

assumes deforestation slows substantially and reforestation occurs at an increasing rate over time. Fossil energy CO_2 declines to 10% of its 2020 value by 2050 and almost to zero by 2070. Emissions of methane and other gases also decrease substantially. N_2O, on the other hand, shows only modest declines over the analysis period. By 2100 net emissions accounting for carbon absorption in the land-use sector are more than 95% lower than in 2020.

When evaluating long-term scenarios like these, it's important not to place too much emphasis on precise numbers, particularly many decades hence. It is valuable, however, to look at broader lessons from such scenarios, and the most important lesson is the rapid rate of change embodied in all scenarios that put a 1.5 °C warming limit in reach. We'll need to build zero-emissions energy and industrial process technologies at high rates and retire existing high-emissions capital on a rapid schedule.

Achieving such speedy emissions reductions will require unprecedented changes in how the global economy generates value. In coming decades, many processes in our economy will need to be re-evaluated and re-designed from scratch to minimize or eliminate emissions.

A.5.3 The folly of delay

To solve the climate problem, we must reduce greenhouse gas emissions to zero as quickly as we can [86]. As William Nordhaus said in 2008 [194], 'There is no case for delay' in starting to reduce emissions.

Delaying mitigation is costly because cumulative emissions are what matter (waiting means we just need to move more quickly later) [87–89, 104, 195]. It is costly because delay means we don't gain the benefits of learning-by-doing, scale, spillovers, and network externalities from deploying zero-emissions technology [154, 161, 196, 197]. It is costly because it adds to the stock of 'stranded assets', fossil fuel-using capital that will need to be retired before the end of its useful life in order to meet emissions targets [100, 104, 198–200]. And it is costly because it adds to climate (and other) damages already being incurred by society if we continue to fail to act [104, 201, 202].

These conclusions are not new. Over a decade ago, the International Energy Agency concluded in its *2009 World Energy Outlook* [203]:

...each year of delay before moving onto the emissions path consistent with a 2 °C temperature increase would add approximately $500 billion to the global incremental investment cost of $10.5 trillion for the period 2010–2030.

In its *2010 World Energy Outlook* [98], IEA increased that estimate of losses for each year of delay to $1 trillion.

The White House Council of Economic Advisors under President Obama declared in 2014 [201]:

...a delay that results in warming of 3° C above pre-industrial levels, instead of 2 °C, could increase economic damages by approximately 0.9 percent of global output. To put this percentage in perspective, 0.9 percent of estimated 2014 US gross domestic product (GDP) is approximately $150 billion [per year].

...net mitigation costs increase, on average, by approximately 40 percent for each decade of delay. These costs are higher for more aggressive climate goals: each year of delay means more CO_2 emissions, so it becomes increasingly difficult, or even infeasible, to hit a climate target that is likely to yield only moderate temperature increases.

The IPCC's special report on 1.5 °C scenarios [104] concluded in 2018

every year's delay before initiating emission reductions decreases by approximately two years the remaining time available to reach zero emissions on a pathway still remaining below 1.5 °C

and

the challenges from delayed actions to reduce greenhouse gas emissions include the risk of cost escalation, lock-in in carbon-emitting infrastructure, stranded assets, and reduced flexibility in future response options in the medium to long term.

Daniel *et al* [153], after applying standard treatments of risk and uncertainty to William Nordhaus's DICE integrated assessment model [118], found

delaying implementation by only 1 y costs society approximately $1 trillion. A 5 y delay creates the equivalent loss of approximately $24 trillion, comparable to a severe global depression. A 10 y delay causes an equivalent loss in the order of $10 trillion per year, approximately $100 trillion in total.

This analysis also found that the cost of delay increases quadratically over time, so that a five year delay costs twenty four times as much as a one year delay and delaying action by ten years instead of five years increases societal costs four-fold.

Delay is costly, and the faster we move to reduce emissions, the easier and cheaper it will be to do so. Conversely, the longer we delay, the more it will cost to fix the problem, and the longer society will incur the huge and avoidable societal costs of pollution from fossil fuel combustion [135, 137, 138, 204]. Delay also means we need to move faster later to stay under a fixed warming limit.

As an example, consider fossil energy carbon dioxide emissions in Grubler *et al*'s LED intervention case [53], which we show in figure A.21. We chose to highlight this scenario because it has no carbon capture, simplifying the story related to delaying emissions reductions.

Energy sector carbon dioxide emissions in this scenario decline 5.6% per year (as a % of 2020 emissions) from 2020 to 2030, reaching 56% below 2020 emissions by

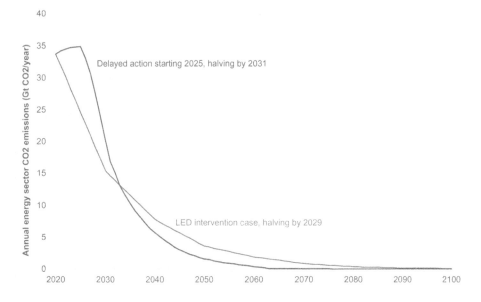

Figure A.21. Fossil energy CO_2 emissions in the low energy demand intervention scenario compared to a delayed action case. Source: Grubler *et al* [53], calculations from Koomey *et al* [205].

2030 (it reaches the 'carbon law' goal of halving emissions a year early, by 2029). The area under that emissions curve equals the cumulative emissions through 2100, which is the 'budget' that we need to meet if we're to keep temperatures from rising more than 1.5 °C.

Figure A.21 also shows what happens if we don't start reducing emissions until 2025 (the delayed action case). The area under this curve for the delayed action case is the same as for the LED intervention case, so these two cases emit the same amount of carbon from the energy sector through 2100.

Delay makes it harder to solve the problem. Starting later eats up more of the carbon budget, and the needed rate of emissions decline goes up to 8.8% per year from 2025 to 2030 (as a % of 2020), if emissions reductions begin right after emissions peak in 2025. This emissions path halves emissions from 2020 by 2031, representing about a two-year delay compared to the LED intervention case.

A.5.4 Can it be done?

The question of whether a modern society can achieve such rapid reductions is one we'll explore in the rest of this book, but it cannot be answered precisely by modeling or analysis. We'll only actually know how far we can reduce emissions once we start trying to do so in earnest, and we really haven't started yet.

Every tenth of a degree matters. If we overshoot 1.5 °C, so be it, but 1.6 °C is much better than 1.7 °C, which is much better than 1.8 °C, and by getting to a climate-positive state we can start to bring temperatures back down and reduce damages from temporarily overshooting climate targets. Even if we are daunted by the challenge of keeping warming below 1.5 °C, we need to try, and we need to act as quickly as we can.

For climate change, 'moving in the right direction' isn't enough, as Solomon Goldstein-Rose points out in his excellent book *The 100% Solution*:

> There's a lot of rhetoric about 'moving in the right direction' on climate change. But because of its difference from other issues (impacts being caused not by each year's emission but by cumulative emissions until we start removing them from the atmosphere), there's not really such a thing as 'moving in the right direction.' Climate change impacts get exponentially worse until we solve the problem 100%. That's why it is so much scarier and more urgent than other problems...

> The idea of 'doing what we can' is dangerous when it comes to climate change because it implicitly accepts that the maximum viable action is less than the minimum needed action.

'Can it be done?' is therefore the wrong question and worrying about feasibility is the wrong framing. The right question is 'How can we change society to do what is necessary?'

Nobody knows what's likely or even possible until we start down the path of aggressively reducing emissions by deploying technology, capital, communications, and institutional innovations at the requisite scale. If we choose to do so, many things will become possible that wouldn't be possible if we didn't.

Feasibility also depends on context, and on what we are willing to pay and prioritize to minimize risks. What if we finally decide (as we should) that it's a real emergency (like World War II)? In that case we'd make every effort to fix the problem, and what would be possible then is far beyond what we could imagine today.

It is therefore a mistake for analysts to impose an informal feasibility judgment when considering a problem like this one, and instead we should aim for what we think is the best outcome from a risk-minimization perspective, and if we don't quite get there, then we'll have to deal with the consequences. But if we aim too low, we might miss possibilities that we'd otherwise be able to capture.

History shows that under the right conditions, societies and industries can move quickly. In the beginning of World War II, the US retooled much of its heavy industry over the span of about 6 months [114], and some other nations engineered similarly rapid change. We now have some technology advantages over industrial firms of that era [115], especially information technology, which is our 'ace in the hole' [116].

We also know that existing technology offers opportunities to reduce emissions substantially right now, and maximizing immediate emissions reductions is where we should be focused, not on worrying about whether we'll be able to get to zero emissions by 2040. Do the obvious things: shut down fossil fuel power plants (starting with coal), mandate electrification where possible, install much more wind, solar, and energy storage, and deploy all other existing emissions reduction technology at scale everywhere we can. Our choices now create our options later because deployment drives costs down, creating new opportunities for reducing emissions elsewhere.

A.5.5 Keeping carbon in the ground

Another way to think about how fast emissions need to come down is by comparing fossil fuel consumption in the SSP-2 reference and LED intervention cases to the latest estimates of proved reserves of fossil fuels. In geology parlance, proved reserves are those stocks of fossil fuels that are known to exist with high confidence and that can be extracted using current technology at current prices. Resources are fossil deposits known with less confidence and/or are not extractable using current technology and current prices.

The first bar in figure A.22 shows proven fossil reserves for 2018 from Bundesanstalt für Geowissenschaften und Rohstoffe (BGR) [117]. They total about 900 billion tonnes of carbon (not carbon dioxide). The SSP-2 reference case implies about 1300 billion tonnes of carbon consumed from 2020 to 2100, which means that a small fraction of the remaining resources (which are more than ten-fold bigger than reserves) would need to be converted to reserves through exploration or using new technology to meet that demand.

The more important bar for our narrative, however, is the one for the LED intervention case. To keep global temperatures from increasing no more than 1.5 °C from pre-industrial times, the world can burn less than one tenth of the fossil carbon implied in the reference case, and only one eighth of the proved reserves. That means we'll either need to keep seven-eighths of the fossil reserves in the ground unburned or identify some way to sequester the carbon from burning it, which will be a heavy lift at the required scale.

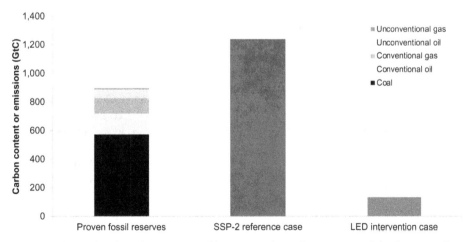

Figure A.22. Comparing the reference case and low energy demand case to proved fossil reserves. Source: BGR [117], Fricko *et al* [44], Grubler *et al* [53].

These exact numbers are dependent on some key assumptions in this scenario, but the main point is the same for all aggressive mitigation scenarios that keep the 1.5 °C warming limit in reach: *We'll need to keep a substantial fraction of proved fossil reserves in the ground or find another safe way to sequester that carbon.*

A.5.6 The economics of climate action

In the early years of studying the economics of climate, economists were concerned about the economic costs of moving too quickly [94, 95, 118–120]. As the problem became better understood, a different picture emerged, indicating that the benefits of at least modest climate action should significantly exceed costs from the societal perspective. Subsequent analyses showed that substantially reducing emissions would come at gross societal costs of *at most* a few percent of GDP in coming decades, resulting in the loss of roughly a single year's growth in GDP [121–128]. Further, these analyses showed that delaying action had serious downside risks, as discussed below.

Over time, economic assessments became more sophisticated and started to include important factors that the earliest simple models omitted—factors whose inclusion generally showed that achieving lower emissions would be cheaper, easier, and more beneficial for society than the earlier assessments indicated [129–133]. Models often omitted air quality and other health co-benefits, which are often big enough to fully offset the gross costs of reducing emissions, justifying significant climate action even before considering greenhouse gas externalities [134–139]. Many models omitted bottom-up analysis of market reforms and technology programs, which are important sources of negative net cost (i.e. societally profitable) emissions reduction options [123, 130, 131, 140–145].

Many **benefit–cost models** ignored climate damages for reference cases, assuming economic growth rates in those cases would be unaffected by unrestricted climate change [146]. Many also relied on flawed estimates of climate damages that vastly underestimate the benefits of reducing emissions [147–150]. Standard calculations of discount rates by Nordhaus and others also included a term for business-as-usual economic growth unencumbered by potential climate damages [151], and that assumption led to higher discount rates that make future benefits of climate action appear less valuable than in reality.

Virtually all models ignored the 'long-tail' risks of catastrophic climate change, which lay bare the weaknesses of the benefit–cost framing like no other issue [81, 152]. Benefit–cost models also often ignored the lessons from financial economics on pricing risk and uncertainty [153] and almost always ignored potential cost reductions associated with economies of scale, learning effects, spillovers, network externalities, and irreversibilities [132, 154–161], although sometimes these effects were studied in specific cases [156, 162]. They also almost always ignore people's asymmetrical treatment of losses versus gains (prospect theory), which again biased the results toward delaying mitigation [150].

Some models also omitted benefits from using carbon tax revenues to reduce inefficiencies in the tax system [163], failed to distinguish between endogenous and directed or induced technical change [160, 164, 165], and included only local or regional emissions trading and focused only on carbon dioxide rather than all greenhouse gases [129, 166]. In almost every case, including these factors in the analysis would have shown emissions reductions to be cheaper for society than in the original analysis.

Prematurely excluding cost-competitive mitigation options from analyses such as these can raise the apparent cost of reducing emissions. For example, in many places excluding the option of extending the useful life of existing nuclear plants would make achieving emissions goals appear to be more expensive than they would be if this option was kept on the table. How this constraint nets out for a particular technology, policy, or scenario depends on the cost for each option.

Conversely, including options as 'backstop' technologies with optimistic costs and potentials, as has happened in the past, for example, in the case of biomass energy with carbon capture, can make achieving emissions reduction goals appear cheaper than they really would be in practice. On balance, however, the models historically have erred more on the side of making the costs of emissions reductions appear to be more expensive than they really are.

A.5.7 Climate action, equity, and justice

While the costs of climate change fall on all of us, a disproportionate share of those costs will continue fall on the poorest countries, the poorest people, and generations yet to be born [167–170]. The pursuit of equity and justice is central to action on climate, a fact that is uncomfortable for many economists. The economics field has traditionally focused on economic efficiency, not on disparities in income and power relationships, but these issues cannot be ignored in facing the climate challenge.

...Climatologist Michael E Mann, writing in his book, *The New Climate War*, wrote social justice is intrinsic to climate action. Environmental crises, including climate change, disproportionately impact those with the least wealth, the fewest resources, and the least resilience. So simply acting on the climate crisis is acting to alleviate social injustice. It's another compelling reason to institute the systemic changes necessary to avert the further warming of our planet [171].

Some climate policies are more justice-enhancing than others. Unless policies are designed with existing inequities in mind, they risk exacerbating inequality and injustice, so it is not always true that 'acting on the climate crisis is acting to alleviate social injustice' [169, 170, 172]. What is true is that those with the fewest resources will be hit the hardest by climate change [169, 170], climate change is already increasing inequality [170, 173, 174], and society has an obligation to reduce those harms by rapidly reducing emissions and structuring climate policies to address (or at a minimum not exacerbate) existing inequities.

There is some empirical work supporting the differential effects of climate change and other environmental issues on less advantaged groups, for example [170, 175–

177], but it's also an intuitively reasonable conclusion. Marginalized people have fewer resources to manage unexpected crises and often live in places susceptible to extreme weather events and environmental pollution. Structural racism exacerbates these inequalities [178, 179].

For example, when extreme temperatures hit, economically disadvantaged communities have less access to well-conditioned private or public spaces. When flooding or hurricanes hit, poorer families have less ability to move to safer ground, either because they are tied to low wage jobs for survival or don't have access to affordable transportation. Communities of color and low-income communities are exposed to more outdoor air pollution than less diverse and more affluent neighborhoods [180], and exposure to air pollution may also be related to increased mortality from COVID-19 and other infections [181]. Many poorer countries are also significantly affected by air and water pollution, and correctly accounting for co-benefits to greenhouse gas emissions reductions should feature prominently in international negotiations on climate targets and commitments [139]. Air pollution from fossil fuels kills millions of people every year [135, 182], and those deaths are borne disproportionately by poorer people.

Climate change is also deeply intertwined with intergenerational justice [63, 183–185]. Those likely to be most affected by a world with a rapidly changing climate are not yet born, while powerful economic interests make their preference for delayed action known loudly in public proceedings and news media around the world.

Another dimension to the question of justice and the climate is the disproportionate effects the wealthy have on emissions [186–188]. The primary beneficiaries of fossil fuel wealth have been rich countries, and rich people everywhere, but they've privatized benefits while socializing costs. That disparity makes it incumbent upon the wealthy to take the lead on climate action. They can afford it, they are historically more to blame for the climate problem than those people who are less well off, and their prominence in society gives them greater ability to influence the opinions of others [189].

As discussed above, humanity's choices for response are threefold: mitigation, adaptation, and suffering. Our decisions on balancing climate mitigation with adaptation and suffering are at their core *moral choices*. How fast and how far we mitigate are not solely questions of economics and are inseparable from questions about equity and justice [167]. The faster and more deeply we reduce emissions and the more we center the correction of existing inequities in our solutions, the less adaptation and suffering we'll do, and the more rapidly justice will be served.

A.5.8 Speed trumps perfection in pandemic response and climate solutions

The seriousness of the climate problem has been obvious to informed observers for at least three decades, but our delay in facing it has virtually guaranteed that little about humanity's response will be optimal. The most important lesson is that we need to get started on rapid emissions reductions as soon as possible. There's no more time to waste.

The IPCC [43], writing in early 2022, emphasized the urgency of responding to the climate problem without delay:

> The cumulative scientific evidence is unequivocal: Climate change is a threat to human well-being and planetary health. Any further delay in concerted anticipatory global action on adaptation and mitigation will miss a brief and rapidly closing window of opportunity to secure a livable and sustainable future for all.

The IPCC assigned *Very High Confidence* to this statement, meaning there is little doubt that the statement is true, with multiple, independent, and consistent lines of evidence supporting it [190]. The quotation itself is scientist-speak for 'It's a bloody emergency!' and 'It's warming, it's us, we're sure, it's bad, we can fix it (but we'd better hurry)'.

There are lessons for climate solutions from responding to other kinds of emergencies. Dr Michael Ryan, Executive Director of the World Health Organization, in talking about pandemic response in early 2020, said this:

> 'Be fast. Have no regrets. You must be the first mover... In emergency response, if you need to be right before you move, you will never win... Speed trumps perfection...The greatest error is not to move. The greatest error is to be paralyzed by the fear of failure.'[7]

We conclude from this advice:

- Don't obsess about optimality, just move as quickly as you can.
- Don't obsess about feasibility, just move quickly as you can.
- Don't obsess about obstacles, just move quickly as you can.

For climate, just like for pandemic response, speed trumps perfection, and that's the attitude we need to take in addressing this problem now, because it's a real emergency [191, 192].

A.6 Greenhouse gas concentrations and temperatures with rapid climate action

Figure A.23 shows greenhouse gas equivalent concentrations from 2020 to 2100 if we manage to follow something such as the path laid out in the LED intervention case (see figure A.16 above for the comparable reference case graph). Those concentrations peak in the 2020s and begin a slow decline to 2100 as the Earth system starts absorbing carbon from the atmosphere more rapidly than we're emitting it and as methane (with its much shorter residence time) exits the atmosphere as emissions decline.

---

[7] https://twitter.com/drericding/status/1340997408503853058?s=10

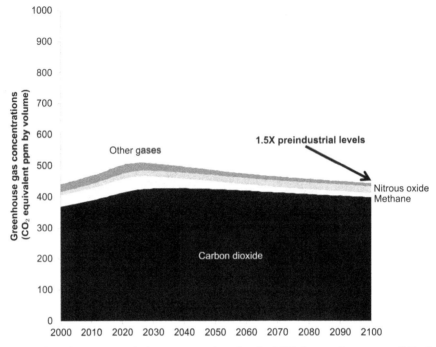

Figure A.23. Greenhouse gas equivalent concentrations for the LED intervention case to 2100. Source: Grubler *et al* [53] and calculations by Koomey.

Figure A.24 shows the temperature implications of the trend in concentrations, showing a temperature peak in the 2040s and a slow decline to 2100. The intervention scenario is a far safer world for humanity, and it's one that substantially reduces the damages to people and ecosystems compared to the reference case. It also is a world that is better, cleaner, and cheaper for people and the biosphere than continuing business as usual.

A.7 Appendix conclusions

We are in a climate emergency, but most people and institutions aren't acting like it. We need to treat climate like the crisis that it is. That means moving more quickly than in normal times, getting started on rapid emissions reductions immediately. There's no more time to waste.

Keeping global surface temperatures from exceeding 1.5 °C above pre-industrial times will not be easy, nor will drawing down GHGs into climate-positive territory. On the contrary, climate change is the biggest collective challenge modern humanity has ever faced.

We'll need to cut absolute global greenhouse emissions in half by 2030, reaching net zero emissions no later than 2040, then remove emissions with climate-positive processes for many decades thereafter. Industrialized nations, including China, should move even more quickly. Aggressive early action in these nations will drive

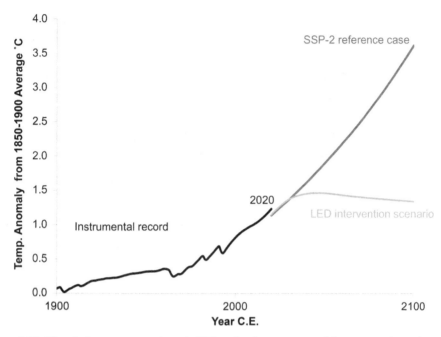

Figure A.24. Historical temperatures through 2020 and reference case and low energy demand scenario temperatures to 2100. Source: Grubler *et al* [53]. Instrumental record taken from figure A.1.

the cost of zero-emissions and climate-positive technologies down substantially, benefitting the entire world.

The vast majority of proved fossil fuel reserves will need to be kept in the ground to stabilize the climate. No fossil fuel companies' business plans currently reflect this reality [193], creating what Al Gore called a 'carbon asset bubble' [107]. When this bubble bursts, as it inevitably will, fossil investors will be left holding the bag.

For climate, just like for pandemic response, speed trumps perfection, and that's the attitude we need to take in addressing this problem now. There will always be social and environmental tradeoffs as we scale up climate solutions, but the direct benefits and co-benefits of scaling solutions generally far outweigh the costs, and we already have techniques for minimizing harm while maximizing benefits for nearly everyone. The rest of this book explores tools needed to speed up climate action and truly face the climate challenge.

Further reading

Burke K D, Williams J W, Chandler M A, Haywood A M, Lunt D J and Otto-Bliesner B L 2018 Pliocene and Eocene provide best analogs for near-future climates *Proc. Natl Acad. Sci.* **115** 13288 https://doi.org/10.1073/pnas.1809600115

Dessler A E 2022 *Introduction to Modern Climate Change* (Cambridge: Cambridge University Press)

Duane T, Koomey J, Belyeu K and Hausker K 2016 *From Risk to Return: Investing in a Clean Energy Economy* (New York: Risky Business) http://riskybusiness. org/fromrisktoreturn/

Goldstein-Rose S 2020 *The 100% Solution: A Plan for Solving Climate Change* (New York: Melville House)

Grübler A *et al* 2018 A low energy demand scenario for meeting the 1.5 °C target and sustainable development goals without negative emission technologies *Nat. Energy* **3** 515–27 https://doi.org/10.1038/s41560-018-0172-6

IPCC 2018 *Global Warming of 1.5 °C. An IPCC Special Report on the Impacts of Global Warming of 1.5 °C Above Pre-industrial Levels and Related Global Greenhouse Gas Emission Pathways, in the Context of Strengthening the Global Response to the Threat of Climate Change, Sustainable Development, and Efforts to Eradicate Poverty* (Geneva: IPCC) https://www.ipcc.ch/sr15

IPCC 2021 *Climate Change 2021: The Physical Science Basis. Contribution of Working Group I to the Sixth Assessment Report of the Intergovernmental Panel on Climate Change* ed V Masson-Delmotte *et al* (Cambridge: Cambridge University Press) https://www.ipcc.ch/report/sixth-assessment-report-working-group-i/

IPCC 2022 *Climate Change 2022: Mitigation of Climate Change. Contribution of Working Group III to the Sixth Assessment Report of the Intergovernmental Panel on Climate Change Due Out March 2022* (Cambridge: Cambridge University Press) https://www.ipcc.ch/report/sixth-assessment-report-working-group-3/

Mann M E and Toles T 2016 *The Madhouse Effect: How Climate Change Denial Is Threatening Our Planet, Destroying Our Politics, and Driving Us Crazy* (New York: Columbia University Press)

Rockström J, Gaffney O, Rogelj J, Meinshausen M, Nakicenovic N and Schellenhuber H J 2017 A roadmap for rapid decarbonization *Science* **355** 1269 https://doi.org/10.1126/science.aah3443

Sovacool B K, Burke M, Baker L, Kotikalapudi C K and Wlokas H 2017 New frontiers and conceptual frameworks for energy justice *Energy Policy* **105** 677–91 https://doi.org/10.1016/j.enpol.2017.03.005

References

[1] Krause F, Bach W and Koomey J 1989 *From Warming Fate to Warming Limit: Benchmarks to a Global Climate Convention* (El Cerrito, CA: International Project for Sustainable Energy Paths) http://mediafire.com/file/pzwrsyo1j89axzd/Warmingfatetowarminglimitbook.pdf

[2] Krause F, Bach W and Koomey J G 1992 *Energy Policy in the Greenhouse* (New York: Wiley)

[3] Nicholas K 2021 *Under the Sky We Make: How to be Human in a Warming World* (New York: Putnam) https://penguinrandomhouse.com/books/665274/under-the-sky-we-make-by-kimberly-nicholas-PhD/

[4] Haustein K, Allen M R, Forster P M, Otto F E L, Mitchell D M, Matthews H D and Frame D J 2017 A real-time global warming index *Sci. Rep.* **7** 15417

[5] Kaufman D, McKay N, Routson C, Erb M, Dätwyler C, Sommer P S, Heiri O and Davis B 2020 Holocene global mean surface temperature, a multi-method reconstruction approach *Sci. Data* **7** 201

[6] Kaufman D *et al* 2020 A global database of Holocene paleotemperature records *Sci. Data* **7** 115

[7] Fogt R L, Sleinkofer A M, Raphael M N and Handcock M S 2022 A regime shift in seasonal total Antarctic sea ice extent in the twentieth century *Nat. Clim. Change* **12** 54–62

[8] Zemp M *et al* 2019 Global glacier mass changes and their contributions to sea-level rise from 1961 to 2016 *Nature* **568** 382–6

[9] Slater T, Hogg A E and Mottram R 2020 Ice-sheet losses track high-end sea-level rise projections *Nat. Clim. Change* **10** 879–81

[10] Slater T, Lawrence I R, Otosaka I N, Shepherd A, Gourmelen N, Jakob L, Tepes P, Gilbert L and Nienow P 2021 Earth's ice imbalance *The Cryosphere* **15** 233–46

[11] Dangendorf S, Hay C, Calafat F M, Marcos M, Piecuch C G, Berk K and Jensen J 2019 Persistent acceleration in global sea-level rise since the 1960s *Nat. Clim. Change* **9** 705–10

[12] Sweet W V *et al* 2022 Global and regional sea level rise scenarios for the United States: updated mean projections and extreme water level probabilities along US coastlines *NOAA Technical Report* NOS 01 National Oceanic and Atmospheric Administration, National Ocean Service, Silver Spring, MD https://oceanservice.noaa.gov/hazards/sealevelrise/sea-levelrise-tech-report.html

[13] Lumpkin R *et al* 2020 Global oceans *Bull. Am. Meteorol. Soc.* **101** S129–84

[14] Kramer R J, He H, Soden B J, Oreopoulos L, Myhre G, Forster P M and Smith C J 2021 Observational evidence of increasing global radiative forcing *Geophys. Res. Lett.* **48** e2020GL091585

[15] McSweeney R 2020 Jet stream: is climate change causing more 'blocking' weather events? *Carbon Brief* https://carbonbrief.org/jet-stream-is-climate-change-causing-more-blocking-weather-events/

[16] Burke K D, Williams J W, Chandler M A, Haywood A M, Lunt D J and Otto-Bliesner B L 2018 Pliocene and Eocene provide best analogs for near-future climates *Proc. Natl Acad. Sci.* **115** 13288

[17] Haq B U 2014 Cretaceous eustasy revisited *Global Planetary Change* **113** 44–58

[18] IPCC 2021 *Climate Change 2021: The Physical Science Basis. Contribution of Working Group I to the Sixth Assessment Report of the Intergovernmental Panel on Climate Change* ed V Masson-Delmotte *et al* (Cambridge: Cambridge University Press) https://ipcc.ch/report/sixth-assessment-report-working-group-i/

[19] Friedlingstein P *et al* 2022 Global carbon budget 2021 *Earth Syst. Sci. Data* **14** 1917–2005

[20] Gütschow J, Jeffery M L, Gieseke R, Gebel R, Stevens D, Krapp M and Rocha M 2016 The PRIMAP-hist national historical emissions time series *Earth Syst. Sci. Data* **8** 571–603

[21] Siegenthaler U *et al* 2005 Stable carbon cycle–climate relationship during the late Pleistocene *Science* **310** 1313–7

[22] Petit J R *et al* 1999 Climate and atmospheric history of the past 420,000 years from the Vostok ice core, Antarctica *Nature* **399** 429–36

[23] IPCC 2021 *Summary for Policy Makers. Climate Change 2021: The Physical Science Basis. Contribution of Working Group I to the Sixth Assessment Report of the Intergovernmental Panel on Climate Change* ed V Masson-Delmotte *et al* (Cambridge: Cambridge University Press) https://ipcc.ch/report/sixth-assessment-report-working-group-i/

[24] NAS 2010 *Advancing the Science of Climate Change* (Washington, DC: National Academy of Sciences) http://nap.edu/catalog.php?record_id=12782

[25] Cook J 2016 Consensus on consensus: a synthesis of consensus estimates on human-caused global warming *Environ. Res. Lett.* **11** 048002

[26] IPCC 2007 *Climate Change 2007: The Physical Science Basis–Contribution of Working Group I to the Fourth Assessment Report of the Intergovernmental Panel on Climate Change* ed S Solomon *et al* (Cambridge: Cambridge University Press) http://ipcc.ch/publication-s_and_data/publications_and_data_reports.shtml

[27] IPCC 2013 *Climate Change 2013: The Physical Science Basis. Contribution of Working Group I to the Fifth Assessment Report of the Intergovernmental Panel on Climate Change* ed T F Stocker *et al* (Cambridge: Cambridge University Press) http://climatechange2013.org

[28] WMO 2021 State of the Global Climate 2020 *Report* WMO-No. 1264 World Meteorological Association, Geneva https://public.wmo.int/en/media/press-release/climate-change-indicators-and-impacts-worsened-2020

[29] Basu S, Scott J L, Miller J B, Andrews A E, Sweeney C, Gurney K R, Xu X, Southon J and Pieter P T 2020 Estimating US fossil fuel CO_2 emissions from measurements of 14 °C in atmospheric CO_2 *Proc. Natl Acad. Sci.* **117** 13300

[30] Broecker W S 1975 Climatic change: are we on the brink of a pronounced global warming? *Science* **189** 460–3

[31] Exxon 1982 CO2 greenhouse effect *Technical Review* 1 April Exxon Corporation https://insideclimatenews.org/wp-content/uploads/2015/09/1982-Exxon-Primer-on-CO2-Greenhouse-Effect.pdf

[32] Jackson R 2020 Eunice Foote, John Tyndall and a question of priority *Notes Records: R. Soc. J. History Sci.* **74** 105–18

[33] Weart S R 2008 *The Discovery of Global Warming* (Cambridge, MA: Harvard University Press) https://history.aip.org/climate/

[34] Rahmstorf S *et al* 2007 Recent climate observations compared to projections *Science* **316** 709

[35] Harries J E, Brindley H E, Sagoo P J and Bantges R J 2001 Increases in greenhouse forcing inferred from the outgoing longwave radiation spectra of the Earth in 1970 and 1997 *Nature* **410** 355–7

[36] Philipona R, Dürr B, Marty C, Ohmura A and Wild M 2004 Radiative forcing—measured at Earth's surface—corroborate the increasing greenhouse effect *Geophys. Res. Lett* **31** L03202

[37] Murphy D M, Solomon S, Portmann R W, Rosenlof K H, Forster P M and Wong T 2009 An observationally based energy balance for the Earth since 1950 *J. Geophys. Res.* **114** D17107

[38] Domingues C M, Church J A, White N J, Gleckler P J, Wijffels S E, Barker P M and Dunn J R 2008 Improved estimates of upper-ocean warming and multi-decadal sea-level rise *Nature* **453** 1090–3

[39] Strong A, Levin K and Tirpak D 2011 *Climate Science: Major New Discoveries* (Washington, DC: World Resources Institute) http://wri.org/publication/climate-science

[40] Marshall B, Driscoll A, Heft-Neal S, Xue J, Burney J and Wara M 2021 The changing risk and burden of wildfire in the United States *Proc. Natl Acad. Sci.* **118** e2011048118

[41] Henson R 2011 *The Rough Guide to Climate Change* 3rd edn (London: Rough Guides)

[42] Zeppetello L R V, Raftery A E and Battisti D S 2022 Probabilistic projections of increased heat stress driven by climate change *Commun. Earth Environ.* **3** 183

[43] IPCC 2022 *Climate Change 2022: Impacts, Adaptation and Vulnerability—The Working Group II Contribution to the Sixth Assessment Report* (Cambridge: Cambridge University Press) https://ipcc.ch/report/sixth-assessment-report-working-group-ii/

[44] Fricko O *et al* 2017 The marker quantification of the shared socioeconomic pathway 2: a middle-of-the-road scenario for the 21st century *Global Environ. Change* **42** 251–67

[45] Hausfather Z and Peters G P 2020 Emissions—the 'business as usual' story is misleading *Nature* **577** 618–20

[46] Hausfather Z and Forster P 2021 Analysis: do COP26 promises keep global warming below 2 °C? *Carbon Brief* https://carbonbrief.org/analysis-do-cop26-promises-keep-global-warming-below-2c

[47] Sognnaes I *et al* 2021 A multi-model analysis of long-term emissions and warming implications of current mitigation efforts *Nat. Clim. Change* **11** 1055–62

[48] Moore F C, Lacasse K, Mach K J, Shin Y A, Gross L J and Beckage B 2022 Determinants of emissions pathways in the coupled climate–social system *Nature* **603** 103–111

[49] Koomey J G 2012 *Cold Cash, Cool Climate: Science-Based Advice for Ecological Entrepreneurs* (El Dorado Hills, CA: Analytics)

[50] Ritchie J and Dowlatabadi H 2017 The 1000 GtC coal question: are cases of vastly expanded future coal combustion still plausible? *Energy Economics* **65** 16–31

[51] Quiggin D, Meyer K D, Hubble-Rose L and Froggatt A 2021 *Climate Change Risk Assessment 2021: The Risks Are Compounding, and without Immediate Action the Impacts Will Be Devastating* (London: Chatham House, Environment and Society Programme) https://www.chathamhouse.org/2021/09/climate-change-risk-assessment-2021

[52] IPCC 2021 Weather and climate 1 extreme events in a changing climate *Climate Change: The Physical Science Basis. Contribution of Working Group I to the Sixth Assessment Report of the Intergovernmental Panel on Climate Change* ed V Masson-Delmotte *et al* (Cambridge: Cambridge University Press) ch 11 https://ipcc.ch/report/sixth-assessment-report-working-group-i/

[53] Grübler A *et al* 2018 A low energy demand scenario for meeting the 1.5 °C target and sustainable development goals without negative emission technologies *Nat. Energy* **3** 27

[54] US Climate Change Science Program 2008 *Weather and Climate Extremes in a Changing Climate—Regions of Focus: North America, Hawaii, Caribbean, and US Pacific Islands* (Washington, DC: US Climate Change Science Program and the Subcommittee on Global Change Research) https://www.globalchange.gov/browse/reports/sap-33-weather-and-climate-extremes-changing-climate

[55] Dai A 2011 Drought under global warming: a review *WIREs Clim. Change* **2** 45–65

[56] Rahmstorf S and Coumou D 2011 Increase of extreme events in a warming world *Proc. Natl Acad. Sci* **108** 17905–9

[57] Romm J 2011 The next dust bowl *Nature* **478** 450–1

[58] Singh J, Ashfaq M, Skinner C B, Anderson W B, Mishra V and Singh D 2022 Enhanced risk of concurrent regional droughts with increased ENSO variability and warming *Nat. Clim. Change* **12** 163–70

[59] Hansen J, Sato M and Ruedy R 2012 *The New Climate Dice: Public Perception of Climate Change* (New York: National Aeronautics and Space Administration, Goddard Institute for Space Studies) https://www.giss.nasa.gov/research/briefs/2012_hansen_17/

[60] Ceballos G, Ehrlich P R and Dirzo R 2017 Biological annihilation via the ongoing sixth mass extinction signaled by vertebrate population losses and declines *Proc. Natl Acad. Sci.* **114** E6089

[61] IPBES 2019 Summary for policymakers of the global assessment report on biodiversity and ecosystem services *Intergovernmental Science-Policy Platform on Biodiversity and Ecosystem Services, IPBES Plenary at Its Seventh Session (Paris, 29 April –4 May)*

[62] Balint M, Domisch S, Engelhardt C H M, Haase P, Lehrian S, Sauer J, Theissinger K, Pauls S U and Nowak C 2011 Cryptic biodiversity loss linked to global climate change *Nat. Clim. Change* **1** 313–8

[63] Thompson A 2022 How climate change will hit younger generations *Sci. Am.* **326** 76

[64] Anderegg W R L, Abatzoglou J T, Anderegg L D L, Bielory L, Kinney P L and Ziska L 2021 Anthropogenic climate change is worsening North American pollen seasons *Proc. Natl Acad. Sci.* **118** e2013284118

[65] Carlson C J, Albery G F, Merow C, Trisos C H, Zipfel C M, Eskew E A, Olival K J, Ross N and Bansal S 2022 Climate change increases cross-species viral transmission risk *Nature* **607** 555–62

[66] Yong E 2022 We created the 'Pandemicene': by completely rewiring the network of animal viruses, climate change is creating a new age of infectious dangers *The Atlantic* 28 April https://theatlantic.com/science/archive/2022/04/how-climate-change-impacts-pandemics/629699/

[67] Grinsted A and Christensen J H 2021 The transient sensitivity of sea level rise *Ocean Sci.* **17** 181–6

[68] Rahmstorf S 2007 A semi-empirical approach to projecting future sea-level rise *Science* **315** 368–70

[69] Vermeer M and Rahmstorf S 2009 Global sea level linked to global temperature *Proc. Natl Acad. Sci.* **106** 21527–32

[70] Ridgwell A and Schmidt D N 2010 Past constraints on the vulnerability of marine calcifiers to massive carbon dioxide release *Nat. Geosci.* **3** 196–200

[71] Sale P 2011 *Our Dying Planet: An Ecologist's View of the Crisis We Face* (Berkeley, CA: University of California Press)

[72] Dias B B, Hart M B, Smart C W and Hall-Spencer J M 2010 Modern seawater acidification: the response of foraminifera to high-CO_2 conditions in the Mediterranean Sea *J. Geol. Soc.* **167** 843–6

[73] Rantanen M, Karpechko A Y, Lipponen A, Nordling K, Hyvärinen O, Ruosteenoja K, Vihma T and Laaksonen A 2022 The Arctic has warmed nearly four times faster than the globe since 1979 *Commun. Earth Environ.* **3** 168

[74] Lenton T M, Rockström J, Gaffney O, Rahmstorf S, Richardson K, Steffen W and Schellnhuber H J 2019 Climate tipping points—too risky to bet against *Nature* **575** 592–5

[75] Kemp L *et al* 2022 Climate endgame: exploring catastrophic climate change scenarios *Proc. Natl Acad. Sci.* **119** e2108146119

[76] Simpson R D 2022 How do we price an unknowable risk? *Issues Sci. Technol.* **38** 47–50 https://issues.org/social-cost-of-carbon-economics-simpson/

[77] Orihuela-Pinto B, England M H and Taschetto A S 2022 Interbasin and interhemispheric impacts of a collapsed Atlantic overturning circulation *Nat. Clim. Change* **12** 558–65

[78] Caesar L, McCarthy G D, Thornalley D J R, Cahill N and Rahmstorf S 2021 Current Atlantic meridional overturning circulation weakest in last millennium *Nat. Geosci.* **14** 118-20

[79] Rahmstorf S, Box J E, Feulner G, Mann M E, Robinson A, Rutherford S and Schaffernicht E J 2015 Exceptional twentieth-century slowdown in Atlantic Ocean overturning circulation *Nat. Clim. Change* **5** 475–80

[80] Jackson L C, Biastoch A, Buckley M W, Desbruyères D G, Frajka-Williams E, Moat B and Robson J 2022 The evolution of the North Atlantic meridional overturning circulation since 1980 *Nat. Rev. Earth Environ.* **3** 241–54

[81] Weitzman M L 2009 On modeling and interpreting the economics of catastrophic climate change *Rev. Econ. Stat.* **91** 1–19

[82] Weitzman M L 2010 What is the 'damages function' for global warming—and what difference might it make? *Clim. Change Econ.* **1** 57–69

[83] Weitzman M L 2011 Fat-tailed uncertainty in the economics of catastrophic climate change *Rev. Environ. Econ. Policy* **5** 275–92

[84] Wagner G and Weitzman M L 2016 *Climate Shock: The Economic Consequences of a Hotter Planet* (Princeton, NJ: Princeton University Press)

[85] IPCC 2022 *Climate Change 2022: Mitigation of Climate Change. Contribution of Working Group III to the Sixth Assessment Report of the Intergovernmental Panel on Climate Change* (Cambridge: Cambridge University Press) https://ipcc.ch/report/sixth-assessment-report-working-group-3/

[86] Allen M R, Friedlingstein P, Girardin C A J, Jenkins S, Malhi Y, Mitchell-Larson E, Peters G P and Rajamani L 2022 Net zero: science, origins, and implications *Annu. Rev. Environ. Resour.* **47** 849–87

[87] Matthews H D, Gillett N P, Stott P A and Zickfeld K 2009 The proportionality of global warming to cumulative carbon emissions *Nature* **459** 829–32

[88] Zickfeld K, Eby M, Matthews H D and Weaver A J 2009 Setting cumulative emissions targets to reduce the risk of dangerous climate change *Proc. Natl Acad. Sci.* **106** 16129

[89] Ricke K L and Caldeira K 2014 Maximum warming occurs about one decade after a carbon dioxide emission *Environ. Res. Lett.* **9** 124002

[90] Matthews H D and Caldeira K 2008 Stabilizing climate requires near-zero emissions *Geophys. Res. Lett.* **35** L04705

[91] Lahn B 2020 A history of the global carbon budget *WIREs Clim. Change* **11** e636

[92] Koomey J 2013 Moving beyond benefit–cost analysis of climate change *Environ. Res. Lett.* **8** 041005

[93] Baumol W J 1972 On taxation and the control of externalities *Am. Econ. Rev.* **62** 307–22 http://jstor.org/stable/1803378

[94] Nordhaus W D 1977 Economic growth and climate: the carbon dioxide problem *Am. Econ. Rev.* **67** 341–46 https://www.jstor.org/stable/1815926

[95] Nordhaus W D 1979 *The Efficient Use of Energy Resources* (New Haven, CT: Yale University Press)

[96] Caldeira K, Jain A K and Hoffert M I 2003 Climate sensitivity uncertainty and the need for energy without CO_2 emission *Science* **299** 2052–4

[97] Meinshausen M, Meinshausen N, Hare W, Raper S C B, Frieler K, Knutti R, Frame D J and Allen M R 2009 Greenhouse-gas emission targets for limiting global warming to 2 °C *Nature* **458** 1158–62

[98] IEA 2010 *World Energy Outlook 2010* (Paris: International Energy Agency, Organization for Economic Cooperation and Development) http://worldenergyoutlook.org/

[99] IEA 2011 *World Energy Outlook 2011* (Paris: International Energy Agency, Organization for Economic Cooperation and Development) http://worldenergyoutlook.org/

[100] IEA 2012 *World Energy Outlook 2012* (Paris: International Energy Agency, Organization for Economic Cooperation and Development) http://worldenergyoutlook.org/

[101] IEA 2020 *World Energy Outlook 2020* (Paris: International Energy Agency, Organization for Economic Cooperation and Development) http://worldenergyoutlook.org/

[102] Luderer G, Pietzcker R C, Bertram C, Kriegler E, Meinshausen M and Edenhofer O 2013 Economic mitigation challenges: how further delay closes the door for achieving climate targets *Environ. Res. Lett.* **8** 034033

[103] Rockström J, Gaffney O, Rogelj J, Meinshausen M, Nakicenovic N and Schellnhuber H J 2017 A roadmap for rapid decarbonization *Science* **355** 1269

[104] IPCC 2018 *Global Warming of 1.5 °C. An IPCC Special Report on the Impacts of Global Warming of 1.5 °C Above Pre-industrial Levels and Related Global Greenhouse Gas Emission Pathways, in the Context of Strengthening the Global Response to the Threat of Climate Change, Sustainable Development, and Efforts to Eradicate Poverty* (Geneva: IPCC) https://ipcc.ch/sr15

[105] Rogelj J, Meinshausen M, Sedláček J and Knutti K 2014 Implications of potentially lower climate sensitivity on climate projections and policy *Environ. Res. Lett.* **9** 031003

[106] McKibben B 2012 Global warming's terrifying new math *Rolling Stone Mag.* 19 July https://www.rollingstone.com/politics/politics-news/global-warmings-terrifying-new-math-188550/

[107] Gore A and Blood D 2013 The coming carbon asset bubble *The Wall Street J.* 29 October http://online.wsj.com/news/articles/SB10001424052702304655104579163663464339836?mod=hp_opinion

[108] Matthews H D *et al* 2020 Opportunities and challenges in using remaining carbon budgets to guide climate policy *Nat. Geosci.* **13** 769–79

[109] Rogelj J, Meinshausen M, Schaeffer M, Knutti R and Riahi K 2015 Impact of short-lived non-CO_2 mitigation on carbon budgets for stabilizing global warming *Environ. Res. Lett.* **10** 075001

[110] Rogelj J, Reisinger A, McCollum D L, Knutti R, Riahi K and Meinshausen M 2015 Mitigation choices impact carbon budget size compatible with low temperature goals *Environ. Res. Lett.* **10** 075003

[111] Tokarska K and Matthews D 2021 Refining the remaining 1.5 °C 'carbon budget' *Carbon Brief* https://carbonbrief.org/guest-post-refining-the-remaining-1-5c-carbon-budget

[112] Matthews H *et al* 2021 An integrated approach to quantifying uncertainties in the remaining carbon budget *Communi. Earth Environ.* **2** 7

[113] MacDougall A H, Zickfeld K, Knutti R and Matthews H D 2015 Sensitivity of carbon budgets to permafrost carbon feedbacks and non-CO_2 forcings *Environ. Res. Lett.* **10** 125003

[114] Herman A 2012 *Freedom's Forge: How American Business Produced Victory in World War II* (New York: Random House)

[115] Lovins A B *et al* 2011 *Reinventing Fire: Bold Business Solutions for the New Energy Era* (White River Junction, VT: Chelsea Green) https://rmi.org/insights/reinventing-fire/

[116] Koomey J G, Matthews H S and Williams E 2013 Smart everything: will intelligent systems reduce resource use? *Annu. Rev. Environ. Resour.* **38** 311–43

[117] BGR 2020 *BGR Energy Study 2019—Data and Developments Concerning German and Global Energy Supplies* (Hannover: Federal Institute for Geosciences and Natural Resources) https://www.bgr.bund.de/EN/Themen/Energie/Produkte/energy_study_2019_summary_en.html

[118] Nordhaus W D 1992 An optimal transition path for controlling greenhouse gases *Science* **258** 1315

[119] Wigley T M L, Richels R and Edmonds J A 1996 Economic and environmental choices in the stabilization of atmospheric CO_2 concentrations *Nature* **379** 240–43

[120] Manne A and Richels R 1997 On stabilizing CO_2 concentrations—cost-effective emission reduction strategies *Environ. Model. Assessment* **2** 251–65

[121] IPCC 2014 *Climate Change 2014: Mitigation of Climate Change. Contribution of Working Group III to the Fifth Assessment Report of the Intergovernmental Panel on Climate Change* ed O Edenhofer *et al* (Cambridge: Cambridge University Press) https://www.ipcc.ch/report/ar5/wg3/

[122] Kriegler E *et al* 2014 The role of technology for achieving climate policy objectives: overview of the EMF 27 study on global technology and climate policy strategies *Clim. Change* **123** 353–67

[123] Krause F, Haites E, Howarth R and Koomey J G 1993 *Cutting Carbon Emissions–Burden or Benefit?: The Economics of Energy-Tax and Non-Price Policies* (El Cerrito, CA: International Project for Sustainable Energy Paths) http://mediafire.com/file/y4r0n4vcvv4257s/cuttingCemissionsburdenorbenefitbook.pdf

[124] Duane T, Koomey J, Belyeu K and Hausker K 2016 *From Risk to Return: Investing in a Clean Energy Economy* (New York: Risky Business) http://riskybusiness.org/fromrisktoreturn/

[125] Williams *et al* 2021 Carbon-neutral pathways for the United States *AGU Adv.* **2** e2020AV000284

[126] USGCRP 2018 *Impacts, Risks, and Adaptation in the United States: Fourth National Climate Assessment* vol 2 (Washington, DC: US Global Change Research Program) https://nca2018.globalchange.gov

[127] DeCanio S J 2003 *Economic Models of Climate Change: A Critique* (Basingstoke: Palgrave-Macmillan)

[128] Lovins A B, Lovins H, Krause F and Bach W 1981 *Least-Cost Energy: Solving the CO_2 Problem* (North Andover, MA: Andover)

[129] Laitner J A, DeCanio S J, Koomey J G and Sanstad A H 2003. Room for improvement: increasing the value of energy modeling for policy analysis *Utilities Policy* **11** 87–94

[130] Krause F, Koomey J and Olivier D 2000 *Cutting Carbon Emissions While Making Money: Climate Saving Energy Strategies for the European Union (Executive Summary for Volume II, Part 2 of Energy Policy in the Greenhouse)* (El Cerrito, CA: International Project for Sustainable Energy Paths) http://mediafire.com/file/bxdgkkb2d5rcjh1/ipsepkyotocosts_eu-report.pdf

[131] Krause F, Baer P and DeCanio S 2001 *Cutting Carbon Emissions at a Profit: Opportunities for the US* (El Cerrito, CA: International Project for Sustainable Energy Paths) http://mediafire.com/file/0aro7bj2d7kqk8w/ipsepcutcarbon_us.pdf

[132] IPCC 2007 *Climate Change 2007: Mitigation of Climate Change–Contribution of Working Group III to the Fourth Assessment Report of the Intergovernmental Panel on Climate Change* ed B Metz *et al* (Cambridge: Cambridge University Press) http://ipcc.ch/publications_and_data/publications_and_data_reports.shtml

[133] Williams J H, DeBenedictis A, Ghanadan R, Mahone A, Moore J, Morrow W R, Price S and Torn M S 2012 The technology path to deep greenhouse gas emissions cuts by 2050: the pivotal role of electricity *Science* **335** 53–9

[134] Nemet G F, Holloway T and Meier P 2010 Implications of incorporating air-quality co-benefits into climate change policymaking *Environ. Res. Lett.* **5** 014007

[135] Vohra K, Vodonos A, Schwartz J, Marais E A, Sulprizio M P and Mickley L J 2021 Global mortality from outdoor fine particle pollution generated by fossil fuel combustion: results from GEOS-Chem *Environ. Res.* **195** 110754

[136] Shindell D 2020 *Health and Economic Benefits of a 2 °C Climate Policy: Testimony by Professor Drew Shindell* (Washington, DC: US House of Representatives) https://oversight.house.gov/sites/democrats.oversight.house.gov/files/Testimony%20Shindell.pdf

[137] Roberts D 2020 Air pollution is much worse than we thought: ditching fossil fuels would pay for itself through clean air alone *Vox* 12 August https://vox.com/energy-and-environment/2020/8/12/21361498/climate-change-air-pollution-us-india-china-deaths

[138] Cohen A J *et al* 2017 Estimates and 25-year trends of the global burden of disease attributable to ambient air pollution: an analysis of data from the Global Burden of Diseases Study 2015 *The Lancet* **389** 1907–18

[139] Scovronick N, Anthoff D, Dennig F, Errickson F, Ferranna M, Peng W, Spears D, Wagner F and Budolfson M 2021 The importance of health co-benefits under different climate policy cooperation frameworks *Environ. Res. Lett.* **16** 055027

[140] Brown M A, Levine M D, Short W and Koomey J G 2001 Scenarios for a clean energy future *Energy Policy* **29** 1179–96

[141] Gumerman E, Koomey J G and Brown M A 2001 Strategies for cost-effective carbon reductions: a sensitivity analysis of alternative scenarios *Energy Policy* **29** 1313–23

[142] Sanstad A H, DeCanio S, Boyd G and Koomey J G 2001 Estimating bounds on the economy-wide effects of the CEF policy scenarios *Energy Policy* **29** 1299–312

[143] Lovins A B, Ürge-Vorsatz D, Mundaca L, Kammen D M and Glassman J W 2019 Recalibrating climate prospects *Environ. Res. Lett.* **14** 120201

[144] DeCanio S 1993 Barriers within firms to energy-efficient investments *Energy Policy* **21** 906–14

[145] DeCanio S J 1998 The efficiency paradox: bureaucratic and organizational barriers to profitable energy-saving investments *Energy Policy* **26** 441–54

[146] Bastien-Olvera B A 2019 Business-as-usual redefined: energy systems under climate-damaged economies warrant review of nationally determined contributions *Energy* **170** 862–8

[147] Moore F C and Delavane B D 2015 Temperature impacts on economic growth warrant stringent mitigation policy *Nature Clim. Change* **5** 127–31

[148] Moore F C, Baldos U, Hertel T and Diaz D 2017 New science of climate change impacts on agriculture implies higher social cost of carbon *Nat. Commun.* **8** 1607

[149] Ackerman F and Stanton E A 2012 Climate risks and carbon prices: revising the social cost of carbon *Economics* **6** 1–25

[150] Kulkarni S, Hof A, van der Wijst K-I and van Vuuren D 2022 Pre-print: disutility of climate change damages warrants much stricter climate targets *Research Square* 10.21203/rs.3.rs-1788130/v1 8 August

[151] Mann G 2022 Check your spillover *Lond. Rev. Books* **44** (10 February) https://lrb.co.uk/the-paper/v44/n03/geoff-mann/check-your-spillover

[152] Cai Y and Lontzek T S 2018 The social cost of carbon with economic and climate risks *J. Political Econ.* **127** 2684–734

[153] Daniel K D, Litterman R B and Wagner G 2019 Declining CO_2 price paths *Proc. Natl Acad. Sci.* **116** 20886

[154] Arthur W B 1990 Positive feedbacks in the economy *Sci. Am.* **262** 92–9 https://www.jstor.org/stable/24996687

[155] Arthur W B 1994 *Increasing Returns and Path Dependence in the Economy* (Ann Arbor, MI: The University of Michigan Press)

[156] Gritsevskyi A and Nakicenovic N 2000 Modeling uncertainty of induced technological change *Energy Policy* **28** 907–21

[157] Rao S, Keppo I and Riahi K 2006 Importance of technological change and spillovers in long-term climate policy *Energy J.* **27** 25–42 https://www.jstor.org/stable/23297059

[158] Arrow K J 1962 The economic implications of learning by doing *Rev. Econ. Studies* **29** 155–73

[159] Vogt-Schilb A, Meunier G and Hallegatte S 2018 When starting with the most expensive option makes sense: optimal timing, cost and sectoral allocation of abatement investment *J. Environ. Econ. Manag.* **88** 210–33

[160] Fischer C and Newell R G 2008 Environmental and technology policies for climate mitigation *J. Environ. Econ. Manag.* **55** 142–62

[161] Sharpe S and Lenton T M 2021 Upward-scaling tipping cascades to meet climate goals: plausible grounds for hope *Climate Policy* **21** 421–33

[162] Nordhaus W D 2014 The perils of the learning model for modeling endogenous technological change *Energy J.* **35** 1–13 http://jstor.org/stable/24693815

[163] Hamond M J, DeCanio S J, Duxbury P, Sanstad A H and Stinson C H 1997 Tax waste, not work *Challenge* **40** 53–62 http://jstor.org/stable/40721868

[164] Acemoglu D, Aghion P, Bursztyn L and Hemous D 2012 The environment and directed technical change *Am. Econ. Rev.* **102** 131–66

[165] Acemoglu D, Akcigit U, Hanley D and Kerr W 2016 Transition to clean technology *J. Political Econ.* **124** 52–104

[166] Weyant JP, de la Chesnaye F C and Blanford G F 2006 EMF 21 Multi-Greenhouse Gas Mitigation and Climate Policy *Energy J.* **27** https://www.jstor.org/stable/i23297028

[167] Mohai P, Pellow D and Roberts J T 2009 Environmental justice *Annu. Rev. Environ. Resour.* **34** 405–30

[168] Sovacool B K *et al* 2017 New frontiers and conceptual frameworks for energy justice *Energy Policy* **105** 677–91

[169] OECD 2021 The inequalities-environment nexus *OECD Green Growth Papers* No.2021/01 OECD

[170] Islam S, Nazrul and Winkel J 2017 Climate change and social inequality *DESA Working paper* No. 152 ST/ESA/2017/DWP/152 United Nations, Department of Economic and Social Affairs, New York https://un.org/esa/desa/papers/2017/wp152_2017.pdf

[171] Mann M E 2021 *The New Climate War: The Fight to Take Back Our Planet* (New York: Public Affairs)

[172] Shi L and Moser S 2021 Transformative climate adaptation in the United States: trends and prospects *Science* **372** eabc8054

[173] Diffenbaugh N S and Burke M 2019 Global warming has increased global economic inequality *Proc. Natl Acad. Sci.* **116** 9808–13

[174] Zheng Y, Davis S J, Persad G G and Caldeira K 2020 Climate effects of aerosols reduce economic inequality *Nat. Clim. Change* **10** 220–4

[175] Hsiang S *et al* 2017 Estimating economic damage from climate change in the United States *Science* **356** 1362

[176] Tessum C W *et al* 2021 $PM_{2.5}$ polluters disproportionately and systemically affect people of color in the United States *Sci. Adv.* **7** eabf4491

[177] UNEP 2021 *Neglected: Environmental Justice Impacts of Marine Litter and Plastic Pollution* (Nairobi: United Nations Environment Program) https://unep.org/resources/report/neglected-environmental-justice-impacts-marine-litter-and-plastic-pollution

[178] Bullard R D 2008 *Dumping in Dixie: Race, Class, and Environmental Quality* 3rd edn (Boulder, CO: Westview)

[179] Bullard R D (ed) 1999 *Confronting Environmental Racism: Voices from the Grassroots* (Boston, MA: South End)

[180] Zou B, Peng F, Wan N, Mamady K and Gaines J W 2014 Spatial cluster detection of air pollution exposure inequities across the United States *PLOS ONE* **9** e91917

[181] Petroni M *et al* 2020 Hazardous air pollutant exposure as a contributing factor to COVID-19 mortality in the United States *Environ. Res. Lett.* **15** 0940a9

[182] Errigo I M *et al* 2020 Human health and economic costs of air pollution in Utah: an expert assessment *Atmosphere* **11** 11

[183] Howarth R B 2011 Intergenerational justice *The Oxford Handbook of Climate Change and Society* ed J S Dryzek, R B Norgaard and D Schlosberg (Oxford: Oxford University Press) ch 23 pp 338–52

[184] Robinson K S 2009 Time to end the multigenerational Ponzi scheme *What Matters* https://web.archive.org/web/20120705215631/http://whatmatters.mckinseydigital.com/climate_-change/time-to-end-the-multigenerational-ponzi-scheme

[185] Rezai A, Foley D K and Taylor L 2012 Global warming and economic externalities *Econ. Theory* **49** 329–51 https://www.jstor.org/stable/41408715

[186] Gore T 2020 *Confronting Carbon Inequality: Putting Climate Justice at the Heart of the COVID-19 Recovery* (Nairobi: Oxfam) https://oxfam.org/en/research/confronting-carbon-inequality

[187] Gore T 2015 *Extreme Carbon Inequality: Why the Paris Climate Deal Must Put the Poorest, Lowest Emitting and Most Vulnerable People First* (Nairobi: Oxfam) https://policy-practice.oxfam.org/resources/extreme-carbon-inequality-why-the-paris-climate-deal-must-put-the-poorest-lowes-582545/

[188] Wiedmann T, Lenzen M, Keyßer L T and Steinberger J K 2020 Scientists' warning on affluence *Nat. Commun.* **11** 3107

[189] Nielsen K S, Nicholas K A, Creutzig F, Dietz T and Stern P C 2021 The role of high-socioeconomic-status people in locking in or rapidly reducing energy-driven greenhouse gas emissions *Nat. Energy* **6** 1011–6

[190] Mastrandrea M D *et al* 2010 *Guidance Note for Lead Authors of the IPCC Fifth Assessment Report on Consistent Treatment of Uncertainties, IPCC Cross-Working Group Meeting on Consistent Treatment of Uncertainties Jasper Ridge, CA, USA* (Geneva: Intergovernmental Panel on Climate Change) https://ipcc.ch/site/assets/uploads/2017/08/AR5_Uncertainty_Guidance_Note.pdf

[191] Fischetti M 2021 We are living in a climate emergency, and we're going to say so *Sci. Am.* 12 April https://scientificamerican.com/article/we-are-living-in-a-climate-emergency-and-were-going-to-say-so/

[192] Hickel J 2021 What would it look like if we treated climate change as an actual emergency? *Current Affairs* 15 November https://currentaffairs.org/2021/11/what-would-it-look-like-if-we-treated-climate-change-as-an-actual-emergency

[193] Kühne K, Bartsch N, Tate R D, Higson J and Habet A 2022 'Carbon bombs'—mapping key fossil fuel projects *Energy Policy* **166** 112950

[194] Nordhaus W D 2008 *A Question of Balance: Weighing the Options on Global Warming Policies* (New Haven, CT: Yale University Press)

[195] UNEP 2020 *Emissions Gap Report 2020* (Nairobi: United National Environment Programme) https://unep.org/emissions-gap-report-2020

[196] Rubin E S, Azevedo I M L, Jaramillo P and Yeh S 2015 A review of learning rates for electricity supply technologies *Energy Policy* **86** 198–218

[197] Grübler A, Nakicenovic N and Victor D G 1999 Dynamics of energy technologies and global change *Energy Policy* **27** 247–80

[198] Bos K and Gupta J 2019 Stranded assets and stranded resources: implications for climate change mitigation and global sustainable development *Energy Res. Soc. Sci.* **56** 101215

[199] Semieniuk G, Holden P B, Mercure J-F, Salas P, Pollitt H, Jobson K, Vercoulen P, Chewpreecha U, Edwards N R and Viñuales J E 2022 Stranded fossil-fuel assets translate to major losses for investors in advanced economies *Nat. Clim. Change* **56** 101215

[200] Kemfert C, Präger F, Braunger I, Hoffart F M and Brauers H 2022 The expansion of natural gas infrastructure puts energy transitions at risk *Nat. Energy* **7** 582–7

[201] CEA 2014 *The Cost of Delaying Action to Stem Climate Change* (Washington, DC: White House Council of Economic Advisors) https://obamawhitehouse.archives.gov/sites/default/files/docs/the_cost_of_delaying_action_to_stem_climate_change.pdf

[202] Swiss Re Institute 2021 *The Economics of Climate Change: No Action Not an Option* (Zurich: Swiss Re Institute) https://www.swissre.com/institute/research/topics-and-risk-dialogues/climate-and-natural-catastrophe-risk/expertise-publication-economics-of-climate-change.html

[203] IEA 2009 *World Energy Outlook 2009* (Paris: International Energy Agency, Organization for Economic Cooperation and Development) http://worldenergyoutlook.org/

[204] McDuffie E E *et al* 2021 Source sector and fuel contributions to ambient PM2.5 and attributable mortality across multiple spatial scales *Nat. Commun.* **12** 3594

[205] Koomey J, Schmidt Z, Hausker K and Lashof D 2022 Black boxes revealed: assessing key drivers of 1.5 °C warming scenarios *White paper* Koomey Analytics, Bay Area, CA

[206] Friedlingstein P *et al* 2021 Global carbon budget 2021 *Earth Syst. Sci. Data Discuss* **2021** 1–191

[207] Keith D W, Holmes G, Angelo D S and Heidel K 2018 A process for capturing CO_2 from the atmosphere *Joule* **2** 1573–94

[208] Sanz-Pérez E S, Murdock C R, Didas S A and Jones C W 2016 Direct capture of CO_2 from ambient air *Chem. Rev.* **116** 11840–76

[209] Buck H J 2020 Should carbon removal be treated as waste management? Lessons from the cultural history of waste *Interf. Focus* **10** 20200010

[210] Madhu K, Pauliuk S, Dhathri S and Creutzig F 2021 Understanding environmental trade-offs and resource demand of direct air capture technologies through comparative life-cycle assessment *Nat. Energy* **6** 1035–44

[211] Fuhrman J *et al* 2020 Food–energy–water implications of negative emissions technologies in a +1.5 °C future *Nat. Clim. Change* **10** 920–7

[212] MacFarling Meure C, Etheridge D, Trudinger C, Steele P, Langenfelds R, van Ommen T, Smith A and Elkins J 2006 The law dome CO_2, CH_4 and N_2O ice core records extended to 2000 years BP *Geophys. Res. Lett.* **33** L14810

[213] Strauss B 2021 The carbon skyscraper: a new way of picturing rapid, human-caused climate change *Washington Post* 12 January https://washingtonpost.com/weather/2021/01/12/carbon-skyscraper-rapid-climate-change/

[214] Meinshausen M *et al* 2017 Historical greenhouse gas concentrations for climate modelling (CMIP6) *Geosci. Model Dev.* **10** 2057–116

[215] Knutti R, Rugenstein M A A and Hegerl G C 2017 Beyond equilibrium climate sensitivity *Nat. Geosci.* **10** 727–36

[216] Min S-K, Zhang X, Zwiers F W and Hegerl G C 2011 Human contribution to more intense precipitation extremes *Nature* **470** 378–81

Appendix B

Modeling capital stock growth and turnover

Tracking capital stocks is necessary for accurate modeling of emissions reduction scenarios. The characteristics of buildings, appliances, and industrial equipment change over time, and those changes need to be tracked to create your business-as-usual and your climate-positive scenarios.

A retirement function ('survival curve') is used to estimate the retirement rate of equipment and structures. Sometimes it is as simple as 'when capital reaches its book life it retires', sometimes it's a function of economic conditions, sometimes it is based on average lifetimes, but particularly for buildings and related equipment, explicit representation of retirement functions is necessary when assessing changes in capital stocks over time.

There are two major cases:

1. *Average equipment lifetimes much longer than the analysis period*, such as industrial equipment, power plants, refineries, and buildings.
    a. For industrial equipment and power plants, these are relatively few in number and large in size, so they are often tracked individually. They are also almost always site built rather than mass produced.

        Most analyses assume that this infrastructure continues to operate for its book life and then retires. Of course, there's no reason why this capital couldn't retire early. It is also often subject to refurbishment, which in some cases can change its characteristics significantly (and likely extends its book life).
    b. For buildings, which are more numerous, aggregate statistical treatments of capital stock turnover are usually required. Residential buildings typically number in the millions in a large country, so treating them individually has historically been difficult, but that's changing as real-time billing data and other building characteristics are available at increasing levels of detail. Even commercial buildings are numerous enough to make individual treatment of capital stocks onerous. Major refurbishment is also significant for buildings, and

doi:10.1088/978-0-7503-4032-8ch15

occurs more rapidly in commercial buildings, typically every 15 years or so, versus more like 30 years for residential structures.

2. *Average equipment lifetimes comparable to or shorter than the analysis period.* This category includes appliances and equipment, which are almost always mass produced rather than site built. These devices are so numerous that statistical treatments of capital stock turnover are required for accurate assessments.

The structure and use of a retirement function are in part dictated by the data that are available. For appliances and buildings, it is common to have data on the number of units existing in a base year, as well as estimates of future shipments over time (from the industry that produces the equipment). The best way to derive a retirement rate is to use historical data on stocks and sales of new units, but sometimes all you have is total stocks and estimates of average lifetimes, which can be enough.

The total number of existing homes/appliances is driven by fundamentals like population growth. New home sales are affected by economic conditions. Retirements are affected by fires, accidents, equipment failure, and the decision to build a new home on an existing lot.

The most commonly used retirement function is called 'exponential decay', which will be appropriate for most applications. Take the inverse of the average lifetime as a percentage (this is the retirement rate) and subtract it from 100%, then multiply that percentage by the equipment stock remaining in year N to get the stock remaining in year $N+1$. This function has the advantage of simplicity and is widely used. Exponential retirement rates can also be derived from historical data on stocks and new construction/appliance sales.

We present an example in table B.1 that can be generalized to category 1b (buildings) and 2 (appliances and equipment), based on building stock data by building type from the US Energy Information Administration's Annual Energy Outlook [1]. We create the stock accounting using the building stock data and assumed lifetimes for each building type.

The split between existing and new buildings/appliances is important because different policies are appropriate for affecting efficiency and fuel choices over time for these two cohorts. Tracking new buildings/appliances over time is also important because equipment standards will change over time, and these need to be tracked explicitly.

For appliances it is common to track each annual cohort of new devices over time. For an example of this kind of detailed stock accounting, see [2]. This approach is more complicated, but it allows for more detailed modeling of efficiency changes and electrification.

For more detailed analyses, sometimes a more sophisticated retirement function is needed. This retirement function is taken from [3], and is shown in figure B.1. It is an approximation that seems to match historical retirement behavior well for major appliances, but the available historical data may yield more accurate results for particular equipment types.

Table B.1. Stock accounting example for US residential building stocks.

Stock Category	House Type	Units	2021	2022	2023	2024	2025	2026	2027	2028	2029	2030	2031	2032	2033	2034	2035	2036	2037	2038	2039	2040	2041	2042	2043	2044	2045	2046	2047	2048	2049	2050
Building stocks																																
	Single family	Millions	86.0	86.9	87.8	88.7	89.6	90.4	91.3	92.1	92.9	93.7	94.5	95.3	96.0	96.7	97.5	98.2	98.9	99.7	100.4	101.1	101.8	102.5	103.2	103.9	104.6	105.3	105.9	106.6	107.3	107.9
	Multi family	Millions	32.4	32.7	33.0	33.2	33.5	33.7	34.0	34.2	34.5	34.7	34.9	35.2	35.4	35.6	35.8	36.0	36.2	36.4	36.6	36.9	37.1	37.3	37.5	37.7	37.9	38.1	38.3	38.5	38.7	38.9
	Mobile homes	Millions	6.6	6.6	6.6	6.6	6.6	6.6	6.6	6.6	6.6	6.6	6.6	6.5	6.5	6.5	6.5	6.5	6.5	6.5	6.5	6.5	6.5	6.5	6.5	6.5	6.5	6.5	6.5	6.5	6.5	6.5
	Total	Millions	125.0	126.2	127.4	128.5	129.7	130.8	131.9	132.9	133.9	135.0	136.0	137.0	137.9	138.9	139.8	140.8	141.7	142.6	143.5	144.5	145.4	146.3	147.2	148.0	148.9	149.8	150.7	151.6	152.4	153.3
Stock existing in 2021 still existing in year N																																
	Single family	Millions	86.0	85.1	84.2	83.4	82.6	81.7	80.9	80.1	79.3	78.5	77.7	77.0	76.2	75.4	74.7	73.9	73.2	72.5	71.7	71.0	70.3	69.6	68.9	68.2	67.5	66.9	66.2	65.5	64.9	64.2
	Multi family	Millions	32.4	31.9	31.4	31.0	30.6	30.1	29.7	29.3	28.8	28.4	28.0	27.6	27.2	26.8	26.5	26.1	25.7	25.3	25.0	24.6	24.3	23.9	23.6	23.2	22.9	22.6	22.3	21.9	21.6	21.3
	Mobile homes	Millions	6.6	6.5	6.5	6.3	6.1	6.0	5.9	5.8	5.7	5.5	5.4	5.3	5.2	5.1	5.0	4.9	4.8	4.7	4.6	4.5	4.4	4.3	4.3	4.2	4.1	4.0	3.9	3.9	3.8	3.7
	Total	Millions	125.0	123.5	122.1	120.7	119.2	117.9	116.5	115.1	113.8	112.5	111.2	109.9	108.6	107.4	106.1	104.9	103.7	102.5	101.3	100.2	99.0	97.9	96.7	95.6	94.5	93.5	92.4	91.3	90.3	89.2
Cumulative stock built since 2021																																
	Single family	Millions		1.8	3.5	5.3	7.0	8.7	10.4	12.0	13.6	15.2	16.8	18.3	19.8	21.2	22.8	24.3	25.8	27.2	28.6	30.1	31.5	32.9	34.3	35.7	37.0	38.4	39.8	41.1	42.4	43.7
	Multi family	Millions		0.8	1.5	2.2	2.9	3.6	4.3	5.0	5.6	6.3	6.9	7.5	8.2	8.8	9.3	9.9	10.5	11.1	11.7	12.2	12.8	13.3	13.9	14.4	15.0	15.5	16.0	16.5	17.1	17.6
	Mobile homes	Millions		0.1	0.2	0.4	0.5	0.6	0.7	0.8	0.9	1.0	1.1	1.2	1.3	1.4	1.5	1.6	1.7	1.8	1.9	2.0	2.1	2.2	2.2	2.3	2.4	2.5	2.5	2.6	2.7	2.8
	Total	Millions		2.7	5.3	7.9	10.4	12.9	15.4	17.8	20.1	22.5	24.8	27.1	29.3	31.5	33.7	35.8	38.0	40.1	42.2	44.3	46.4	48.4	50.4	52.4	54.4	56.4	58.3	60.2	62.2	64.1
Implied new home construction in year N																																
	Single family	Millions		1.8	1.7	1.7	1.7	1.7	1.6	1.6	1.6	1.6	1.6	1.5	1.5	1.5	1.5	1.5	1.4	1.4	1.4	1.4	1.4	1.4	1.4	1.4	1.4	1.4	1.3	1.3	1.3	1.3
	Multi family	Millions		0.8	0.7	0.7	0.7	0.7	0.7	0.7	0.7	0.7	0.6	0.6	0.6	0.6	0.6	0.6	0.6	0.6	0.6	0.6	0.6	0.6	0.5	0.5	0.5	0.5	0.5	0.5	0.5	0.5
	Mobile homes	Millions		0.1	0.1	0.1	0.1	0.1	0.1	0.1	0.1	0.1	0.1	0.1	0.1	0.1	0.1	0.1	0.1	0.1	0.1	0.1	0.1	0.1	0.1	0.1	0.1	0.1	0.1	0.1	0.1	0.1
	Total	Millions		2.7	2.6	2.6	2.6	2.5	2.5	2.4	2.4	2.4	2.4	2.3	2.3	2.2	2.2	2.2	2.1	2.1	2.1	2.1	2.1	2.0	2.0	2.0	2.0	2.0	2.0	2.0	1.9	1.9

Sources: Building stocks to 2050 from US EIA Annual Energy Outlook 2022. **Building lifetimes** assumed to be 100 years for SF, 70 years for MF, and 50 years for MHs. Retirement rates = 1/building lifetime by building type. Retirement rates used to calculate stock existing in 2021 still existing in year N. Cumulative stock built since 2021 equals total building stock minus stock existing in 2021 still existing in year N. Implied new construction in year N is the difference between cumulative stock built since 2021 in year $N+1$ and year N.

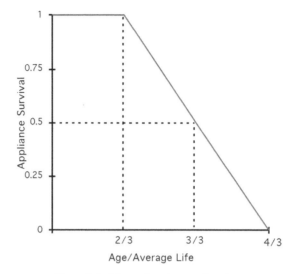

Figure B.1. Schematic retirement function.

For this function no appliances retire in the first 2/3 of their average life, and all units are retired by 4/3 of their average life. Expressed as equations, this function is as follows:

- if age $\leqslant$ [2/3 × (average life)] then 100% survive.
- if age > [2/3 × (average life)] and age < [4/3 × (average life)] then [2 – age × 1.5/(average life)] survive.
- if age $\geqslant$ [4/3 × (average life)] then 0% survive.

All retirement functions are approximations to a complex process of equipment turnover and replacement. Fortunately, results from most analyses are not very sensitive to the form of the retirement function used but ignoring stock turnover entirely is not an option for accurate analyses over several decades.

References

[1] US DOE 2022 *Annual Energy Outlook 2022, with Projections to 2050* (Washington, DC: Energy Information Administration, US Department of Energy) https://eia.gov/aeo
[2] Webber C A, Brown R E, Mahajan A and Koomey J G 2002 *Savings Estimates for the ENERGY STAR Voluntary Labeling Program: 2001 Status Report* (Berkeley, CA: Lawrence Berkeley National Laboratory) LBNL-48496
[3] Koomey J G, Mahler S A, Webber C A and McMahon J E 1999 Projected regional impacts of appliance efficiency standards for the US residential sector *Energy* **24** 69–84

IOP Publishing

Solving Climate Change
A guide for learners and leaders
Jonathan Koomey and Ian Monroe

Appendix C

How we know that much existing fossil capital will need to retire

Grübler's low energy demand (LED) scenario implies about 7%/year carbon emissions reductions for the three decades after 2020 (compounded), with little difference in this rate between the three decades [1]. This rate of emissions reductions is aggressive, but not quite as aggressive as these percentages might imply at first glance, particularly in later years of the scenario.

As emissions actually start declining, a given percentage reduction represents a smaller *absolute amount* of equipment to be retired or replaced each year (because the percentage is relative to a smaller base). That's why it's often helpful to express reduction rates as a percentage of some base year value, in our case 2020.

This approach indicates what fraction of 2020 emissions would have to be eliminated in any year to meet the constraints of the LED mitigation case, and it is proportional to the amount of capital equipment associated with those emissions (as long as there isn't much change in the carbon intensity of energy supply, which is true in the SSP-2 reference case [2]). Using this metric, the reduction rate for total carbon emissions in the first decade after 2020 in the LED scenario is about 5.4% of year 2020 emissions every year. Between 2030 and 2040 it's 2.3% of year 2020 emissions per year, and from 2040 to 2050, it's about 1.2% of year 2020 emissions every year. In absolute terms, the rate of equipment retired in the LED intervention case from 2040 to 2050 drops to about one quarter of the retirement rate from 2020 to 2030.

Figure C.1 shows an illustrative calculation about energy-related carbon dioxide emissions to give you a feel for the magnitudes. We've taken the liberty of vastly simplifying the story to make a few key points. First, we made a y-axis that shows carbon emissions as a fraction of 2020 emissions, so the lines on the graph are shown as an index with 2020 = 1.0. The topmost line is the SSP-2 reference case emissions from 2020 to 2050. That scenario shows average annual growth in emissions of 1.5% of year 2020 emissions.

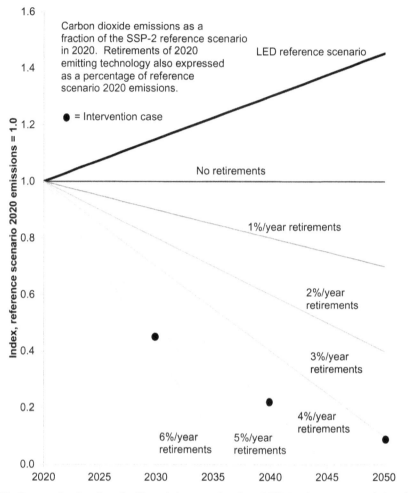

Figure C.1. Energy-related carbon dioxide emissions as a fraction of SSP-2 reference case emissions in 2020, assuming different retirement rates of 2020 capital stock and full replacement of retired stock and new growth with zero emission resources.

Then we plotted a horizontal line to represent carbon emissions from the 2020 capital stock assuming there are no retirements of that equipment (or equivalently that it is replaced when retired with capital equipment that has same emissions intensity as the 2020 stock). This line also corresponds to the emissions path that would prevail if all incremental energy service demand growth is met instead with energy technologies that emit no CO_2 starting in 2020, but equipment existing in 2020 continues to emit the same amount to 2050.

Finally, we plotted emissions pathways assuming different retirement rates for the 2020 capital stock, and assuming that all growth in emissions is met with zero-emissions energy technologies, as is all energy service demand for retiring equipment that is displaced (retirement rates are expressed as a percentage of 2020 equipment stock). This thought experiment allows us to assess the rate of equipment retirement embodied in the LED intervention case emissions path, as shown below.

The retirement rates are related to the lifetimes of capital equipment. In the simplest case, a 1%/year absolute retirement rate means that the average lifetime of the capital stock is 100 years. Retirement rates of 2%, 3%, 4%, 5%, and 6% imply lifetimes of 50, 33, 25, 20, and 16.7 years, respectively, and are also expressed as a percentage of emissions in 2020 (this makes these retirement rates linear and absolute, as opposed to exponential, which is another simplification). In the real world there is great complexity in lifetimes and retirement rates of capital equipment, but for the high-level calculation here, this rough approximation is good enough.

It's important to distinguish between capital stocks on the supply and demand sides because their lifetimes are so different. Supply side equipment, such as power plants and refineries, typically lasts 25 to 50 years, while most end-use equipment is replaced in 10 to 20 years. Building shells last longer, typically 100 years for houses and about 50 years for commercial buildings, but these usually undergo major retrofits every 20 to 30 years. A weighted average lifetime of 33 years corresponds to the 3%/year retirement case, which happens to roughly mimic the average annual retirement rate to 2050 of the emissions path for the LED intervention case.

This graph repays careful study. The average required rate of emissions reductions from 2020 to 2050 in the LED intervention case is near the limit of what can be expected by taking maximum advantage of natural stock turnover when replacing the year 2020 infrastructure (assuming a 33 year average lifetime). The retirements are front-loaded in this scenario, however. From 2020 to 2030, retirements are more than 5% of year 2020 infrastructure every year, with retirements really slowing down after 2030. That means that some existing fossil capital will need to be scrapped to meet the emissions goals of the LED intervention case, particularly in the years to 2030.

Figure C.1 indicates the scope of the challenges we face. On average, every year between 2020 and 2050 we'll need to build the equivalent of about 4.5% of year 2020 energy infrastructure but do it using zero emitting technologies[1]. From 2020 to 2030, we'll need to build the equivalent of 9% of year 2020 infrastructure in zero-emissions technologies.

Of course, we would have had to build that infrastructure anyway, we'll just need to do it with zero-emissions technologies instead of standard ones, which is likely to be somewhat more expensive in the beginning. In later years economies of scale will take hold and the net direct cost of the energy system is unlikely to cost more than a few percent of GDP relative to the business-as-usual case and have significantly lower non-climate related pollution costs as well as much lower climate risks [3].

Energy-related carbon dioxide emissions in the reference case track energy related capital stocks because the emissions intensity of primary energy supply doesn't vary much in this case. The no-policy case emissions grow at 1.5% of 2020 emissions

---

[1] The 4.5% is the sum of 3%/year retirements and about 1.5% per year growth in emissions, all expressed as a percentage of year 2020 emissions.

every year over this period (that corresponds to about a 1.3% compounded annual growth rate). The 'no retirements' case represents emissions from the 2020 capital stock assuming there are no retirements of that equipment (or equivalently that it is replaced when retired with capital equipment that has exactly the same emissions characteristics as the 2020 stock). Finally, we plotted emissions pathways assuming different retirement rates and assuming that all growth in emissions is met with zero-emissions technologies, as is all demand for replacement equipment (retirement rates also expressed as a percentage of 2020 equipment stock).

Figure C.2 is the same graph but for all warming agents, not just carbon dioxide.

One important finding these graphs make clear is that the rate of emissions reductions in the LED intervention case slows down drastically after 2030. If we're able to build 9% of 2020 fossil capital in the years before 2030, wouldn't we keep

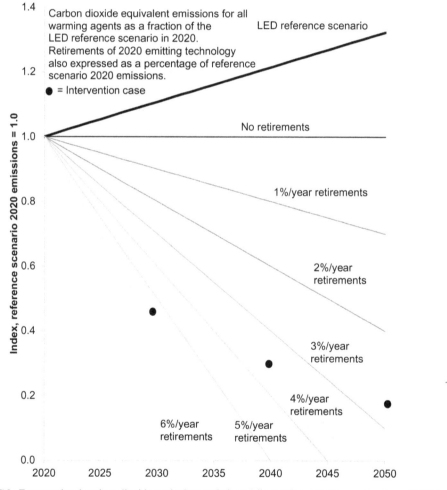

Figure C.2. Energy-related carbon dioxide equivalent emissions (all warming agents) as a fraction of SSP-2 reference case emissions in 2020, assuming different retirement rates of 2020 capital stock and full replacement of retired stock and new growth with zero emission resources.

doing that? If we retire fossil capital at 5.4%/year after 2030 we'd get to zero emissions well before 2050. Why would retirements slow down after 2030?

References

[1] Arnulf G *et al* 2018 A low energy demand scenario for meeting the 1.5 °C target and sustainable development goals without negative emission technologies *Nat. Energy* **3** 515–27

[2] Fricko O *et al* 2017 The marker quantification of the shared socioeconomic pathway 2: a middle-of-the-road scenario for the 21st century *Global Environ. Change* **42** 251–67

[3] IEA 2020 *World Energy Outlook 2020* (Paris: International. Energy Agency, Organization for Economic Cooperation and Development) http://worldenergyoutlook.org/

Appendix D

Expanded Kaya decomposition

The Kaya identity illustrates the key drivers for fossil carbon dioxide emissions from the energy sector. This identity decomposes carbon emissions as a product of aggregate wealth, energy intensity of economic activity, and carbon intensity of the energy supplied. Professor Kaya presented this equation to help understand the implications of history and future scenarios in a simple 'back of the envelope' way.

We show the familiar 'four-factor' Kaya identity in equation (D.1):

$$\text{Carbon dioxide emissions} = P \cdot \frac{\text{GNP}}{P} \cdot \frac{\text{PE}}{\text{GNP}} \cdot \frac{C}{\text{PE}} \qquad \text{(D.1)}$$

where

- P is population in any year;
- GNP is gross national product per year, a measure of economic activity;
- PE is primary energy consumption per year, including conversion and energy transmission losses;
- C is total net carbon dioxide emitted per year from the primary energy resource mix;
- $\frac{\text{GNP}}{P}$ is the average income per person per year;
- $\frac{\text{PE}}{\text{GNP}}$ is the primary energy intensity of the economy; and
- $\frac{C}{\text{PE}}$ is the net carbon dioxide intensity of supplying primary energy.

The Kaya identity reflects a more general identity that expresses impact (I) as a product of human population (P), affluence (A), and technology (T) [1, 2]. Population is the same in both the Kaya and IPAT identities, GNP/person represents affluence, and the other two terms characterize technology.

This formulation implies that a larger number of people with a higher income and more extensive use of certain technologies will have a greater impact on the environment. The role of technology can be ambiguous—technologies that produce and combust fossil fuels are the primary anthropogenic source of carbon dioxide,

doi:10.1088/978-0-7503-4032-8ch17

while technologies for harnessing renewable energy and nuclear power, sequestering carbon, and improving efficiency can reduce net anthropogenic carbon emissions.

This appendix relies on methods developed for previous work [3], enhanced and updated, as explored in a recent white paper [4]. We summarize key drivers of emissions scenarios in the energy sector using an **expanded Kaya identity**, in which we disaggregate key terms to address energy supply losses, the fraction of primary energy delivered by fossil fuels, and fuel switching among fossil fuels (this disaggregation is explained in more detail in [3]). We supplement the expanded Kaya identity with additional graphs that tell the complete high-level emissions story for each scenario.

The expanded Kaya identity, as described in [3], reads as show in equation (D.2):

$$C_{\text{Fossil Fuels}} = P \cdot \frac{\text{GNP}}{\text{P}} \cdot \frac{\text{FE}}{\text{GNP}} \cdot \frac{\text{PE}}{\text{FE}} \cdot \frac{\text{PE}_{\text{FF}}}{\text{PE}} \cdot \frac{\text{TFC}}{\text{PE}_{\text{FF}}} \cdot \frac{\text{NFC}}{\text{TFC}} \qquad (D.2)$$

where

$C_{\text{Fossil Fuels}}$ represents carbon dioxide (CO_2) emissions from fossil fuels combusted in the energy sector in any year;

P is population;

GNP is gross national product per year (measured consistently using purchasing power parity or market exchange rates);

FE is final energy consumed per year;

PE is total primary energy consumed per year, calculated using the direct equivalent (D equation) method, as discussed in Koomey *et al* [3];

PE_{FF} is primary energy consumed per year associated with fossil fuels;

TFC is total fossil energy CO_2 emitted by the primary energy resource mix per year;

NFC is net fossil CO_2 emitted to the atmosphere per year after accounting for fossil sequestration;

$\frac{\text{GNP}}{\text{P}}$ represents annual economic activity per person;

$\frac{\text{FE}}{\text{GNP}}$ represents final energy intensity of economic activity;

$\frac{\text{PE}}{\text{FE}}$ represents the energy system loss factor (ESLF) which is a measure of total losses throughout the energy system supply chain;

$\frac{\text{PE}_{\text{FF}}}{\text{PE}}$ the ratio PE_{FF}/PE we call the fossil fuel fraction, which is the fraction of primary energy supplied by fossil fuels;

$\frac{\text{TFC}}{\text{PE}_{\text{FF}}}$ the ratio TFC/PE_{FF} we call the emissions intensity of fossil fuel production, changes in which measures fuel switching among fossil fuels (such as switching power plants from being fired by coal to being fired by fossil gas, or switching from oils with higher life-cycle emissions to those with lower life-cycle emissions, as described in [5–7]); and

$\frac{\text{NFC}}{\text{TFC}}$ is an index characterizing the fraction of energy-sector emissions that reach the atmosphere, which is a measure of how much energy-sector fossil sequestration a scenario contains.

This identity allows us to disentangle key drivers affecting scenario results in the energy sector, and to show graphically which of these drivers are most important.

Because we care about all emissions that cause warming, we also need the more comprehensive relationship summarized in equation (D.3), which includes all emissions in terms of carbon dioxide equivalent:

$$C_{Total}^{eq} = C_{Fossil\ Fuels} + C_{Industry} + C_{Land\text{-}use} + C_{Non\text{-}CO_2 gases}^{eq} - CS_{Biomass} \qquad (D.3)$$

where

$C_{Fossil\ Fuels}$ is defined in equation (D.2);

$C_{Industry}$ represents carbon dioxide emissions from industrial processes (non-energy uses of fossil fuels that result in emissions, such as cement and aluminum production), some models combine these emissions with fossil fuel combustion emissions, but they should be split out for clarity and internal consistency checks;

$C_{Land\text{-}use}$ represents net carbon dioxide emissions from changes in agriculture and land-use that are not associated with emissions reductions from biomass CCS, this term can be negative if there is significant reforestation;

$C_{Non\text{-}CO_2 gases}^{eq}$ represents emissions of other greenhouse gases converted to CO_2 equivalent using relative factors of global warming potential[1]; and

$CS_{Biomass}$ represents net negative emissions from sequestering carbon emissions associated with biomass combustion (in effect, such sequestration removes carbon from the biosphere), the emissions reductions from this source must be carefully distinguished from those of land-use changes.

If direct air capture of CO_2 is present in future scenarios (as seems likely) an additional term would be needed in equation (D.3).

Substituting equation (D.2) into equation (D.3) we obtain equation (D.4), which we refer to as our *fully expanded decomposition*:

$$C_{Total}^{eq} = P \cdot \frac{GWP}{P} \cdot \frac{FE}{GWP} \cdot \frac{PE}{FE} \cdot \frac{PE_{FF}}{PE} \cdot \frac{TFC}{PE_{FF}} \cdot \frac{NFC}{TFC} + C_{Industry} + C_{Land\text{-}use}$$
$$+ C_{Non\text{-}CO_2 gases}^{eq} - CS_{Biomass} \qquad (D.4)$$

Equation (D.4) allows us to compare emissions savings in every sector from scenario modeling runs, assuming that those modeling exercises release sufficient data to calculate all terms in our fully expanded decomposition.

---

[1] We convert emissions of the two major non-CO_2 greenhouse gases (methane and nitrous oxides) to CO_2 equivalents using 100 year global warming potentials (including climate feedbacks) from the IPCC's *Sixth Assessment Report* [8: table 7.SM.7]. For both models we calculate total F-gas emissions in CO_2 equivalent using GWPs from the same source using the three major categories of such gases reported by the models: PFCs, HFCs, and SF_6.

References

[1] Ehrlich P R and Holdren J P 1971 Impact of population growth *Science* **171** 1212–7

[2] Ehrlich P R and Holdren J P 1972 One-dimensional ecology *Bull. Atomic Sci.* **28** 18–27

[3] Koomey J, Schmidt Z, Hummel H and Weyant J 2019 Inside the black box: understanding key drivers of global emission scenarios *Environ. Model. Softw.* **111** 268–81

[4] Koomey J, Schmidt Z, Hausker K and Lashof D 2022 Black boxes revealed: assessing key drivers of 1.5 °C warming scenarios *White paper* Koomey Analytics, Bay Area, CA

[5] Koomey J, Gordon D, Brandt A and Bergeson J 2016 *Getting Smart about Oil in a Warming World* (Washington, DC: Carnegie Endowment for International Peace) http://carnegieendowment.org/2016/10/04/getting-smart-about-oil-in-warming-world-pub-64784

[6] Gordon D, Brandt A, Bergeson J and Koomey J 2015 *Know Your Oil: Creating a Global Oil-Climate Index* (Washington, DC: Carnegie Endowment for International Peace) http://goo.gl/Jly9Op

[7] Brandt A R, Masnadi M S, Englander J G, Koomey J and Gordon D 2018 Climate-wise choices in a world of oil abundance *Environ. Res. Lett.* **13** 044027

[8] IPCC 2021 *Climate Change 2021: The Physical Science Basis. Contribution of Working Group I to the Sixth Assessment Report of the Intergovernmental Panel on Climate Change* ed V Masson-Delmotte *et al* (Cambridge: Cambridge University Press) https://www.ipcc.ch/report/sixth-assessment-report-working-group-i/

Appendix E

Proper treatment of primary energy

E.1 Primary energy conventions can be misleading

For electricity generation using combustion sources, half to two-thirds of the primary energy (PE) input is lost as heat, but if we replace that generation using non-combustion sources, we can simply eliminate that waste [1, 2]. Some widely used models use conventions for estimating primary energy for non-combustion sources that (i) are inconsistent and (ii) mask this important benefit of non-combustion generation.

Some energy models use a convention called 'the substitution method', in which every kilowatt-hour of electricity generated by non-combustion sources such as wind, solar, geothermal, and nuclear power is assigned an amount of primary energy equivalent to the amount of primary energy that would have been required if the generation came from fossil-fired power plants, or are based on the efficiency of converting natural energy flows to electricity. The method of imputing average system losses to non-combustion sources using the proxy of fossil fuel generation has some justification when there is a significant amount of final energy delivered by fuel-based generation sources.

There are issues with the substitution approach, however. Imputing losses for non-combustion resources in essence creates 'fictional' PE losses that aren't evident in the actual energy supply system or are losses that are not consequential for an emissions perspective (such as geothermal energy or nuclear heat not converted to electricity). If non-combustion resources displace combustion sources with real conversion losses, those losses are eliminated, and primary energy use *should* go down. Using the substitution approach masks that contribution.

E.1.1 Disentangling energy intensity and supply chain losses

The most widely used metric for energy intensity of economic activity (PE/GWP) refers to primary energy (PE), which is the total energy input to the economy from all sources, measured as the energy potential in fossil fuels and biomass at the point of extraction [3]. The PE/GWP metric is sensitive to four types of changes in an

Table E.1. Energy-economy dynamics that affect the ratio of primary energy to GWP.

Category	Cause	Example
Energy supply losses in the supply chain from primary to final energy	Technological improvements in the efficiency of energy supply conversion	A shift toward cogeneration of heat and electricity
	Changes in the balance of demand for final energy sources	A rising share of final energy from electricity
	Interfuel substitution among primary energy sources supplying each final energy type	A rising share of electricity generated using natural gas
End-use efficiency in the conversion of final energy to end-uses	Technological improvements in the efficiency of end-use energy conversion	More efficient lights or motors
	Interfuel substitution among final energy sources	A shift from all-gasoline to plug-in electric hybrid vehicles
Structural change in the economy	Changes in the modes of energy service delivery	Globalized trade patterns; urban development patterns
	Changes in the types of economic activity	A rising share of economic activity from the services sector
Conservation	Reduction in non-productive energy uses	Carpooling; changes in personal behavior or lifestyle

energy-economy system, each of which is affected by specific dynamics outlined with examples in table E.1.

Although the definition of energy intensity as a ratio of primary energy to economic activity (PE/GWP) is widely used in the literature, there are long-standing arguments for separating final energy to better assess trends in end-use demands and to isolate the first effect (energy supply losses) in table E.1 [4]. Final energy is the energy that is actually delivered to the customer's meter or gas tank, and it can include electricity, gasoline, hydrogen, or direct uses of natural gas, coal and biomass [3]. The Special Report on Emissions Scenarios (SRES) reports final energy (FE) in its detailed appendices, allowing a more accurate assessment of the energy intensity of the economy [5], and a few analysts have disaggregated FE/GWP and PE/FE in previous scenario decomposition studies [6–8].

Following those authors, the PE/GWP metric can be further disaggregated as shown in equation (E.1):

$$\frac{PE}{GWP} = \frac{FE}{GWP} \cdot \frac{PE}{FE} \qquad (E.1)$$

where

$\dfrac{FE}{GWP}$ is the final energy intensity of economic activity, and

$\dfrac{PE}{FE}$ is a measure of total energy system supply losses for delivering final energy to users.

The ratio of primary energy to final energy delivered at an economy-wide scale indicates the portion of potential energy lost in the supply chain. A value of 1.0 indicates zero conversion losses in delivering final energy to users, so this ratio will be greater than 1.0 for all real energy systems.

In the context of a single technology type (e.g. an electric power plant), the ratio of final energy to primary energy (FE/PE) is a metric that captures both the efficiency of conversion and efficiency of energy transport/transmission. At an aggregate data level, however, it is not accurate to represent the ratio of primary energy to final energy (PE/FE) as the inverse of the conversion efficiency, as is done, for example, in Kawase *et al* [7]. Table E.1 lists three different dynamics in an energy system that can affect supply losses, and a change in technical efficiency is only one.

Energy supply losses do occur as a result of some inefficiency along the way, but the value of the metric itself varies even if conversion efficiencies in the system are held constant. Changes in the PE/FE ratio can result from changes in the balance between fuels supplying a single type of final energy (e.g. natural gas versus coal for electricity generation) or the balance between final energy types (e.g. supplying water heating using electricity or natural gas). For the sake of precision and clarity this article departs from the system efficiency terminology used by some other authors in favor of the more precise term energy supply loss factor (ESLF).

E.1.2 Converting non-combustion energy production to primary energy

One of the key issues in understanding energy systems is assessing the total energy consumed by the system, including all the losses in making energy available to consumers. This assessment is complicated because of variations in how different energy sources produce fuels or electricity that allow us to do useful work.

Primary energy is the energy contained in fossil and biomass fuels, measured as (for example) the heat content of coal that goes into a power plant's boiler [3]. The difference between primary energy and secondary energy is the conversion loss in converting coal to electricity. The secondary energy is the amount of electricity injected into the grid at the busbar (measured in kWh), which also called net generation (after accounting for on-site use of electricity to run the plant). Final energy is the electricity delivered to the customer's meter, which is lower than that injected by the power plant into the grid because of transmission and distribution (T&D) losses.

Nuclear, hydroelectric, solar electric, wind power, and other non-combustion sources of electricity (or hydrogen or process heat created using these fuels) do not have losses that result in additional emissions like fossil fuel generation does. What should define the quantity of primary energy for these sources? There is the energy embodied in the nuclear fuel and the solar flux hitting a photovoltaic panel, but what does it mean to 'consume' that energy from the perspective of the emissions calculated by the Kaya identity?

To fully account for global energy use in emissions scenarios, all non-thermal sources of electricity generation, hydrogen, and process heat have traditionally been assigned a primary energy value based on some measure of the amount of fuel needed to generate equivalent amounts of secondary energy, plus the associated transmission and distribution (T&D) losses to transport the secondary energy to the customer's meter. This approach assumes that the alternative to the non-combustion energy is fossil fuel-fired combustion/generation.

For many years, this method (termed the substitution method) was considered in the scenario analysis community to be the 'customary convention'. The standard prescription for efficiency of conversion of primary to final energy in electricity generation was a constant 38.6% [9, 10: p 90]. This convention implies a final to primary energy factor of 9.33 MJ kWh^{-1} (kWh measured at the customer's meter). For direct heat treated with the substitution method, a different efficiency of conversion may be used—for example, 85%, as found in IIASA's *Global Energy Assessment* [11: p 1820].

One could also imagine a 'dynamic substitution' approach in which non-combustion sources are assigned energy supply chain losses equal to those of the average losses in the combustion part of the energy system as they change over time. Whereas the original substitution method assumed constant losses over time, this alternative method would assess losses as they evolve in the energy systems being modeled. That means it would capture the shift from (for example) older inefficient plants to newer efficient ones.

This method of imputing average system losses to non-combustion sources has some justification when there is a significant amount of final energy delivered by fuel-based energy sources, a situation that holds now and into the near future for many energy scenarios. It also allows for accurate comparison of the contribution of both combustion and non-combustion resources to the generation mix.

There are issues with the substitution approach, however, even if using the more accurate 'dynamic' version. Imputing losses for non-combustion resources in essence creates 'fictional' primary energy losses that aren't evident in the actual energy supply system. If non-combustion resources displace combustion sources with real conversion losses, those losses are eliminated, and primary energy use *should* go down. Using the substitution approach masks that contribution.

In 1998 modelers participating in the landmark *Special Report on Emissions Scenarios* prepared for the Intergovernmental Panel on Climate Change adopted an alternative convention for non-combustion electricity generation based on the heat content of the electricity power plants delivered to the busbar. This convention equates primary energy of electricity generation to the secondary energy at the busbar, using a conversion factor of 3.6 MJ kWh^{-1}. It then subtracts T&D losses to get to final energy. The modelers adopted this method, termed *direct equivalence*, as their common convention to harmonize assumptions and facilitate the comparison of results.

SRES designated nuclear power to be treated with the direct equivalent method along with solar power, wind power, hydropower, geothermal power, and other renewable sources of electricity and hydrogen [5][1]. Hundreds of mitigation scenarios based upon the reference scenarios developed for SRES have inherited the direct equivalence assumption, and aside from cautionary notes buried deep in the SRES report itself (on pages 216 and 221) and a sidebar treatment in Nakicenovic *et al* [10, p 90], it has seldom been mentioned in the literature.

If more direct equivalent sources enter the supply mix, primary energy use will decline because conversion losses from combustion are eliminated. The substitution approach would instead indicate that total primary energy and losses in the system would change more modestly as a function of the efficiency of combustion plants remaining in the system after existing plants are displaced, a counterintuitive result. Primary energy calculated using the direct equivalent approach correctly characterizes energy system losses over time (in the form of the energy supply loss factor).

Scenario modelers, who rely heavily on the 'four-factor' Kaya identity, have often compared historical changes in the ratio of primary energy to GNP to the results of model projections, failing to distinguish quantitatively between changes attributable to the shift to non-combustion direct equivalent resources and those due to changes in final energy intensity. One example is Loftus *et al* [12], which relies on aggregate trends in primary energy to GNP for its otherwise rigorous scenario comparisons. Another is Peters *et al* [13], which notes the possibility of splitting out the effects of these two factors in their 'methods' discussion but still shows a graph of historical and projected primary energy use over time in their figure 3.

Even if the scenario modelers understand this distinction, our experience is that policy makers can be easily misled by this way of presenting the data, thinking that scenarios with large reductions in the ratio of PE to GNP demonstrate significant end-use efficiency when in many scenarios significant savings come from increasing penetration of non-combustion resources. *This conceptual confusion is avoided by splitting those two key drivers in our expanded Kaya identity*, and we strongly caution against relying on the ratio of PE to GNP in almost all cases.

It is also important to note that two of the most important energy data agencies, the US Energy Information Administration (EIA) and the International Energy Agency (IEA), have adopted conventions about primary energy that can lead to confusion. EIA uses the dynamic substitution method for all non-combustion resources, with non-biomass renewables assigned the annual average conversion efficiency of fossil fuel plants, and nuclear power assigned the annual average thermal efficiency of nuclear plants. Electricity imports are treated using direct equivalence[2]. IEA treats renewables like solar, wind, and hydro using direct equivalence, geothermal energy as a thermal power plant with 10% efficiency, and

---

[1] It is important to clarify that engineering-economic models used to produce global energy scenarios *do* consider the technical efficiency of the engineered systems that harness non-thermal renewable resources and nuclear power. Indeed, technical efficiency is a vital characteristic of cost and performance parameters for each technology type. However, the *primary energy* data calculated for each technology type by models using SRES terms is reported in terms of direct equivalence as described above.

[2] https://www.eia.gov/tools/glossary/index.php?id=P

nuclear power with the thermal efficiency of 33%.[3] These choices should be reconsidered given the wide acceptance of the direct equivalent method in the scenario modeling community and the need to avoid inconsistencies in how primary energy conversions are treated.

More research is clearly needed on methods for assessing trends in primary energy. As Nakicenovic *et al* [10: p 90] point out, 'The very concept of primary energy becomes increasingly problematic, particularly as renewable energy forms gain importance'. The equations in appendix D show how correctly accounting for the convention of direct equivalence fits into the expanding Kaya identity.

References

[1] Ehrlich P R and Holdren J P 1971 Impact of population growth *Science* **171** 1212–7

[2] Eyre N 2021 From using heat to using work: reconceptualising the zero carbon energy transition *Energ. Effic.* **14** 77

[3] Grübler A, Nakicenovic N, Pachauri S, Rogner H-H and Smith K R 2015 *Energy Primer: Based on Chapter 1 of the Global Energy Assessment* (Laxenburg: International Institute for Applied Systems Analysis) https://iiasa.ac.at/web/home/research/Flagship-Projects/Global-Energy-Assessment/Chapter1.en.html

[4] Schipper L, Meyers S, Howarth R and Steiner R 1992 *Energy Efficiency and Human Activity: Past Trends, Future Prospects* (New York: Cambridge University Press)

[5] Nakicenovic N *et al* 2000 *Special Report on Emissions Scenarios (SRES), A Special Report of Working Group III of the Intergovernmental Panel on Climate Change* (Cambridge: Cambridge University Press)

[6] Grübler A, Messner S, Schrattenholzer L and Schafer A 1993 Emission reductions at the global level *Energy* **18** 539–81

[7] Kawase R, Matsuoka Y and Fujino J 2006 Decomposition analysis of CO_2 emission in long-term climate stabilization scenarios *Energy Policy* **34** 2113–22

[8] Price L, de la Rue du Can S, Sinton J, Worrell E, Zhou N, Sathaye J and Levine M 2006 *Sectoral Trends in Global Energy Use and Greenhouse Gas Emissions* (Berkeley, CA: Lawrence Berkeley National Laboratory) LBNL-56144

[9] Grübler A and Nakicenovic N 1996 Decarbonizing the global energy system *Technol. Forecast. Soc. Change* **53** 97–110

[10] Nakicenovic N, Grübler A and McDonald A (ed) 1998 *Global Energy Perspectives* (Cambridge: Cambridge University Press)

[11] IIASA 2012 *Global Energy Assessment* (Cambridge: Cambridge University Press)

[12] Loftus P J, Cohen A M, Long J C S and Jenkins J D 2015 A critical review of global decarbonization scenarios: what do they tell us about feasibility? *WIREs Clim. Change* **6** 93–112

[13] Peters, G P, Andrew R M, Canadell J G, Fuss S, Jackson R B, Korsbakken J I, Le Quéré C and Nakicenovic N 2017 Key indicators to track current progress and future ambition of the Paris Agreement *Nat. Clim. Change* **7** 118–22

---

[3] https://www.iea.org/reports/world-energy-balances-overview

Appendix F

Estimated annual revenues from fossil fuel companies and tobacco companies in 2019

This appendix documents estimated annual revenues from fossil fuel companies and tobacco companies in 2019, using methods similar to those found in the appendices to Koomey [1]. Table F.1 shows revenues for top oil and gas companies, table F.2 shows revenues for coal companies, and table F.3 shows revenues for the top tobacco companies. Fossil fuel revenues totaled at least $5 trillion in 2019, with 90% of that revenue attributable to oil and gas, only 10% to coal. Fossil fuel industry revenues were more than 13-fold bigger than tobacco revenues in total in that year. Other data for 2022 show oil and gas production revenues of $5 trillion/year[1] and coal production revenues of $600 billion/year, so both methods yield comparable results[2].

Table F.1. Oil and gas industry global revenue for top companies in 2019.

Company	Revenue Billion 2019 US$	% of total	Notes
Sinopec Group	435	9.7%	2
China Petroleum and Chemical Corp.	429	9.6%	2
CNPC	401	9.0%	2
PetroChina	364	8.1%	2
Royal Dutch Shell	345	7.7%	3
Saudi Arabian Oil Company	297	6.6%	2
BP	278	6.2%	3
Exxon Mobil	265	5.9%	3
Total	176	3.9%	3

(Continued)

---

[1] https://www.ibisworld.com/global/market-size/global-oil-gas-exploration-production/
[2] https://www.ibisworld.com/global/market-size/global-coal-mining/

Table F.1. (*Continued*)

Company	Revenue Billion 2019 US$	% of total	Notes
Chevron	147	3.3%	3
OAO Rosneft	137	3.1%	2
Gazprom	118	2.6%	2
Valero Energy	108	2.4%	3
Petroleo Brasileiro	85	1.9%	2
ENI	80	1.8%	3, 4
Pemex	79	1.8%	2
PTT	73	1.6%	2
ENGIE	67	1.5%	3, 4
Equinor	66	1.5%	3
ONGC	62	1.4%	2
Petronas	59	1.3%	2
Pertamina	56	1.3%	2
SOCAR	50	1.1%	2
E.ON	46	1.0%	3, 4
ConocoPhillips	37	0.8%	3
CNOOC Limited	34	0.8%	2
National Iranian Oil Company	30	0.7%	7
Ecopetrol	22	0.5%	2
KazMunayGas	21	0.5%	2
Repsol YPF	16	0.4%	2
Nigerian National Petroleum Corporation	16	0.4%	2
NNPC	16	0.4%	2
RWE AG	15	0.3%	3, 4
Sonangol	15	0.3%	2
Lukoil	12	0.3%	3, 5
Indian Oil	8	0.2%	3, 6
Naftogaz	6	0.1%	2
Marathon Oil	5	0.1%	3
TAQA	5	0.1%	2
Total	4480	100%	

Notes:

1. Data compiled by Zachary Schmidt (zacharym.schmidt@gmail.com), April–June 2021.
2. Source: National Oil Company Database (https://www.nationaloilcompanydata.org).
3. Source: Company Annual Report.
4. Revenues converted to US dollars using 1.12 dollar/euro.
5. Revenues converted to US dollars using 0.0155 dollar/Russian rubles.
6. Revenues converted to US dollars using 0.0142 dollar/Indian rupee.
7. Source: https://www.bp.com/content/dam/bp/business-sites/en/global/corporate/pdfs/energy-economics/statistical-review/bp-stats-review-2020-full-report.pdf. Oil revenues estimated using total company oil production (3.535 million barrels/day, p 16) and the average spot crude oil price for 2019 (p 26). Gas revenues estimated using total company natural gas production (8.79 exajoules/year, p 35) and the average natural gas price for 2019 (p 39).

Table F.2. Coal industry global revenue for 2019.

Company	Revenue Billion 2019 US$	% of total	Notes
Glencore	215	42.9%	2
Peabody Energy Corporation	46	9.2%	2
BHP Billiton	44	8.8%	2
Rio Tinto	43	8.6%	2
Vale SA	38	7.5%	2
Shenhua Group	35	7.0%	2, 5
Anglo American	30	6.0%	2
RWE Group	15	2.9%	2, 3
Teck Resources	12	2.4%	2
Yanzhou Coal Mining Company Limited	10	2.0%	2, 5
Bumi Resources	5	0.9%	2
Adaro Indonesia	3	0.7%	2
Banpu	3	0.6%	2
Coal India	1	0.3%	2, 4
Arch Coal	1	0.2%	2
Total	501	100	

Notes:
1. Data compiled by Zachary Schmidt (zacharym.schmidt@gmail.com), April–June 2021.
2. Source: Company Annual Report.
3. Revenues converted to US dollars using 1.12 dollar/euro.
4. Revenues converted to US dollars using 0.0142 dollar/Indian rupee.
5. Revenues converted to US dollars using 0.1448 dollar/Chinese yuan.

Table F.3. Tobacco manufacturing industry global revenue for 2019.

Company	Revenue Billion 2019 US$	% of total
China National Tobacco Corp.	150	40.8%
Philip Morris International Inc.	78	21.2%
Imperial Brands	40	11.0%
British American Tobacco PLC	33	9.0%
Altria Group	25	6.8%
Japan Tobacco Inc.	20	5.4%
PT Gudang Garam Tbk.	8	2.2%
ITC Limited	7	1.8%
KT&G	4	1.2%

(Continued)

Table F.3. (*Continued*)

Company	Revenue Billion 2019 US$	% of total
Universal Corporation	2	0.6%
Eastern Co SAE	1	0.2%
Total	368	100%

Notes:
1. Data compiled by Zachary Schmidt (zacharym.schmidt@gmail.com), April–June 2021.
2. Sources: https://vape.hk/china-tobacco-1205-6-billion for China National Tobacco Corp., Company Annual reports for all others.

Reference

[1] Koomey J G 2012 *Cold Cash, Cool Climate: Science-Based Advice for Ecological Entrepreneurs* (El Dorado Hills, CA: Analytics)

IOP Publishing

Solving Climate Change
A guide for learners and leaders
Jonathan Koomey and Ian Monroe

Appendix G

The effect of carbon prices on existing coal-fired electricity generation and retail gasoline prices

Table G.1 shows the calculation of the effect of a $10/tonne carbon dioxide charge for existing coal-fired power plants, assuming a higher heating value (HHV) efficiency of 33% (from [1]) and the carbon content of typical US bituminous coal (based on HHV, from the US Energy Information Administration [2]).

Table G.1. Effect of a $10/tonne tax on carbon dioxide emissions for an existing coal plant.

	Units	Bituminous Coal steam existing
HHV efficiency	%	33
Carbon content	g C/kWh.f	86.8
Emissions	g C/kWh.e at busbar	263
Emissions	g CO$_2$/kWh.e at busbar	965
Cost at busbar for $10/tonne CO$_2$ tax	2020 ¢/kWh	1.0

Notes
1. g C/kWh.f = grams of carbon contained in fuel (coal) equivalent to 1 kWh (3412 Btus or 3.6 MJ) of heat value, taken from https://www.eia.gov/environment/emissions/co2_vol_mass.php.
2. g C/kWh.e = grams of carbon emitted per kWh of electricity generated from burning coal at 33% efficiency, measured at the busbar.
3. g CO$_2$/kWh.e = grams of carbon dioxide emitted per kWh of electricity generated from burning coal at 33% efficiency, measured at the busbar.
4. CO$_2$ emissions calculated assuming 100% combustion and the ratio of carbon dioxide to carbon molecular weights (44/12).
5. Price per gallon calculated for $10 per tonne CO$_2$ charge (2020 $).

Table G.2 shows the calculation of the effect of a $10/tonne carbon dioxide charge on retail gasoline prices, using the carbon content of typical US motor gasoline (based on HHV, from the US Energy Information Administration [2]).

Table G.2. Effect of a $10/tonne tax on carbon dioxide emissions for US motor gasoline.

	Units	Gasoline
Carbon content per energy content	g C/kWh.f	66.3
CO_2 emissions per energy content	g CO_2/kWh.f	243
Energy content in kWh per gallon	kWh/gallon	35.25
CO_2 emissions per gallon	kg CO_2/gallon	8.58
Price effect of CO_2 tax	2020 US ¢/gallon	0.09

Notes

1. g C/kWh.f = grams of carbon contained in fuel (gasoline) equivalent to 1 kWh (3412 Btus or 3.6 MJ) of heat value, taken from https://www.eia.gov/environment/emissions/co2_vol_mass.php.

2. CO_2 emissions calculated assuming 100% combustion and the ratio of carbon dioxide to carbon molecular weights (44/12).

3. Energy content in kWh/gallon calculated from data at https://www.eia.gov/energyexplained/units-and-calculators/british-thermal-units.php.

4. Price per gallon calculated for $10 per tonne CO_2 charge (2020 $).

References

[1] Koomey J *et al* 2010 Defining a standard metric for electricity savings *Environ. Res. Lett.* **5** 014017

[2] EIA 2016 *Carbon Dioxide Emissions Coefficients* (Washington, DC: Energy Information Administration, US Department of Energy) https://eia.gov/environment/emissions/co2_vol_-mass.php

Printed in the USA
CPSIA information can be obtained
at www.ICGtesting.com
JSHW050626070224
56719JS00003B/29

9 780750 340304